Sediment Dynamics in Artificial Nourishments

Sediment Dynamics in Artificial Nourishments

Editor

Carlos Daniel Borges Coelho

MDPI

Basel • Beijing • Wuhan • Barcelona • Belgrade • Novi Sad • Cluj • Manchester

Editor
Carlos Daniel Borges Coelho
Department of Civil Engineering
University of Aveiro
Aveiro
Portugal

Editorial Office
MDPI
St. Alban-Anlage 66
4052 Basel, Switzerland

This is a reprint of articles from the Special Issue published online in the open access journal *Journal of Marine Science and Engineering* (ISSN 2077-1312) (available at: www.mdpi.com/journal/jmse/special_issues/bz_sediment_artificial_nourishments).

For citation purposes, cite each article independently as indicated on the article page online and as indicated below:

Lastname, A.A.; Lastname, B.B. Article Title. *Journal Name* **Year**, *Volume Number*, Page Range.

ISBN 978-3-0365-8521-5 (Hbk)
ISBN 978-3-0365-8520-8 (PDF)
doi.org/10.3390/books978-3-0365-8520-8

Contents

About the Editor

Carlos Daniel Borges Coelho

Carlos Coelho is an Associated Professor with Habilitation in the Civil Engineering Department of the Aveiro University, where he completed a PhD. in Civil Engineering, in the field of Coastal Engineering, in 2005. He is responsible for several Curricular Units in the area of Hydraulics in Civil Engineering at Aveiro University. He was the President of the Board of the Portuguese Association for Water Resources (2021-2023) and he was the President of the Specialized Commission for Coastal and Marine Zones of the Portuguese Association for Water Resources (2018-2021). He is one of the Portuguese delegates at PIANC, The World Association for Waterborne Transport Infrastructure. His research has been published in several international and national journals and international and national conference proceedings papers. He has supervised several PhD and dozens of master's theses, as well as supervising several ongoing projects. Carlos Coelho was a member of the Commission that followed the Littoral Working Group and also collaborates on consultancy work, developing coastal management plans and sediment dynamics and hydrologic studies.

Preface

Sediment dynamics in artificial nourishments is an important topic in the current worldwide research regarding coastal erosion mitigation and climate change adaptation. Artificial nourishments allow negative budgets to decrease by adding sediment to the coastal system. However, due to the complexity of coastal processes, sediments dynamics after the intervention are difficult to evaluate. This is evident in the works being presented and discussions going on in the main international coastal conferences and forums. Therefore, when this Special Issue was announced, very positive reactions were received from colleagues developing research on the topic, who praised the idea and registered their interest in publishing in the proposed Special Issue, which confirmed the relevance of the theme. The twelve manuscripts published in this Special Issue reinforce the importance of research into understanding sediment dynamic processes related to artificial nourishments.

The existence of a volume that compiles research in a specific topic makes literature reviews easier for the next researchers. This is an increasing trend in science and research that also allows communities of authors and readers to be built, who will discuss the latest information and data on sediment dynamics and develop new ideas and research directions. National institutions, governmental agencies and entities responsible for coastal management and planning will certainly benefit from this collection of papers. Decision makers will be able to more effectively support their strategies with adequate information and knowledge. Consequently, this publication is directed to all coastal agents interested in sediment dynamics in artificial nourishments.

In this way, as Guest Editor, I would like to thank MDPI and the *Journal of Marine Science and Engineering* for the opportunity to call for manuscripts that describe some of the most updated results on the topic of sediment dynamics in artificial nourishments and to thank all the authors that contributed with their work and knowledge to improve the performance of coastal interventions and preserve sustainable worldwide coastal systems. I hope readers will enjoy this Special Issue.

Carlos Daniel Borges Coelho
Editor

Journal of
Marine Science and Engineering

MDPI

Editorial

Sediment Dynamics in Artificial Nourishments

Carlos Coelho

RISCO & Department of Civil Engineering, University of Aveiro, Campus Universitário de Santiago, 3810-193 Aveiro, Portugal; ccoelho@ua.pt

Worldwide, coasts present increasing erosion trends, regardless of the investments made to mitigate them. In fact, serious erosion problems relating to significant negative sediment budgets in coastal systems have been identified. Artificial nourishments are a coastal erosion mitigation strategy that allows for these negative budgets to be decreased by adding sediment to the coastal system. However, due to the complexity of coastal processes, after the intervention, sediment dynamics present difficult evaluations. Important technical questions remain without an adequate answer. For instance, guidance for nourishment designs should consider the time needed for the cross-shore profile to reach its equilibrium configuration after the intervention. Other designing characteristics should consider the nourished sediments' longshore transport velocity and the time sand takes to benefit the neighbor beaches under different morphological and hydrodynamic conditions to better define the re-nourishment frequency. It is also important to control the mass center of the nourishment interventions under different longshore sediment transport conditions, understanding the nourished morphological shape and the degree of diffusion of the sediments over time.

The social perceptions of artificial nourishments also depend on the technician's ability to explain what is intended with these coastal interventions. A subaerial reinforcement of a beach may lead to larger recreative areas, with quick positive effects on their recreative use, but the movement of sediments to the submerged bar causes a negative idea of rapid losses and inadequate intervention. On the other hand, reinforcing a submerged bar may increase the beach berm over time due to cross-shore dynamic processes, causing a positive reaction in coastal populations that did not realize that the sediments were previously deposited in the subaqueous portion of the beach profile. The location of the deposition in the profile also depends on the costs of the intervention, the equipment, used and the sediment sources. Sand from maritime sources is simpler to deposit in the submerged part of the profile, while terrestrial sediments are easier to use when reinforcing dunes or beaches. Knowing the existing volumes available and the distances of the sources is also fundamental to define the nourishment strategies over time.

Citation: Coelho, C. Sediment Dynamics in Artificial Nourishments. *J. Mar. Sci. Eng.* **2023**, *11*, 1433. https://doi.org/10.3390/jmse11071433

Received: 27 June 2023
Accepted: 14 July 2023
Published: 18 July 2023

Another relevant topic relating to artificial nourishment interventions is the effective benefit they represent in decreasing the maintenance needs of existing coastal structures or mitigating wave overtopping. In fact, the maintenance of coastal structures and overtopping and flooding events may correspond to high costs that can be diminished by the nourishments, but these positive impacts are scarcely quantified. The engineering, technical, and social aspects of nourishment interventions require scientific knowledge and continuous research. Considering this, the present Special Issue compiles the most updated scientific knowledge on understanding the processes of sediment dynamics following artificial nourishment. This Special Issue promotes discussions on cross-shore and longshore nourished sediment distribution, monitoring works, the impacts of shoreline evolution after nourishments, the longevity of the nourishments, ecological impacts, and the interactions of artificial nourishments with other coastal structures.

Artificial nourishments are presently some of the most applied coastal erosion mitigation interventions [1]. Nearshore nourishments are constructed for shoreline protection from waves, to provide sediment nourishment to the beach profile, and to beneficially use sediment dredged from navigation channel maintenance [2]. According to Kroon et al. [3], sandy nourishments can provide additional sediment to a coastal system to maintain its recreational or safety function under rising sea levels. These nourishments can be implemented at sandy beach systems as valid measures of mitigating coastal erosion in some erosional hot spots; they are also considered measures of adaptation under the present climate change scenario, including the impacts of an increasing sea level [4,5].

This Special Issue provides new insights into the time scales of beach responses to high-magnitude shoreface interventions on different sandy coasts. Monitoring results are shown and discussed in four manuscripts [3–6]. At the Hondsbossche Dunes, the Netherlands, a combination of shoreface, beach, and dune nourishment of 35 million m^3 sand was built to replace the flood protection function of an old sea dike while creating additional space for nature and recreation. Over a five-year period, net volume losses from the project area were less than 5% of the initial nourished sand volume. The dune volume has increased and the dune foot migrated seaward at the entire nourished site, regardless of whether the subaqueous profile gained or lost sediment. Natural forces in the years after implementation provided a significant contribution to the growth in dune volume and related safety against flooding [3]. At the Aveiro coast (Costa Nova-Ílhavo), Portugal, the first monitoring results of a $\approx 2.4 \times 10^6$ m^3 shoreface nourishment, the largest performed in Portugal until now, are presented by Pinto et al. [4] and Mendes et al. [5]. The morphological development, impacts on adjacent beaches due to alongshore spreading and cross-shore redistribution, and the contribution to the sediment budget of the nourished sediment cell were evaluated. The results show rapid morphological change over the placement area, with a decrease of about 40% of the initial volume. Sediment spreading also induced the accretion of the subaerial section of Costa Nova beaches in front of the placement area, reversing their long-term erosive trend [4]. These two study sites show the specificity of each location in the performance and evaluation of nourishments.

A nearshore nourishment project was completed during the summer of 2021 in Harvey Cedars, NJ, USA, with 67,500 m^3 of dredged sediment from Barnegat Inlet placed along approximately 450 m of beach at a depth of 3–4 m [6]. Altering the cross-shore profile geometry due to the introduction of new sediments induces a non-equilibrium situation with respect to the local wave dynamics [6]. In fact, hydrodynamics are changed by nourishment interventions that alter the bottom characteristics, inducing different wave propagation conditions, namely refraction and breaking. These types of effect need to be better understood, and potential increases in the concentrations of suspended sediments, mainly during intervention works, which may represent relevant environmental impacts, must be evaluated.

This Special Issue also represents an important step in compiling updated knowledge about monitoring works and example situations in which the interactions with coastal structures are described. Two manuscripts [7,8] describe the relationships between nourishments and existing coastal structures, and they also combine the monitored information and the different characteristics of the materials used in the nourishment operation (sand and gravel). The capacity to distinguish the different effects of sediment grain sizes is highlighted as an important research topic due to the complexity of the processes relating to different sand characteristics and the ability to model those processes.

This Special Issue also contributes to numerical modelling research, which is also applied in cost–benefit assessments, ecological impacts, and laboratory works. It is poorly understood how the placed sediments' morphology and depth influence nearshore processes operating on wave-dominated coasts [2]. Ferreira et al. [9] also state that it is essential to understand and adequately model the shoreline response after a nourishment operation in order to support the definition of the best intervention scenarios. Numerical modelling can help in understanding processes and anticipating behaviours. Johnson et al. [2] investigated the wave fields, sediment transport, and morphological responses to three common nearshore nourishment shapes, a nearshore berm (elongated bar), undulated nearshore berm, and small discrete mounds, using numerical experiments. The simulation results indicate that shallower, more continuous berms attenuate the most wave energy, while deeper, more diffuse placements retain more sediment. Sancho et al. [10] aimed to compare and identify the most effective intervention in terms of reducing beach erosion or even promoting beach accretion forced by local wave conditions, supported by a shoreline evolution model calibrated with in situ field data. Ferreira et al. [9] studied the effects of artificial nourishment on the longshore sediment transport and consequently on the morphological evolution at the intervention site and nearby areas over a time span of 5 years. In a different approach, Guimarães et al. [11] used 3D movable bed physical modelling to test the impacts of beach nourishment on hydrodynamics, sediment transport, and morphodynamics. Nourishments may also have ecological consequences. This Special Issue presents a study referring to the possible interaction between nourishments and the success of invading species [12]. It is suggested that direct communication between environmental regulators and scientists is crucial for improving both scientific research and environmental management policies [12].

Measures to mitigate coastal erosion usually present negative aspects and thus, when a coastal intervention is performed, is it desirable to define a solution that presents low levels of negative physical impacts while being simultaneously economically attractive. Supported by the results of models, an integrated cost–benefit methodology is presented to analyse the performances of artificial nourishments [1]. The approach presented by Coelho et al. [1] encompasses a shoreline evolution model (to estimate maintained, gained, or lost coastal areas over time) and a cost–benefit evaluation (combining the monetary benefits of land use and the ecosystem services of the territory with the costs of the artificial nourishment interventions, depending on their sand volumes along time). Thus, the performance of an artificial nourishment should be analysed by assessing the effectiveness of different scenarios from physical, social, environmental, and economic perspectives.

In general, the sediment dynamics of nourished sediments are deeply evaluated in this Special Issue, which aims to contribute to scientific knowledge about the permanence of sediment at the deposition site and the frequency required for new nourishments. The variety of research topics presented in this Special Issue are demonstrative of the importance of artificial nourishment as a coastal erosion mitigation strategy and the complexity of the processes involved in this type of intervention. The monitoring works highlight the diversity of parameters to monitor, and the site-specific conditions reveal the difficulty of defining common behaviours while also demonstrating the interactions with other coastal structures and the importance of the materials considered in the intervention. Monitoring is also supporting in modelling works, which are increasingly relevant in projecting future scenarios for adequate long-term planning. The obtained modelling results improve the understanding of nearshore nourishment shapes and can support decision makers in identifying the most appropriate construction technique or intervention scenarios for future nearshore nourishment projects. Consequently, the research presented in this Special Issue can support decision makers in identifying the most proper management action where coastal erosion problems persist and nourishment interventions are required.

Conflicts of Interest: The author declares no conflict of interest.

References

1. Coelho, C.; Lima, M.; Ferreira, A.M. A cost-benefit approach to discuss artificial nourishments to mitigate coastal erosion. *J. Mar. Sci. Eng.* **2022**, *10*, 1906. [CrossRef]
2. Johnson, C.L.; McFall, B.C.; Krafft, D.R.; Brown, M.E. Sediment Transport and Morphological Response to Nearshore Nourishment Projects on Wave-Dominated Coasts. *J. Mar. Sci. Eng.* **2022**, *9*, 1182. [CrossRef]
3. Kroon, A.; de Schipper, M.; de Vries, S.; Aarninkhof, S. Subaqueous and Subaerial Beach Changes after Implementation of a Mega Nourishment in Front of a Sea Dike. *J. Mar. Sci. Eng.* **2022**, *10*, 1152. [CrossRef]
4. Pinto, C.A.; Taborda, R.; Andrade, C.; Baptista, P.; Silva, P.A.; Mendes, D.; Pais-Barbosa, P. Morphological Development and Behaviour of a Shoreface Nourishment in the Portuguese Western Coast. *J. Mar. Sci. Eng.* **2022**, *10*, 146. [CrossRef]

5. Mendes, D.; Pais-Barbosa, J.; Baptista, P.; Silva, P.A.; Bernardes, C.; Pinto, C. Beach Response to a Shoreface Nourishment (Aveiro, Portugal). *J. Mar. Sci. Eng.* **2022**, *9*, 1112. [CrossRef]
6. McGill, S.P.; Harris, B.D.; McFall, B.C.; Krafft, D.R.; Bain, R.L.; Olsen, N.R.; Conery, I.W.; Chasten, M.A. Morphological Analysis of a Nearshore Nourishment along the Atlantic Coast of New Jersey, USA. *J. Mar. Sci. Eng.* **2022**, *10*, 1622. [CrossRef]
7. Kondrat, V.; Šakurova, I.; Baltranaitė, E.; Kelpšaitė-Rimkienė, L. Natural and Anthropogenic Factors Shaping the Shoreline of Klaipėda, Lithuania. *J. Mar. Sci. Eng.* **2022**, *9*, 1456. [CrossRef]
8. Cinelli, I.; Anfuso, G.; Bartoletti, E.; Rossi, L.; Pranzini, E. The Making of a Gravel Beach (Cavo, Elba Island, Italy). *J. Mar. Sci. Eng.* **2022**, *9*, 1148. [CrossRef]
9. Ferreira, A.M.; Coelho, C. Artificial Nourishments Effects on Longshore Sediments Transport. *J. Mar. Sci. Eng.* **2022**, *9*, 240. [CrossRef]
10. Sancho, F. Evaluation of Coastal Protection Strategies at Costa da Caparica (Portugal): Nourishments and Structural Interventions. *J. Mar. Sci. Eng.* **2023**, *11*, 1159. [CrossRef]
11. Guimarães, A.; Coelho, C.; Veloso-Gomes, F.; Silva, P.A. 3D Physical Modeling of an Artificial Beach Nourishment: Laboratory Procedures and Nourishment Performance. *J. Mar. Sci. Eng.* **2022**, *9*, 613. [CrossRef]
12. Zviely, D.; Zurel, D.; Edelist, D.; Bitan, M.; Spanier, E. Does Sand Beach Nourishment Enhance the Dispersion of Non-Indigenous Species?—The Case of the Common Moon Crab, Matuta victor (Fabricius, 1781), in the Southeastern Mediterranean. *J. Mar. Sci. Eng.* **2022**, *9*, 911. [CrossRef]

Journal of
*Marine Science
and Engineering*

MDPI

Article

A Cost–Benefit Approach to Discuss Artificial Nourishments to Mitigate Coastal Erosion

Carlos Coelho [1], Márcia Lima [1,2] and Margarida Ferreira [1,3,*]

1 RISCO & Civil Engineering Department, University of Aveiro, Campus Universitário de Santiago, 3810-193 Aveiro, Portugal
2 Dreams & Faculty of Natural Sciences, Engineering and Technologies, Lusophone University of Porto, Rua de Augusto Rosa 24, 4000-098 Porto, Portugal
3 CESAM & Civil Engineering Department, University of Aveiro, Campus Universitário de Santiago, 3810-193 Aveiro, Portugal
* Correspondence: margarida.ferreira@ua.pt

Abstract: Worldwide, artificial nourishments are being considered as one of the main coastal erosion mitigation measures. However, this solution is not permanent, since the natural removal of sediments that occurs after the sand deposition leads to the need of re-nourishment projects; thus, its performance and longevity dependent on several design parameters (placement site and extension alongshore, frequency, and volume, etc.) In this work, a methodological approach for cost–benefit assessment is applied to analyze the performance of artificial nourishments from a physical and economical point of view, by analyzing the effectiveness of different scenarios. The study was developed considering two study areas: a hypothetical situation (generic study area) and a real coastal stretch (Barra-Vagueira, located in the Portuguese west coast). The findings show the complexity in defining the best nourishment option, being dependent on the wave climate, site specific conditions, and main goal of the intervention. The proposed cost–benefit approach allows one to obtain and compare the physical and economic performance of artificial nourishments to mitigate coastal erosion, aiding the decision-making processes related to coastal planning and management.

Keywords: LTC; COAST; nourishment scenarios; Barra-Vagueira; coastal management

Citation: Coelho, C.; Lima, M.; Ferreira, M. A Cost–Benefit Approach to Discuss Artificial Nourishments to Mitigate Coastal Erosion. *J. Mar. Sci. Eng.* **2022**, *10*, 1906. https://doi.org/10.3390/jmse10121906

Academic Editor: Felice D'Alessandro

Received: 7 November 2022
Accepted: 24 November 2022
Published: 5 December 2022

1. Introduction

Worldwide, coastal erosion is a major issue in coastal zones, being estimated that 24% of the world's sandy beaches are eroding at rates exceeding 0.5 m/year [1]. In the last decades, artificial nourishments have become one of the main coastal intervention measures to mitigate the shoreline retreat and climate change effects around the world. This measure acts directly on the cause of erosion, trying to circumvent the deficit of sediments in coastal areas. However, artificial nourishments are not a permanent solution, since the removal of sediments by the wave and current actions, which occurs naturally from the deposition site, requires re-nourishment interventions over time to maintain its designed function. Periodic re-nourishment intervals range from 2 to 10 years and depend on several factors related to the initial design, namely wave climate, frequency and type of storms, and sand characteristics [2–4].

Artificial nourishments present positive and negative aspects, with different impacts, costs, and benefits; therefore, this evaluation is dependent upon technical and local variables [5–8]. Thus, to mitigate coastal erosion based on artificial nourishments in a sustainable long-term perspective, it is necessary to maximize the benefits and reduce the costs of the intervention. The nourishment design defines the volume, frequency of intervention, and placement site and depends on the sediment dynamics at the placement site, etc. An overview of beach nourishment practices around the world shows an increasing tendency to perform more recurrent nourishment interventions with increasing sediments volumes [9–11].

Cost–benefit analyses are tools that can be applied to compare costs and benefits, allowing one to take into consideration the economic feasibility of each strategy [12]. To obtain the balance between costs and benefits, it is necessary to quantify all the effects of the intervention during its life cycle and to convert them to the same unit, obtaining a unique unit that allows for the comparison of costs and benefits [13].

This work aims to present contributions to analyze and discuss the physical and economic feasibility of artificial nourishments interventions, considering their costs and benefits. For that, COAST software develop at University of Aveiro (version 1) [7] is applied, allowing an integrated, well-defined, and sequential cost–benefit approach in order to support decision-making for planning and coastal management. The software is applied for two study cases, one considering a hypothetical study area and the other corresponding to a real domain, the coastal stretch Barra-Vagueira, in Portugal. For both study areas, a 20-year time horizon was defined, and the physical and economic performance of artificial nourishment scenarios was evaluated. The obtained modelled results allow for the discussion of the cost–benefit of the nourishments with regard to the functions of different design parameters, such as volume, placement site, placement alongshore extension, and frequency.

2. Methodology

Cost–benefit analysis can be applied to compare nourishment costs and benefits, allowing one to take into consideration the economic feasibility of different scenarios [12]. In this study, COAST software, developed in the University of Aveiro by Lima, in 2018 [7], was applied to evaluate the economic performance of the artificial nourishments. This software applies a sequential and integrated methodology to compare the benefits and costs of coastal erosion mitigation measures and is based on three main stages [7,14,15]:

1st—Assessment of the physical performance of the coastal erosion mitigation measures, evaluating the shoreline evolution along time, allowing for the estimation of the benefits of the intervention;

2nd—Definition of the nourishment volume and consequent costs of the intervention, considering its life cycle;

3rd—Cost–benefit assessment of each intervention scenario, based on the analysis of the evolution of the economic parameters NPV (net present value), BCR (benefit–cost ratio) and break-even year (the year where the costs are compensated for by the benefits).

In the first stage is applied a long-term shoreline evolution numerical model (LTC—long-term configuration [5]) to obtain the physical impacts of the nourishment scenarios. The coastal areas gained and/or not lost due to the intervention are evaluated along time to estimate the benefits. In the second stage, the costs of the nourishments and re-nourishments are considered. Finally, in the third stage, costs and benefits obtained through time in the previous stages are summed (considering a discount rate to address the monetary values to the reference year) and compared, obtaining the net present value (NPV), benefit-cost ratio (BCR), and break-even year.

2.1. Benefits

The main benefits of coastal erosion mitigation measures are linked to their capacity to slow down the erosion trend and, consequently, to reduce the irreversible loss of territory, economic activities, infrastructures, and goods during the life cycle defined for the project [12,16]. In the COAST software, the benefits of coastal erosion mitigation measures are obtained as a result of the territory maintained, gained, or lost along time, plus the territory value. Following this approach, the software applies a one-line model (LTC) to obtain the areas of territory maintained or gained, by comparing the results with a non-intervention scenario. Within a specific study area, the software allows the user to define different monetary values for the territory alongshore.

LTC [5] is a shoreline evolution numerical model for sandy beaches that combines a simple classical one-line model with a rule-based model for erosion/accretion volumes distribution along the beach profile [17,18]. The model was designed to support coastal management and planning, allowing one to obtain shoreline projections in a medium- to long-term perspective (decades), considering different coastal erosion mitigation measures without a high computational effort and without a complex calibration and validation process [18–22].

To apply LTC, the user defines the bathymetry and topography of the study area, the wave climate offshore, the domain boundary conditions, water level, and sediment's properties [5]. To simulate artificial nourishments, the users define the nourishment volume, starting and ending instants of the intervention, and the deposition area [5,22].

A key parameter to obtain the benefits of coastal erosion mitigation measures is the value of the coastal territories that are maintained due to the interventions. Coastal territories provide multiple functions and services, as they merge frequent urban uses and natural ecosystems functions, making it possible to identify, within a range of few kilometers, a high variety of land uses. Those land use values should consider social, environmental, and cultural aspects and require sensitivity analysis prior to its application to adequately characterize the provided services of the territory [7].

2.2. Costs

The costs of artificial nourishments are related to the means employed to collect, transport, and deposit the sediments on the coast (dredges, trucks, pipelines, boats, etc.) and mainly, the distance between the source and deposition site (beach, dune, or littoral drift).

The sources of nourishment sediments may include extraction from quarries, reprocessing of quarry/mining waste, fossil beach deposits, port and harbor dredging activities, lagoon or barrier island coasts, offshore deposits, or derived from downdrift or updrift sediments and fluvial supplies [23]. According to Dean [24], the main advantage of using offshore sources is the ease of being able to find large stretches of adequate sand, usually 1 to 20 km away from the deposition site. However, offshore dredging must be carried out outside the active beach profile, otherwise, the sediments will just be moved from one place to the other, in the same dynamic coastal system. The material dredged from ports and navigation channels represents an economically interesting sediment source, since dredging is imperative in these places to guarantee navigation conditions. In this case, those sediments can be reused downdrift for beach nourishment, which will bring the advantages of relatively short distances and related costs for transportation [25].

In the present study, it was assumed a unitary sediments nourishment cost equal to 2 EUR/m^3, considering all the nourishment operations (sediments dredging, transport, and deposition). This value was defined on the basis of the unitary costs of sand transposition studies carried out for the Portuguese west coast [26,27].

2.3. Cost–Benefit Analysis

Benefit–cost ratio (BCR), net present value (NPV) and break-even year are three economic parameters that compare costs and benefits, allowing one to make considerations about the economic feasibility of a project. The break-even year corresponds to the year in which the benefits are compensated for by the costs.

The NPV evaluation criterion is given as the sum of discounted benefits minus the sum of discounted costs that occur in each period, t, over the lifetime of the project, T [13], and is given by:

$$NPV = \sum_{t=0}^{T} \frac{B_t}{(1+r)^t} - \sum_{t=0}^{T} \frac{C_t}{(1+r)^t} \tag{1}$$

where r is the time discount rate. The investment is considered economically viable when the NPV > 0, i.e., when the benefits' present value (first term on right-hand side of Equation (1)) exceeds the costs' present value (second term on right-hand side).

The BCR evaluation criterion is given by the sum of discounted benefits relative to the sum of discounted costs that occur in each period, t, over the lifetime of the project, T [13], and is given by:

$$BCR = \sum_{t=0}^{T} \frac{B_t}{(1+r)^t} \bigg/ \sum_{t=0}^{T} \frac{C_t}{(1+r)^t} \quad (2)$$

The investment is considered economically viable when the BCR > 1, i.e., when the benefits' present value (numerator on right-hand side of Equation (2)) exceeds the costs' present value (denominator on right-hand side). Note that the BCR = 1 when the NPV = 0.

3. Case Studies and Nourishment Scenarios

The cost–benefit of artificial nourishments interventions was discussed using two study areas. The first study focused on discussing the physical and economic performance of nourishment interventions considering a hypothetical study area to better control all the considered parameters. Afterwards, the cost–benefit methodology was applied to study nourishment interventions performed in a real coastal stretch located in the NW coast of Portugal (Barra-Vagueira). For each study area, first, the shoreline projection was obtained for a reference scenario, corresponding to the natural shoreline evolution, without new coastal interventions during the simulation period. Secondly, assuming a baseline scenario for nourishment intervention, alternative scenarios were tested through changes in one of the factors at a time (extension alongshore, placement site, frequency, and volume).

For each nourishment scenario, the benefits were obtained annually by comparison with the shoreline position, considering the reference scenario and the ones in which the artificial nourishments are performed. All the analyses were carried out for a time horizon of 20 years. This time frame was defined in line with the period usually defined by the Portuguese coastal management authorities for the life cycle of coastal interventions projects. Furthermore, the analyzed period is in line with the numerical model assumptions for medium- to long-term projections, and the period of 20 years avoids the effect of additional unpredicted long-term processes and/or anthropogenic and social actions.

3.1. Generic Study Area

The calculation domain of the generic study area in LTC was defined by regular topohydrography with an area of 8 × 10 km^2, Figure 1, represented by a regular square grid (20 m spaced), with 401 × 501 points (respectively, in the cross-shore and longshore directions). Dean profile, $h = Ax^m$ [28], was considered to define the bathymetry of the study area in which the sediment-dependent scale parameter (A) was defined to be equal to 0.127, according to Portuguese beaches sediment grain size, and the parameter related to the beach exposure to wave energy (m) was defined to be equal to 2/3 (value generally considered for intermediate beaches). The topography was defined with a constant slope equal to 2%. The cross-shore active profile was considered constant throughout the simulation period, with a height of 10 m defined by the depth of closure (DoC = 8 m) and the wave run-up limit (R_u = 2 m). At the northern boundary, it was defined to be a closed condition, where there are no sedimentary exchanges with the adjacent cell, and at the south boundary, the extrapolation of the longshore sediment transport of nearby conditions was adopted.

The wave characteristics were considered constant along all the numerical modelling simulations, with offshore wave height (H_0) of 2 m, wave period of 9.34 s (T), and 10 degrees wave direction with the west, clockwise (α_0). The alongshore sediment transport was defined to be calculated according to CERC formula [29].

The longshore extension of the generic study area was divided into three different zones. As shown in Figure 2, from north to south, the northern longshore extension of 1.0 km was defined with beach uses (beach provides coastal protection and recreational uses). Then, an urban area that can support several different activities and uses (restaurants, hotels, economic services, etc.) is defined, measuring 1.5 km in length. The remaining longshore extension was defined as a forest, providing this land use regulation for timber, habitat for biodiversity, erosion control, and many others [30–32]. The monetary value attributed to each land use is presented in Table 1. The time discount rate (r) was considered to be 3% [32].

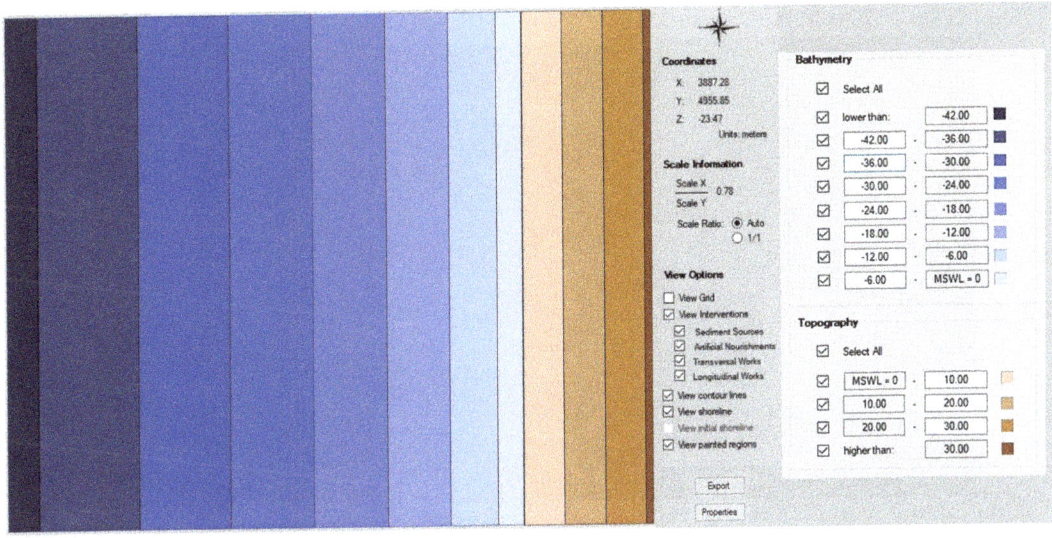

Figure 1. Schematization of the study area: land uses and artificial nourishment baseline scenario.

Table 1. Economic land value defined in the case study based on [32].

	Description (km)	Location	Extension (km)	Value (€/m^2/Year)
Zone 3	Beaches	North limit	1.0	2.00
Zone 2	Urban area	Intermediate	1.5	10.00
Zone 1	Forests	South limit	7.5	0.20

The artificial nourishment baseline scenario considers the nourishment of 1 million m^3 of sediment, every 5 years, at an average deposition rate of 10 thousand m^3 per day. The nourished area is characterized by a longshore extension of 500 m, centered in the urban area, and covering the entire cross shore active profile width, approximately 600 m (Figure 2). Table 2 summarizes the nine alternative nourishment scenarios evaluated.

3.2. Barra-Vagueira Coastal Stretch

Barra-Vagueira coastal stretch (Figure 3), NW coast of Portugal, with an extension of approximately 12 km, is an erosion hotspot, presenting, in the last 50 years, erosion rates that reached 8 m/year [34,35]. In the last decades, artificial nourishment has become one of the main coastal interventions to mitigate coastal erosion, being performed in the coastal stretch low frequency interventions of thousands cubic meters of sediments [36,37].

The numerical domain of this coastal stretch was defined on the basis of the shoreline position identified through the analysis of the digital elevation model provided by COSMO [38]. On the basis of the shoreline position, the bathymetry and topography of the stretch was represented by a regular grid of points spaced 20 m in both directions (west–east and north–south), with an extension of 6 × 12 km^2. On the basis of the COSMO surveys [38], the bathymetry was adjusted to Dean profile's shape, being considered regular and parallel. The topography was approximated to a constant slope equal to 3%. The numerical domain includes the 19 coastal structures existing on the stretch Barra-Vagueira (8 groins and 11 longitudinal revetments).

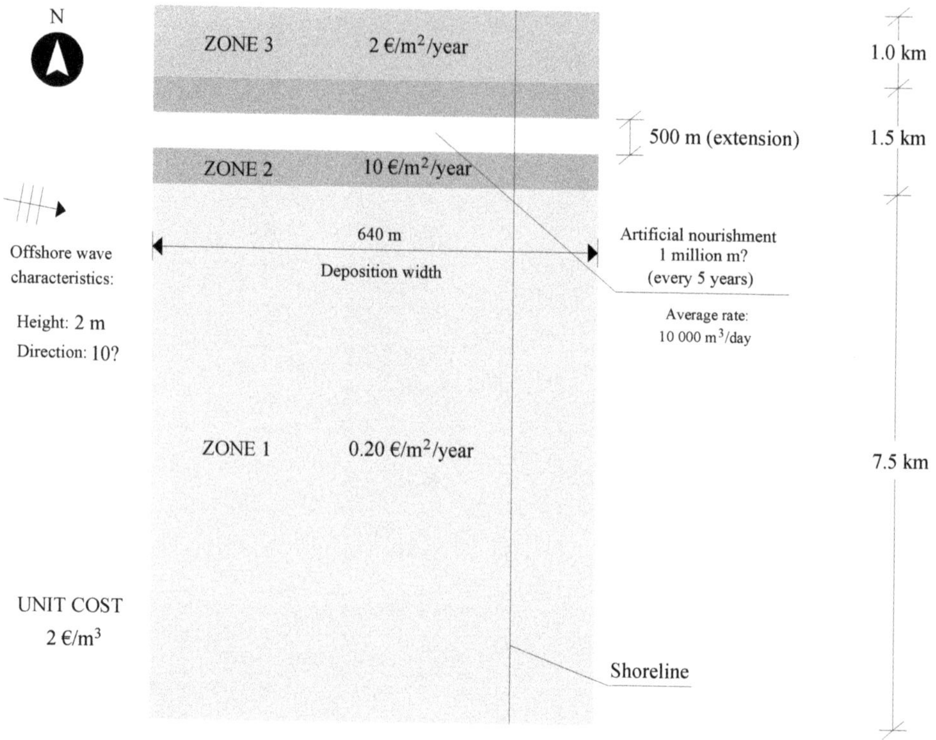

Figure 2. LTC numerical model representation of the study area obtained with COAST software interface, developed by Lima, in 2018 [33].

Table 2. Artificial nourishments scenarios characteristics.

Design Parameter		1	2	3
Extension	*i*	1000 m (north)	1000 m (south)	1500 m
Location	*ii*	500 m (north)	500 m (south)	-
Frequency	*iii*	400 thousand m^3 every 2 years	2 million m^3 every 10 years	-
Volume	*iv*	one-half million m^3	2 million m^3	-

The wave climate used as input of the model considered the wave series generated in the scope of the MarRisk research project [39], and the sediment transport was defined to be calculated according to the CERC formula [29]. The model was calibrated through the values of the shoreline change rate, comparing the mean shoreline retreat obtained numerically with values described in the bibliography.

The economic value of the land use was based on several studies [27,40,41]. In these studies, the authors divided the longshore extension of Barra-Vagueira into 12 sectors (Figure 4), obtaining in each sector the territory value through a methodology that combines land uses, based on the analysis of the Portuguese land cover map [42] and benefits transfer approaches.

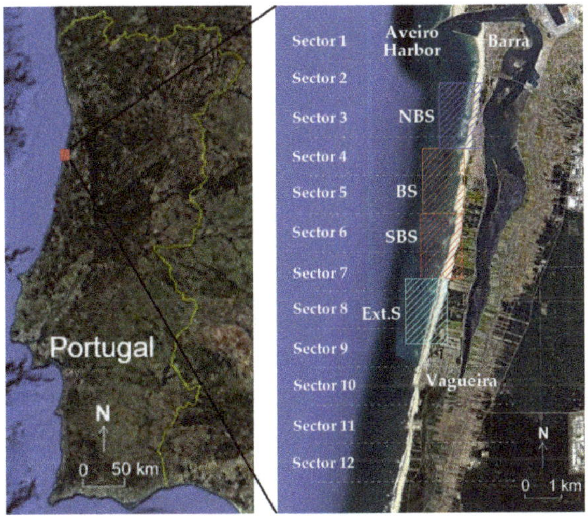

Figure 3. Coastal stretch Barra-Vagueira (the rectangles represent the nourishment location scenarios: BS—baseline; NBS—North of BS; SBS—South of BS; and Ext.S—Extreme South).

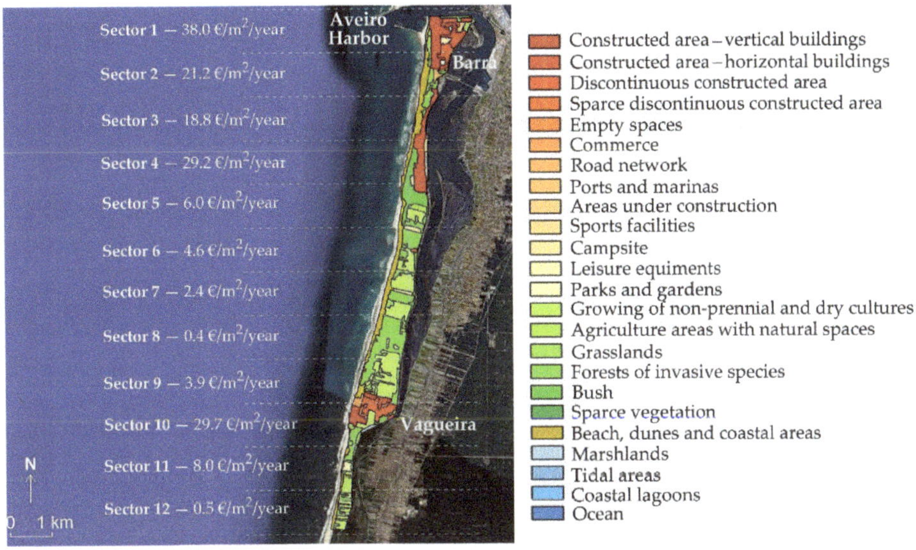

Figure 4. Monetary value of the coastal sectors [27,40,41] with overlay of the land use map on the study area [42].

A nourishment baseline scenario (BS) was adopted, considering the annual deposition of 5×10^5 m^3 of sediments, in front of a groin field, referred to as BS in Figure 3. This baseline scenario was defined on the basis of the historical data of artificial nourishments interventions that have been performed in the study area [36,37]. Nine alternative scenarios have been defined through changes in one of the factors at a time, considering three alternatives for each of the analyzed parameters, namely volume, frequency, and placement site alongshore (Table 3).

Table 3. Alternative nourishment scenarios for the coastal stretch Barra-Vagueira (nourishment placement sites are shown in Figure 3).

Design Parameter	1	2	3
Volume (m^3/year)	1×10^5	3×10^5	7×10^5
Placement site	NBS	SBS	Ext.S
Frequency	2.5×10^5 m^3 each half year	1×10^6 m^3 each 2 years	2×10^6 m^3 each 4 years

4. Results

The main results of all the analyzed scenarios are presented and discussed here. For both of the study areas, the results for the reference scenario are presented firstly, following by the physical and economic performance of nourishment interventions scenarios tested. Final remarks highlight the major outlines of the obtained results.

4.1. Generic Study Area

4.1.1. Reference Scenario

The evolution of the shoreline position after 5, 10, and 20 years is shown in Figure 5, where substantial erosion problems are observed. Without interventions to mitigate erosion effects, the results suggest a shoreline retreat close to 230 m in the northern boundary of the calculation domain, after 20 years of simulation. In addition, the extension of the stretch defined as urban uses is affected by erosion. In economic terms, through the balance between the areas of territory lost over time (LTC results in the reference scenario) and the valuation of the territory (land use and ecosystems services), it is expected that the shoreline retreat represents a loss of approximately 0.8 million euros, after five years of simulation. After ten years, the costs exceed 3 million euros and at the end of the defined time horizon (20 years), the land loss represents approximately 12 million euros (all values updated to Year 0).

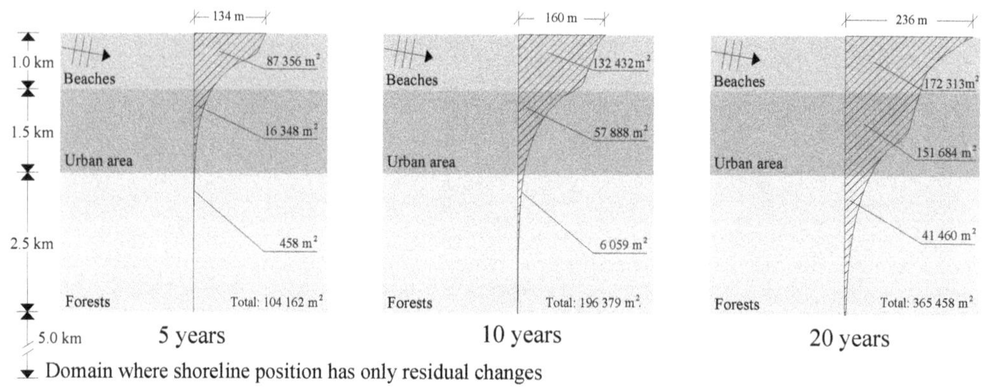

Figure 5. Shoreline position in the reference scenario, along the time (cross-shore scale equal to 10 times the longitudinal scale).

4.1.2. Artificial Nourishment Scenarios

Table 4 summarizes the physical and economic results obtained for the artificial nourishment scenarios. The results show that performing a nourishment project (according to the parameters defined as the baseline scenario) allows for a decrease in the erosion areas in approximately 23 ha over the 20 years of simulation (in comparison with the reference scenario). The total addition of 4 million m^3 of sediments to the coastal system over the 20 years of simulation (four nourishment interventions) represents a total investment of approximately 6.5 million euros (at present values, considering the discount rate). The break-even is achieved after 13 years, and after 20 years, it is verified that the benefits are approximately 70% higher than the total investment costs (BCR = 1.73). Despite the positive physical and economic aspects related to the intervention, based on Figure 6, it is possible to observe that the nourishment baseline scenario does not ensure the total protection of the urban water front.

Table 4. Summary of the physical and economic results of the artificial nourishments scenarios.

	Scenario	Territory Area (ha)			$BCR_{20\ yr}$	$NPV_{20\ yr}$	Costs		Break-Even
		Accretion	Erosion	Impact	(–)	(€)	Initial (€)	Total * (€)	(Years)
BS	Figure 2	2.6	16.6	22.6	1.73	4,733,343	2,000,000	6,497,129	13
i.1	1000 m (north)	2.1	15.0	23.7	1.82	5,356,332			12
i.2	1000 m (south)	3.5	17.5	22.5	1.68	4,441,918	2,000,000	6,497,129	13
i.3	1500 m	2.9	16.0	23.4	1.76	4,917,737			13
ii.1	500 m (north)	1.3	13.4	24.5	1.80	5,167,493	2,000,000	6,497,129	13
ii.2	500 m (south)	3.6	19.0	21.2	1.51	3,311,101			14
iii.1	400 thousand m^3 every 2 years	2.9	15.5	24.0	1.75	4,643,467	800,000	6,220,104	12
iii.2	2 million m^3 every 10 years	3.3	16.9	23.0	1.98	6,856,026	4,000,000	6,976,376	8
iv.1	one-half million m^3	0.4	23.8	13.2	2.11	3,606,909	1,000,000	3,248,565	9
iv.2	2 million m^3	11.9	8.9	39.6	1.68	8,842,361	4,000,000	12,994,259	12

* Values updated to the initial simulation instant, according to the discount rate (*r*).

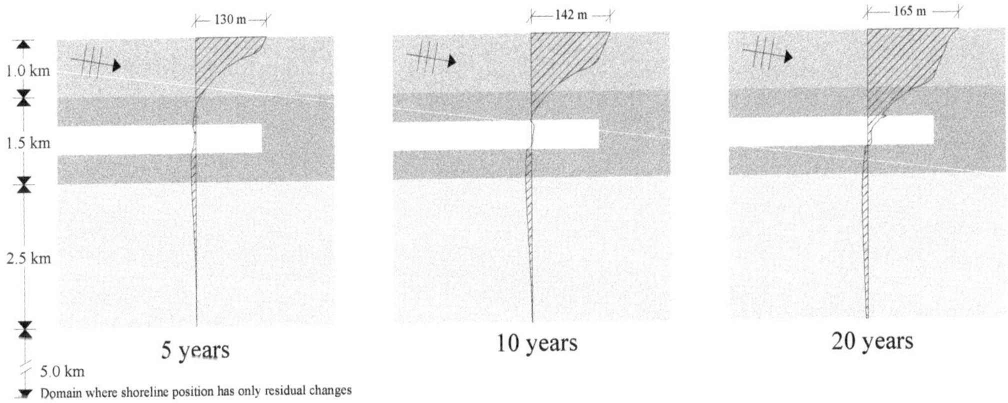

Figure 6. Shoreline position in the artificial nourishment baseline scenario along time (cross-shore scale equal to 10 times the longitudinal scale).

In the following sections, changes in artificial nourishment baseline scenario characteristics are analyzed.

Extension

To assess the impact of the alongshore nourishment extension, three alternatives to the baseline scenario were considered, corresponding to extended nourished areas: (*i.1*) extending the nourishment area in 500 m in the north direction; (*i.2*) extending the nourishment area in 500 m in the south direction; and (*i.3*) 1500 m extension covering all the urban waterfront.

Based on the results presented in Table 4 and Figure 7, it is slightly positive to extend the nourishment area 500 m to the north. The three scenarios present similar results, but Scenario *i.1* presents simultaneously better physical and economic performances, reaching the break-even year one year earlier than the remaining scenarios (12 years instead of 13 years).

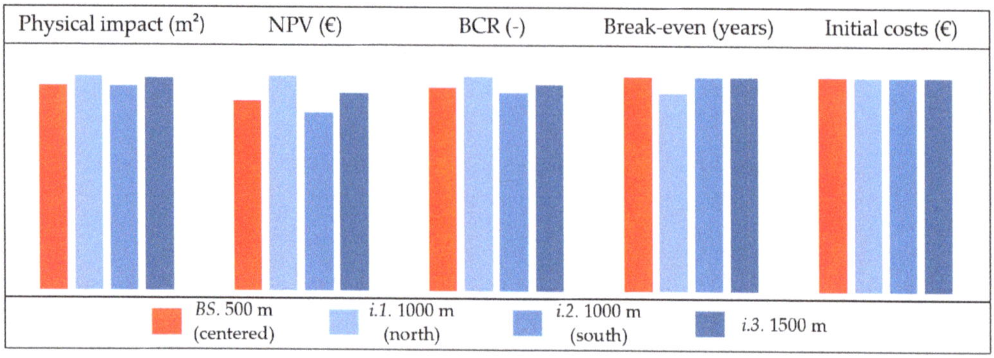

Figure 7. Normalized representation of physical and economic impacts of the artificial nourishment extension scenarios.

Location

In an attempt to optimize the nourishment impacts in the shoreline evolution, two alternative locations to deposit the sediments were tested: (*ii.1*) move the nourishment area in 500 m to the north and (*ii.2*) move the nourishment area in 500 m to the south.

As can be observed from Table 4 and Figure 8, the results suggest that the northern nourishment location results in a smaller loss of territory and, simultaneously, it represents the most positive economic result (BCR = 1.80, with the break-even reached after 13 years, as in the baseline scenario).

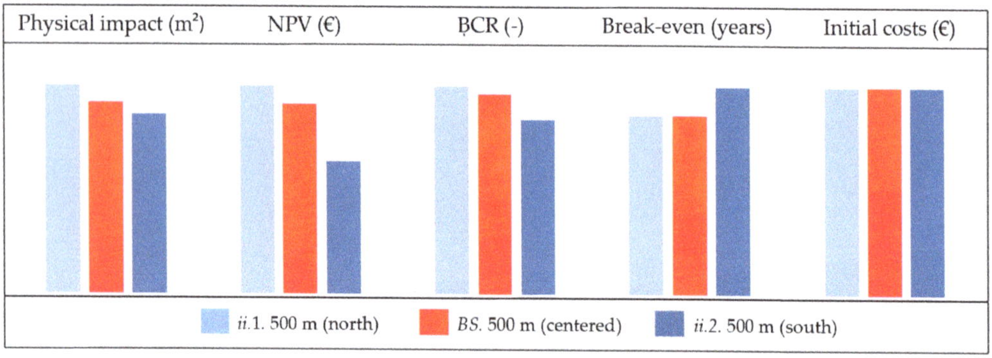

Figure 8. Normalized representation of physical and economic impacts of the artificial nourishment location scenarios.

Frequency

The costs and benefits of nourishments frequency were assessed by comparison of two alternative scenarios: (*iii.1*) 400 thousand m^3 of sediments deposited every 2 years; and (*iii.2*) 2 million m^3 of sediments every 10 years. Despite the different volumes considered in each intervention, the cost per m^3 was considered always the same.

The scenario with the best physical impact is not the same as the one with the best economic performance (Table 4 and Figure 9). Artificial nourishments performed in shorter time intervals (Scenario *iii.1*) presents smaller land losses over the 20 years, but the economic benefit is higher if only two interventions of 2×10^6 m^3 of sediments each are performed (Scenario *iii.2*). Although in Scenario *iii.2* the break-even point is lower and the NPV is higher (after 20 years), this scenario requires a higher initial investment (4 million euros, while in Scenario *iii.1*, only 800 thousand euros are needed), which may be a constraint, as the initial financial availability is often a constraining factor in the choice of the coastal intervention.

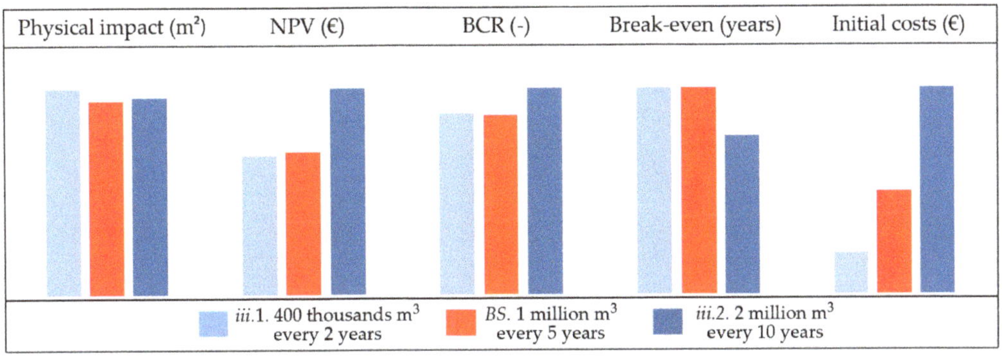

Figure 9. Normalized representation of physical and economic impacts of the artificial nourishment frequency scenarios.

Volume

Regarding the total volume of sediments deposited, two additional artificial nourishment scenarios were analyzed: (*iv.1*) 500 thousand m^3 of sediments every 5 years and (*iv.2*) 2 million m^3, again, every five years. Naturally, in physical terms, Scenario *iv.2* is more attractive, decreasing the territory loss by approximately 40 ha, in comparison with the reference scenario. After 20 years, this scenario leads to a global accretion when compared with the initial instant (Year 0). Despite the physical benefits of higher nourishment volumes, the BCR shows that Scenario *iv.1* presents a better economic performance (Table 4 and Figure 10). Thus, it is highlighted that although Scenario *iv.2* represents a higher initial and total investment cost, it also presents important shoreline evolution benefits.

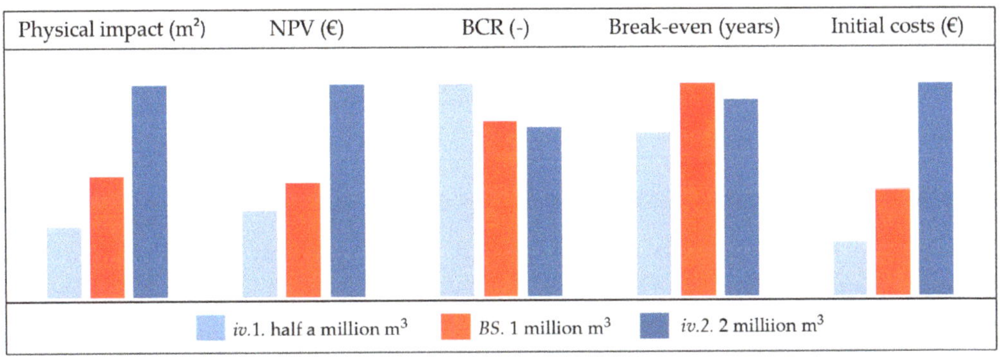

Figure 10. Normalized representation of physical and economic impacts of the artificial nourishment volume scenarios.

4.2. Generic Study Area

4.2.1. Reference Scenario

Shoreline projection results show that if no actions are taken to mitigate coastal erosion, the Barra-Vagueira shoreline will continue to retreat (Figure 11). Considering a time horizon of 20 years, the annual mean shoreline retreat approaches 1.45 m/year, representing a loss of territory of approximately 35 ha. Considering the land uses, this erosion effects represent a loss of approximately 37 million € (Table 5).

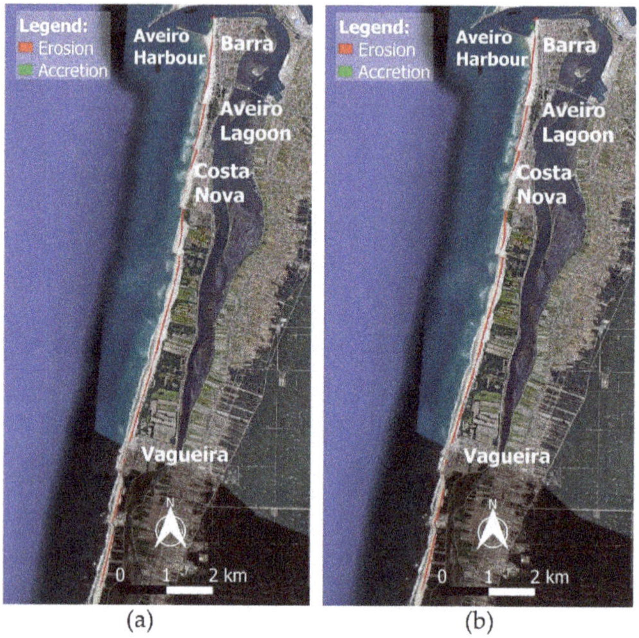

Figure 11. Barra-Vagueira shoreline evolution if no new coastal interventions are adopted: (**a**) 10 years; (**b**) 20 years.

Table 5. Physical and economic losses at Barra-Vagueira coastal stretch, after 10 and 20 years of simulation.

	Area Lost (ha)	Mean Shoreline Retreat (m/Year)	Economic Loss * (Million €)
10 years	27.05	2.25	16.67
20 years	34.71	1.45	36.96

* Values updated to the initial simulation instant, according to the discount rate (*r*).

4.2.2. Artificial Nourishments

The nourishments allow for the mitigation of the negative erosion effects expected in the reference scenario, decreasing the shoreline retreat rate. However, the economic viability of the nourishments was achieved for only one of the tested scenarios. Table 6 summarizes the physical impact by presenting the percentage of territory not lost at the end of 20 years, the economic performance through the benefit–cost ratio (BCR), and net present value (NPV).

Regarding the nourishment baseline scenario, it is observed that 5×10^5 m^3/year of sediment decreases the loss of territory by approximately 15%, but the economic viability of the intervention is not achieved during the 20 years evaluated. The total estimated costs are 15.32 million euros, and the benefits approach 3.79 million euro (Figure 12).

Results of nourishment volume scenarios suggest that both physical and economic performance increases with the deposited volume (Figure 13). The annual deposition of 1×10^5 of sediment presents the lower total costs (3.06 million euros), but the area not lost is also lower, leading to a benefit–cost ratio of approximately 0.03. The deposition of 7×10^5 m^3/year of sediment is the scenario that presents the higher total cost (21.45 million euros) but is also the scenario that presents the higher total benefits (7.72 million euros), with a benefit–cost ratio of 0.36.

Table 6. Physical and economic performance at the end of 20 years: percentage of coastal area not lost; benefit–cost ratio and net present value.

Scenario		Physical Impact (%)	$BCR_{20\,yr}$ (–)	$NPV_{20\,yr}$ (€)
Volume (m^3/year)	1×10^5	0.20	0.03	−3,145,364
	3×10^5	8.99	0.23	−7,096,792
	5×10^5	15.66	0.25	−11,530,765
	7×10^5	29.00	0.36	−13,731,977
Placement site	NBS	23.73	1.10	1,583,142
	BS	15.66	0.25	−11,530,765
	SBS	20.22	0.15	−12,987,722
	Ext.S	26.75	0.07	−14,185,279
Frequency	2.5×10^5 every half year	10.59	0.17	−12,773,599
	5×10^5 every year	15.66	0.25	−11,530,765
	1×10^6 every 2 years	17.49	0.32	−10,594,789
	2×10^6 every 4 years	20.07	0.40	−9,557,717

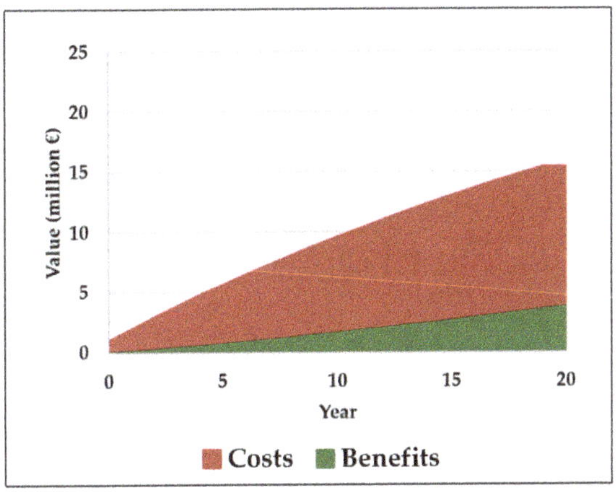

Figure 12. Accumulated costs and benefits evolution for the baseline scenario.

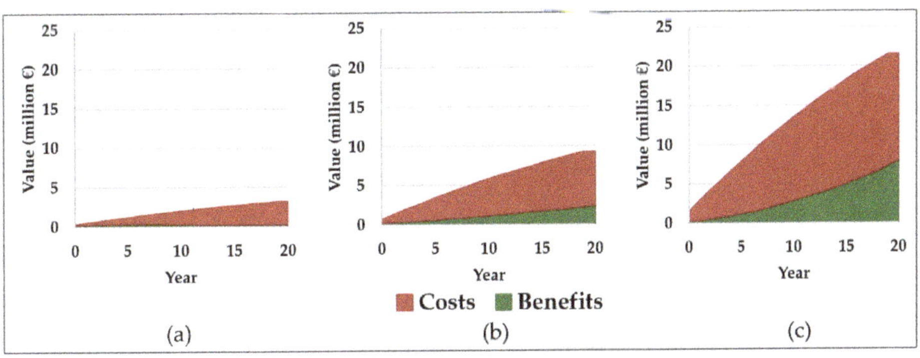

Figure 13. Accumulated costs and benefits evolution for the nourishment volume scenarios: (**a**) 1×10^5 m^3/year; (**b**) 3×10^5 m^3/year; and (**c**) 7×10^5 m^3/year.

Considering the results that assessed the impact of the deposition site (Figure 14), it is observed that the northern area deposition (NBS) is the one with the best economic performance, with the costs outweighing by the benefits (BCR = 1.1, with the break-even reached after 17 years). Note that, despite different distances to the deposition site, the unitary cost of dredging/deposition intervention was considered the same in all the scenarios.

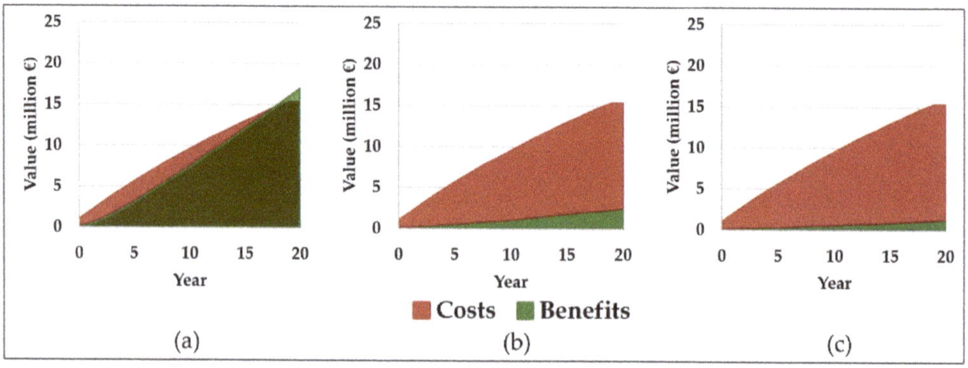

Figure 14. Accumulative costs and benefits evolution, for the placement site scenarios: (**a**) NBS; (**b**) SBS; and (**c**) Ext.S.

The best economic performance is not necessarily connected with the intervention that produces a greater impact on the reduction of the area lost on the coastal stretch. Although the deposition of sediments in the scenario Ext.S is the intervention that led to a greater reduction on the area not lost (26.75%), this scenario is the one which presents the worst economic performance. These results are due to the accumulative evolution of the benefits based on the relationship between the location of the area not lost and its land value. The total costs for these scenarios are the same (15.32 million euros). Sediment deposition in scenario NBS leads to a better performance, where the higher value of territory was estimated (Sectors 2–4), which significantly increases the benefits, outweighing the total costs. For the same reason, when comparing the deposition in SBS and in BS, it is observed that scenario SBS produces a higher reduction in the area lost, but the best economic performance is obtained for the deposition in BS. Finally, the results indicate that it is more efficient to carry out fewer artificial nourishment interventions, with larger volumes (Figure 15). According to the results that evaluated the nourishment frequency, the deposition of 2×10^6 every 4 years is the one that presents the best physical and economic performance (BCR = 0.4).

On the basis of the comparison of costs and benefits of the interventions (Figure 15), it is observed that the deposition of 2×10^6 each 4 years represents an approximately 5 thousand euros higher cost than the remaining frequency scenarios that were tested. However, the benefits are approximately 2.5 times higher than those obtained for the scenarios that considered the deposition of lower volumes (2.5×10^5 every half year and 5×10^5/year).

4.3. Final Remarks

The presented study discusses physical and economic performance of artificial nourishments based on the results at two study areas, showing that the proposed approach is applicable to both conceptual research and applied projects. Comparing both studied areas, the higher the volume, the better the BCR for Barra-Vagueira, but the contrary is observed in the hypothetical study. Location of the deposition of the nourishment at northern areas (updrift) is better in both studies, considering the adopted assumptions. The lower the frequency of the interventions, the better the RBC. In fact, these results show that the main goal of the nourishment needs to be very well defined and is site-specific, because a better scenario in one aspect may be worse in another one. However, the main results show that erosion effects represent important economic losses and artificial nourishments allow for the mitigation of shoreline retreat rates, decreasing the land losses. The physical performance and consequent economic evaluation of the nourishments depends on the design parameters and the valuation of the land uses and ecosystem services.

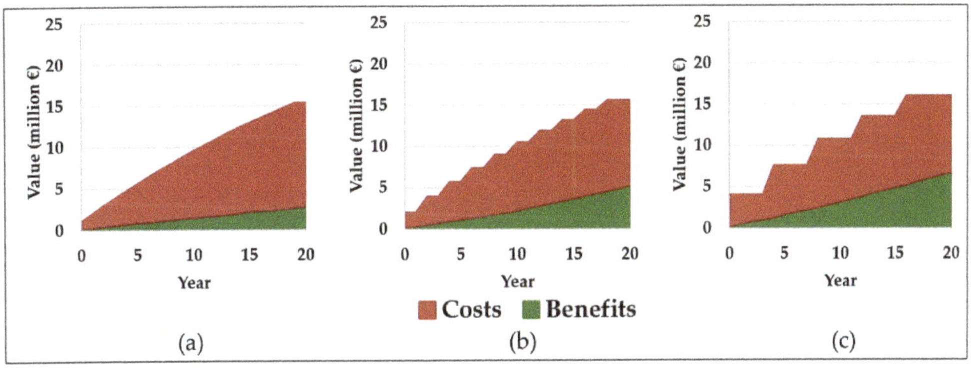

Figure 15. Accumulative costs and benefits evolution for the scenarios that assessed the nourishment frequency: (**a**) 2.5×10^5 m^3 every half year; (**b**) 1×10^6 m^3 every 2 years; and (**c**) 2×10^6 m^3 every 4 years.

In the generic study area, moving the deposition site and extending its area to the north represents physical and economic advantages. Frequent nourishment interventions with lower volumes will induce larger accretion areas over time, but higher economic gains correspond to a larger volume and lower nourishment frequencies. A higher total volume of nourished sediment in the coastal system provides larger accretion areas over the entire 20-year time period that was analyzed and simultaneously results in a higher net present value. At the Barra-Vagueira coastal stretch, the economic viability of the intervention was achieved for only one of the tested scenarios. Thus, the economic feasibility of the interventions is not necessarily connected with the greater physical impact of the nourishment, but it is also strongly dependent on the monetary value attributed to the territory.

5. Discussion

To mitigate the shoreline retreat rate's negative effects, the responsible authorities try to identify which is the most effective measure, with lower costs. For that, it is necessary to develop studies regrading coastal erosion mitigation that compromises two dimensions, namely physical effectiveness and economic performance. In the last decades, economical tools, such cost-effectiveness, cost–benefit, and efficiency analyses, have been applied in the scope of coastal management and planning [43]. These tools provide a more comprehensive and sustainable coastal management. The main outcome is the possibility of comparing solutions considering physical and economical effectiveness, allowing a better management of resources, and the possibility of supporting the planning process. Cost-effectiveness studies provide insight into which interventions achieve coastal protection objectives at the lowest cost [44,45]. Cost–benefit studies provide insight into which interventions provide the largest net benefits [12,46–54]. In short, it can be considered that nourishment design should rely on cost–benefit analyses [55]. Thus, the main relevance of the proposed methodology to design the best nourishment intervention is to follow a well-defined and sequential approach, considering numerical modelling and costs and benefits assessment.

All the stages of the proposed methodology may result in discussion due to the inherent uncertainties of each process and required adopted assumptions.

In this study, a simple one-line shoreline evolution model was applied to characterize the territory areas lost, maintained, or gained. According to [56,57], simplified models allow for the predication of coastal zone evolution in medium- to long-term scales, avoiding the time demands due to both field data collection and computational time, as well as the time and scale limitations of the three-dimensional models. However, as concluded by Coelho et al. [58], there are still gaps in the long-term numerical modelling of artificial nourishments, and it is important to increase modelling capabilities, by coupling cross- and longshore processes, aiding the selection of optimal artificial nourishment schemes and the establishment of more efficient coastal management policies. Larson et al. [59] have highlighted the need for long-term morphological modelling that considered short-term effects, since long-term changes may be influenced by short-term responses. In addition, the long-term impacts of nourishment interventions in the shoreline evolution needs further research. Ferreira and Coelho [22] report that the changes on the local bathymetry due to the nourishments have consequences in the sediment transport patterns and, therefore, in the shoreline evolution.

The benefits of artificial nourishments are considered as the capacity to slow down the shoreline retreat, allowing the territory to maintain its functions and its monetary value over time. However, further research must also be developed to improve the land use and ecosystems valuation, as this monetary value must represent all the economic, social, and cultural aspects related to the activities developed in the coastal zones and surrounding communities. According to Lima et al. [60], nourishment interventions may represent ecosystem preservation and maintenance of natural landscapes, may increase the beach demand that leads to economic gains, may promote new economic and tourist activities, allow for decreased costs related to eventual overtopping events, and also reduce the maintenance costs of existing coastal structures. In fact, Noordegraaf [61] developed a study case in a beach under erosion and reports the intention of people to make more visits when the beach increases; additionally, on the basis of numerical studies, Passeri et al. [62] suggest that during stronger storms, nourishments reduce dune overtopping.

Additionally, since coastal territories provide multiple functions, the different stakeholders have different interests. The proposed methodology allows a comparison of options, enabling one to assess costs and benefits evolution, considering direct costs and benefits for different groups. Furthermore, since the methodology considers the timing of each of the costs and benefits associated with particular options and converts future costs and benefits into today's prices, all impacts can be meaningfully compared, regardless of timing.

Finally, as previously indicated, the territory value is extremely dependent on the land uses, and, thus, an adequate sensitivity analysis should be performed to well characterize the land use value. An important part of the studies regarding land use and economic value of the territories is focused on estimating the demand curve, thus trying to identify changes in the consumers behavior due to a specific intervention. According to Parson et al. [63], artificial nourishment effects in the beach width have consequences on the consumption of the users. On the basis of a study conducted at Delaware beaches (USA) to estimate the value of beach width and artificial nourishments for recreational purposes by its users, the authors concluded that reducing of the beach width to a quarter of the current width would lead to a loss of approximately USD 5.00 per day at the beach, per visitor. Widening to twice the current beach width would lead to an increase per day and per visitor at the beach of USD 2.75. Landry et al. [64] performed a study in North Carolina to estimate the demand curve and the 'willingness to pay' by users and non-users, for coastal management measures. On the basis of the results, it was estimated that the marginal value of incremental beach width for users and non-users ranges from USD 0.23 to USD 0.48 per meter. Thus, the proposed approach makes it easier to assess the benefit evolution of artificial nourishments as a function of the territory value that, as demonstrated by the two study cases discussed previously, can be extremely dependent on the user's evaluation and may be dependent on local aspects.

6. Conclusions

This work aimed to present a well-defined and sequential approach to evaluate the costs and benefits of different artificial nourishment scenarios. A hypothetical study area and a real domain (Barra-Vagueira) were considered, and the nourishment scenarios were defined to discuss the impact of the design parameters that conditioned the costs and benefits of the interventions.

Cost–benefit analysis gives important insights for better design of nourishment interventions. In this evaluation, benefits are related to the shoreline evolution and are dependent on the numerical assumptions defined to develop the study. They are also functions of the valuation of the coastal territory that must include social, environmental, and cultural aspects, which are difficult to estimate. The costs estimation requires the prediction of the evolution of all costs of the project during its life cycle, which should consider the dredging operations, the transport (highly dependent of the oil prices), and the deposition processes along the time.

Overall, the findings show that in coastal areas susceptible to erosion (where the sediments volume available for the littoral drift are below the potential sediment transport capacity), important economic losses will occur if no decisions are adopted. Artificial nourishments allow for the mitigation of erosion problems, but the physical and economic performance of the intervention is site-specific and dependent upon design parameters. The results highlight the importance of a good definition of the objectives to be reached when an artificial nourishment is proposed, as the scenarios that present higher positive physical impacts are not necessarily the same as those that present the best economic performance.

The application of the presented results to other coastal areas is naturally limited by the specific conditions of each site (land use values, wave climate conditions, existing coastal intervention characteristics, nourishment scenarios, etc.) However, the easy approach defined by the methodology allows for a quick sensitivity analysis to those conditions, permitting its general worldwide application. Thus, it is considered that the proposed methodology and the COAST software represents one step toward a well-supported decision-making process, aiding in coastal management and planning.

Author Contributions: Conceptualization, C.C., M.L. and M.F.; methodology, C.C. and M.L.; software, C.C. and M.L.; validation, M.L. and M.F.; formal analysis, M.L. and M.F.; investigation, M.L. and M.F.; resources, C.C.; data curation, C.C., M.L. and M.F.; writing—original draft preparation, M.L. and M.F.; writing—review and editing, C.C., M.L. and M.F.; visualization, M.L. and M.F.; supervision, C.C.; project administration, C.C.; funding acquisition, C.C. All authors have read and agreed to the published version of the manuscript.

Funding: This research received no external funding.

Institutional Review Board Statement: Not applicable.

Informed Consent Statement: Not applicable.

Data Availability Statement: Not applicable.

Acknowledgments: This work was financially supported by the project "Integrated Coastal Climate Change Adaptation for Resilient Communities", INCCA—POCI-01-0145-FEDER-030842, funded by FEDER, through "Programa Operacional Competitividade e Internacionalização" in its ERDF component and by national funds (OE), through FCT/MCTES. Ferreira, M. is supported by the doctoral Grant 2021.07269.BD financed by the Portuguese Foundation for Science and Technology (FCT).

Conflicts of Interest: The authors declare no conflict of interest.

References

1. Luijendijk, A.; Hagenaars, G.; Ranasinghe, R.; Baart, F.; Donchyts, G.; Aarninkhof, S. The state of the world's beaches. *Sci. Rep.* **2018**, *6641*, 11. [CrossRef] [PubMed]
2. NRC. *National Research Council: Beach Nourishment and Protection*; The National Academic Press: Washington, DC, USA, 1995; p. 334.
3. ASBPA. *Shore Protection Assessment: How Beach Nourishment Projects Work*; Shore Protection Assessment; US Army Corps of Engineers: Vicksburg, MA, USA, 2007; p. 9.
4. De Schipper, M.; de Vries, S.; Ruessink, G.; de Zeeuw, R.; Rutten, J.; van Gelder-Maas, C.; Stive, M. Initial Spreading of a Mega Feeder Nourishment: Observations of the Sand Engine Pilot Project. *Coast. Eng.* **2016**, *111*, 23–38. [CrossRef]
5. Coelho, C. Riscos de Exposição de Frentes Urbanas para Diferentes Intervenções de Defesa Costeira. PhD Thesis, University of Aveiro, Aveiro, Portugal, 2005; p. 404. (In Portuguese)
6. ANCORIM. *Overview of Soft Coastal Protection Solutions*; BRGM/ONF: Bordeaux, France, 2017; p. 54.
7. Lima, M. Ferramenta Numérica de Análise do Impacto de Intervenções de Defesa Costeira na Evolução da Linha de Costa: Custos e Benefícios. Ph.D. Thesis, University of Aveiro, Aveiro, Portugal, 2018; p. 336. (In Portuguese)
8. Coelho, C.; Narra, P.; Marinho, B.; Lima, M. Coastal management software to support the decision-makers to mitigate coastal erosion. *J. Mar. Sci. Eng.* **2018**, *8*, 37. [CrossRef]
9. Adapt. European Climate Adaptation Platform Climate. European Commision and European Environment Agency. Available online: https://climate-adapt.eea.europa.eu (accessed on 10 January 2022).
10. Hamm, L.; Capobianco, M.; Dette, H.; Lechuga, A.; Spanhoff, R.; Stive, M. A summary of European experience with shore nourishment. *Coast. Eng.* **2002**, *47*, 237–264. [CrossRef]
11. Zandmotor Monitoring. Rijkswaterstaat Ministerie van Infrastructuur en Waterstaat. Available online: https://dezandmotor.nl/ (accessed on 23 January 2022).
12. Roebeling, P.; Coelho, C.; Reis, E. Coastal erosion and coastal defense interventions: A cost-benefit analysis. *J. Coast. Res.* **2011**, *Special Issue 64*, 1415–1419.
13. Zerbe, R.; Dively, D. *Benefit-Cost Analysis in Theory and Practice*; Harper Collins College Publishers: New York, NY, USA, 1994; p. 557.
14. Lima, M.; Coelho, C.; Veloso-Gomes, F.; Roebeling, P. An integrated Physical and Cost Benefit Approach to Assess Groins as a Coastal Erosion Mitigation Strategy. *Coast. Eng.* **2020**, *156*, 103614. [CrossRef]
15. Lima, M.; Coelho, C.; Cachim, P. Programing methods for pre-design of coastal structures. In Proceedings of the V International Conference on Computational Methods in Marine Engineering, Hamburg, Germany, 29–31 May 2013; pp. 868–879.
16. Penning-Rowsell, E.; Green, C.; Thompson, P.; Coker, A.; Tunstall, S.; Richards, C.; Parker, D. *The Ecomomics of Coastal Management: A Manual of Benefit Assessment Techniques*, 1st ed.; Belhaven Press: London, UK; Boca Raton, FL, USA, 1992; ISBN 1-85293-161-2.
17. Coelho, C.; Taveira Pinto, F.; Veloso Gomes, F.; Pais-Barbosa, J. Coastal Evolution and Coastal Works in the Southern Part of Aveiro Lagoon Inlet, Portugal. In Proceedings of the 29th International Conference on Coastal Enginnering, Lisbon, Portugal, 19–24 September 2004.
18. Coelho, C.; Veloso-Gomes, F.; Silva, R. Shoreline coastal evolution model: Two Portuguese case. *Coast. Eng.* **2007**, *Volume 5*, 3430–3441.
19. Coelho, C.; Lima, M.; Veloso-Gomes, F. Relationship between cross-shore active profile and one-line shoreline evolution models performance. *J. Coast. Res.* **2013**, *2*, 2107–2112. [CrossRef]
20. Guimarães, A.; Lima, M.; Coelho, C.; Silva, R.; Veloso-Gomes, F. Groin impacts on updrift morphology: Physical and numerical study. *Coast. Eng.* **2016**, *109*, 63–75. [CrossRef]
21. Lima, M.; Coelho, C. LTC shoreline evolution model: Assumptions, evolution, validation and application. *J. Integr. Coast. Zone Manag.* **2017**, *17*, 5–17.
22. Ferreira, M.; Coelho, C. Artificial nourishments effects on longshore sediments transport. *J. Mar. Sci. Eng.* **2021**, *9*, 240. [CrossRef]
23. French, P. *Coastal Defences: Processes, Problems and Solutions*; Routledge: London, UK, 2001; 306p, ISBN 0-415-19845-3.
24. Dean, R. *Beach Nourishment Theory and Practise*; Advanced Series on Ocean Engineering 18; World Scientific: Singapore, Singapore, 2002; p. 403.
25. Haney, R.; Kouloheras, L.; Malkoski, V.; Mahala, J.; Unger, Y. *Beach Nourishment MassDEP's Guide to Best Management Practices for Projects in Massachusetts*; Massachusetts Department of Environmental Protection and Massachusetts Office of Coastal Zone Management: Boston, MA, USA, 2007; p. 32.
26. Carvalho, R.; Afonso, C.; Figueira, P.; Pound, M.; Coelho, C.; Silva, P.; Afonso, A. *Estudo de Viabilidade de Transposição Aluvionar das Barras de Aveiro e da Figueira da Foz–Relatório Intercalar 3, Tarefa 6*; University of Aveiro: Aveiro, Portugal, 2021; p. 31. (In Portuguese)
27. Coelho, C.; Silva Afonso, A.; Bernardes, C.; Silva, P.A.; Baptista, P.; Roebeling, P.; Férnadez Fernández, S.; Abreu, T.; Oliveira, T.; Ferreira, M.; et al. *Estudo de Viabilidade de Transposição Aluvionar das Barras de Aveiro e da Figueira da Foz–Sumário Executivo*; University of Aveiro: Aveiro, Portugal, 2021; p. 26. (In Portuguese)
28. Dean, R.G. Equilibrium beach profiles: Characteristics and applications. *J. Coast. Res.* **1991**, *7*, 53–84.
29. CERC. *Shore Protection Manual*; U.S. Army Corps of Engineers. Coastal Engineering and Research Center. U.S. Government Printing Office: Washington DC, USA, 1984; Volume 1.

30. Costanza, R.; d'Arge, R.; De Groot, R.; Farber, S.; Grasso, M.; Hannon, B.; Limburg, K.; Naeem, S.; Oneill, R.V.; Paruelo, J.; et al. The value of the world's ecosystem services and natural capital. *Nature* **1997**, *387*, 253–260. [CrossRef]
31. Costanza, R.; De Groot, R.; Sutton, P.; Van Der Ploeg, S.; Anderson, S.J.; Kubiszewski, I.; Farber, S.; Turner, R.K. Changes in the global value of ecosystem services. *Glob. Environ.* **2014**, *26*, 152–158. [CrossRef]
32. Roebeling, P.; d'Elia, E.; Coelho, C.; Alves, T. Efficiency in the design of coastal erosion adaptation strategies: An environmental-economic modelling approach. *Ocean. Coast. Manag.* **2018**, *160*, 175–184. [CrossRef]
33. Lima, M.; Coelho, C. Improvement of an existing shoreline evolution numerical model. In Proceedings of the VIII International Conference on Computational Methods in Marine Engineering, Goteborg, Sweden, 13–15 May 2019; pp. 502–513.
34. Santos, F.; Lopes, A.; Moniz, G.; Ramos, L.; Taborda, R. Gestão da zona costeira: O desafio da mudança. *Relatório Grupo Trab. Litoral* **2014**, *1*, 237. (In Portuguese)
35. Lira, C.; Silva, A.; Taborda, R.; de Andrade, C. Coastline evolution of Portuguese low-lying sandy coast in the last 50 years: An integrated approach. *Earth Syst. Sci. Data Discuss.* **2016**, *8*, 22.
36. Pinto, C.; Silveira, T.; Teixeira, S. Alimentação artificial de praias na faixa costeira de Portugal continental: Enquadramento e retrospetiva das intervenções realizadas (1950–2017). Relatório Técnico. *APA–Port. Environ. Agency* **2018**, *1*, 66. (In Portuguese)
37. Pinto, C.; Silveira, T.; Teixeira, S. Beach nourishment practice in mainland Portugal (1950-2017): Overview and retrospective. *Ocean. Coast. Manag.* **2020**, *192*, 105211. [CrossRef]
38. COSMO. Coastal Monitoring Programme of Continental Portugal: COSMO. APA–Portuguese Environment Agency. Available online: https://cosmo.apambiente.pt/ (accessed on 20 January 2022).
39. Ferreira, M.; Fernández-Fernández, S.; Coelho, C.; Silva, P.A. Caracterização de tempestades e da deriva litoral no litoral de Aveiro, em cenários de alterações climáticas. In Proceedings of the XV Congresso Da Água, Lisbon, Portugal, 22–26 March 2021; pp. 1–4. (In Portuguese)
40. Ferreira, M.; Coelho, C.; Lima, M.; Bernardes, C.; Baptista, P.; Silva, P.A.; Carvalho, R.; Pound, M.; Pinto, C. Feasibility study of sand bypass at Aveiro and Figueira da Foz inlets. In Proceedings of the 37th International Conference on Coastal Engineering, Sydney, Australia, 4–9 December 2022; p. 3.
41. Ferreira, M.; Lima, M.; Coelho, C.; Nolasco, A.; Cardoso, A.; Rocha, B.; Narra, P.; Silva, P.; Carvalho, R.; Pound, M.; et al. Análise custo-benefício de soluções de transposição de sedimentos. In Proceedings of the 10as Jornadas de Engenharia Costeira e Portuária, Sines, Portugal, 7–8 April 2022; p. 12. (In Portuguese)
42. COS: Carta de Uso e Ocupação do Solo–2018. Direção-Geral do Território. Available online: http://mapas.dgterritorio.pt/DGT-ATOM-download/COS_Final/COS2018_v1/COS2018_v1.zip (accessed on 5 February 2021).
43. Breil, M.; Catenacci, M.; Travisil, C. Impatti del Cambiamento Cimatico Sulle Zone Costiere: Quantificazione Economica di Ompatti e di Misure di Sdattamento–Sintesi di Risultati e Indicazioni Metodologiche per la Ricerca Futura. In In Proceedings of the APAT Workshop on "Cambiamenti Climatici e Ambiente Marino-Costiero: Scenari Futuri per un Programma Nazionale di Adattamento", Palermo, Italy, 27–28 June 2007 ; 27–28 June 2007.
44. Taborda, R.; Magalhães, F.; Ângelo, C. Evaluation of Coastal Defence Strategies in Portugal. In *Environmentally Friendly Coastal Protection*; Springer: Dordrecht, Netherlands, 2005.
45. Chu, M.L.; Guzman, J.A.; Muñoz-Carpena, R.; Kiker, G.A.; Linkov, I. A simplified approach for simulating changes in beach habitat due to the combined effects of long-term sea level rise, storm erosion, and nourishment. *Environ. Model. Softw.* **2014**, *52*, 111–120. [CrossRef]
46. Alexandrakis, G.; Manasakis, C.; Kampanis, N.A. Valuating the effects of beach erosion to tourism revenue. A management perspective. *Ocean. Coast. Manag.* **2015**, *111*, 1–11. [CrossRef]
47. Bosello, F.; Roson, R.; Tol, R. Economy-wide Estimates of the Implications of Climate Change: Sea Level Rise. *Environ. Resour. Econ.* **2007**, *37*, 549–571. [CrossRef]
48. Costa, L.; Tekken, V.; Kropp, J. Threat of sea level rise: Costs and benefits of adaptation in European union coastal countries. *J. Coast. Res.* **2009**, *SI 56*, 223–227.
49. Darwin, R.F.; Tol, R. Estimates of the Economic Effects of Sea Level Rise. *Environ. Resour. Econ.* **2001**, *19*, 113–129. [CrossRef]
50. Martínez, M.L.; Intralawan, A.; Vázquez, G.; Pérez-Maqueo, O.; Sutton, P.; Landgrave, R. The coasts of our world: Ecological, economic and social importance. *Ecol. Econ.* **2007**, *63*, 254–272. [CrossRef]
51. Martino, S.; Amos, C.L. Valuation of the ecosystem services of beach nourishment in decision-making: The case study of Tarquinia Lido, Italy. *Ocean. Coast. Manag.* **2015**, *111*, 82–91. [CrossRef]
52. Neumann, J.; Hudgens, D.; Herter, J.; Martinich, J. The economics of adaptation along developed coastlines. *Wiley Interdiscip. Rev. Clim. Change* **2011**, 2, 89–98. [CrossRef]
53. Turner, R.K.; Burgess, D.; Hadley, D.; Coombes, E.; Jackson, N. A cost–benefit appraisal of coastal managed realignment policy. *Glob. Environ. Change* **2007**, *17*, 397–407.
54. Roebeling, P.C.; Rocha, J.; Alves, F.; Vizinho, A. *Cost-Benefit Analysis of Coastal Erosion Adaptation Pathways along the Barra-Vagueira Coastal Stretch*; Technical Report Prepared in the Context of the Project BASE (Bottom-Up Climate Adaptation Strategies for a Sustainable Europe); Department of Environment and Planning, University of Aveiro: Aveiro, Portugal, 2015.
55. Nicholls, R.J.; Tol, R.S.J. (2006). Impacts and responses to sea-level rise: A global analysis of the SRES scenarios over the twenty-first century. *Philos. Trans. R. Soc. A Math. Phys. Eng. Sci.* **2006**, *364*, 1073–1095. [CrossRef]

56. Larson, M. Numerical modelling. In *Encyclopedia of Coastal Science*; Schwartz, M.L., Ed.; Springer: Dordrecht, Netherlands, 2005; pp. 730–733.
57. Burcharth, H.; Hawkins, S.; Zanuttigh, B.; Lamberti, A. *Environmental Design Guidelines for Low Crested Coastal Structures: Design Tools Related to Engineering*; Elsevier Science: Amsterdam, The Netherlands, 2007; pp. 203–333. ISBN 978-0-08-044951-7.
58. Coelho, C.; Ferreira, M.; Marinho, B. Numerical modelling of artificial sediment nourishment impacts. *J. Coast. Res.* **2020**, *95*, 209–213. [CrossRef]
59. Larson, M.; Palalane, J.; Fredriksson, C.; Hanson, H. Simulating cross shore material exchange at decadal scale. Theory and model component validation. *Coast. Eng.* **2016**, *116*, 57–66. [CrossRef]
60. Lima, M.; Coelho, C.; Alves, F.; Marto, M. *Base de Dados #2–Impactos de Medidas de Mitigação e Adaptação à Erosão Costeira e às Alterações Climáticas*; Technical Report Prepared in the Scope of the Research Project INCCA-INtegrated Coastal Climate Change Adaptation for Resilient Communities (POCI-01-0145-FEDER-030842); University of Aveiro: Aveiro, Portugal, 2021; 74p, (In Portuguese). Avaliable online: http://incca.web.ua.pt/index.php/base-de-dados-2/ (accecced om 31 October 2022).
61. Noordegraaf, I. Economic Valuation of Increased Beach Width at Praia da Vagueira (Central Portugal). Master's Thesis, University of Aveiro, Aveiro, Portugal, 2020; p. 80.
62. Passeri, D.; Bilskie, M.; Hagen, S.; Mickey, R.; Dalyander, P.; Gonzales, V. Assessing the effectiveness of nourishments in decadal barrier island morphological resilience. *Water* **2021**, *13*, 944. [CrossRef]
63. Parson, G.; Chen, Z.; Hidrue, M.; Standing, N.; Lilley, J. Valuing beach width for recreational use: Combining revealed and stated preference data. *Mar. Resour. Econ.* **2013**, *28*, 221–241. [CrossRef]
64. Landry, C.; Shonkwiler, J.; Whitehead, J. Economic values of coastal erosion management: Joint estimation of use and existence values with recreation demand and contingent valuation data. *J. Environ. Econ. Manag.* **2020**, *103*, 102364. [CrossRef]

Journal of
Marine Science and Engineering

MDPI

Article

Sediment Transport and Morphological Response to Nearshore Nourishment Projects on Wave-Dominated Coasts

Cody L. Johnson *, Brian C. McFall, Douglas R. Krafft and Mitchell E. Brown

Coastal and Hydraulics Laboratory, U.S. Army Engineer Research and Development Center, Vicksburg, MS 39180, USA; Brian.C.McFall@usace.army.mil (B.C.M.); Douglas.R.Krafft@usace.army.mil (D.R.K.); Mitchell.E.Brown@usace.army.mil (M.E.B.)
* Correspondence: Cody.L.Johnson@usace.army.mil; Tel.: +1-601-634-2552

Abstract: Nearshore nourishments are constructed for shoreline protection from waves, to provide sediment nourishment to the beach profile, and to beneficially use dredged sediment from navigation channel maintenance. However, it is poorly understood how placement morphology and depth influence nearshore processes operated on wave-dominated coasts. This study investigates the wave fields, sediment transport, and morphological response to three common nearshore nourishment shapes, nearshore berm (elongated bar), undulated nearshore berm, and small discrete mounds, with numerical experiments utilizing the Coastal Modeling System. The nourishments are placed in depths between 3 m and 7 m with a volume of approximately 100,000 m^3 and between 400 m and 1000 m in alongshore length. Numerical experiments are carried out in three distinct coastal settings with representative wave climates and geomorphology. Simulation results indicate that shallower, more continuous berms attenuate the most wave energy, while deeper, more diffuse placements retain more sediment. Results from this study improve the understanding of nearshore nourishment shapes and can support decision makers identifying the most appropriate construction technique for future nearshore nourishment projects.

Keywords: nearshore nourishment; wave dissipation; nearshore sediment transport; coastal hydrodynamic modeling

Citation: Johnson, C.L.; McFall, B.C.; Krafft, D.R.; Brown, M.E. Sediment Transport and Morphological Response to Nearshore Nourishment Projects on Wave-Dominated Coasts. *J. Mar. Sci. Eng.* **2021**, *9*, 1182. https://doi.org/10.3390/jmse9111182

Academic Editor: Carlos Daniel Borges Coelho

Received: 4 October 2021
Accepted: 25 October 2021
Published: 27 October 2021

1. Introduction

Sandy beaches are a natural or nature-based feature (NNBF) that serve a crucial societal, cultural, economic, and environmental role [1] but typically require periodic maintenance through sand nourishment. Nourishment projects can be on the subaerial beach or in the nearshore. Nearshore nourishment projects are commonly completed around the globe. In the United States (U.S.), the U.S. Army Corps of Engineers currently place more than 7.7 million m^3 in the nearshore annually [2]. Nourishments in the U.S. are constructed in a range of dimensions and conditions, with placement depths ranging from 1.5 m to 26 m, nourishment volumes between 18,000 m^3 to 14,300,000 m^3, and heights of 0.5 m to 9 m [2]. Projects that place sediment in the nearshore have many names, including nearshore nourishment, profile nourishment, nearshore placement, littoral placement, or shoreface nourishment. When the sediment is intentionally placed as an artificial sandbar or mound, it is called a nearshore berm. Regardless of the name, the projects nourish the beach profile by placing sediment in the nearshore zone.

Nearshore nourishment projects can be constructed by beneficially using dredged sediment from navigation channels, and they can be used as a flood risk reduction method. Goals for these projects often include dissipating wave energy farther from the shoreline, keeping sediment in the littoral system, and nourishing the beach profile with minimal environmental impacts. Previous nearshore nourishment research initiatives have focused on the transport rate and direction of the sediment placed in the nearshore [3–6], and the morphodynamic response to the placed sediment [7–9]. Different construction methods will place sediment differently in the nearshore and thereby create nourishment features with different geometries. There has been minimal effort applied to understanding how the geometry of the nearshore nourishment features influence the sediment transport and nearshore morphological change.

Constructing a linear nearshore berm is a nourishment strategy that has been researched with field studies [10], physical models [11–13], and numerical tools and models [9,13–22]. Linear berms can be constructed by placing sediment with a hydraulic pipeline or by bottom dumping hoppers or scows [23]. Undulated nearshore berms, i.e., shore parallel berms with sinusoidal variation in berm height, may be used to create a shoreline response of multiple small cusps to minimize the environmental impact to beach dwelling species, e.g., filter feeders such as the Emerita talpoida (mole crab) and Donax variabilis (coquina clam) [24,25]. Many nearshore nourishment projects constructed with hopper dredges beneficially use dredged navigation channel sediment and place the sediment in small discrete mounds in a large placement area within the littoral zone [26]. This typically results in an approximately random distribution of individual mounds in the nearshore. This project investigates how the nearshore nourishment shape, such as these, and placement depth influence the sediment transport and morphodynamics on wave-dominated coasts when forced by different wave climates. Other factors contributing to nearshore circulation, such as wind stress and tides, have been neglected in order to isolate the dominant processes of wave-driven sediment transport.

In this section, a brief introduction to nearshore nourishments and the current work is provided with references to the relevant literature. The manuscript is structured as follows. Section 2 explains the methodology of the study, including the Coastal Modeling System (CMS), creation of numerical experiments, and the data analysis techniques. Section 3 details the results of the numerical experiments, and an analysis of the simulated data is presented. Section 4 provides a discussion of the analysis and model results with emphasis on the practice of nearshore nourishment design. Section 5 summarizes the work with an itemized list of conclusions.

2. Materials and Methods

2.1. Idealized Scenarios

2.1.1. Regional Study Sites

In order to realistically approximate the conditions of actual nearshore nourishments, model inputs were derived from historical project sites. These sites, located within three contrasting coastal regions of the U.S., were selected to represent a range of wave climates and nearshore geomorphologies. The sites selected were South Padre Island, TX (SP); Vilano Beach, FL (VB); Ogden Dunes, IN (OD), which represent the Gulf of Mexico, East Coast, and Great Lakes regions, respectively (see Figure 1). These locations have been the site of nearshore placements in the past, as well as data collection efforts, which further facilitated model set up. Bathymetry and wave climate data were required from each site to create inputs for each model scenario. Overall, OD has less energetic waves and a steeper offshore profile than the other sites and is situated within a large interior lake (Lake Michigan); SP has more energetic waves than OD, but it has shorter wave periods than VB due to wave generation occurring within the Gulf of Mexico; VB has longer period waves originating within the south Atlantic Ocean and is the most energetic of the three sites. Additional wave climate details are described in Section 2.1.4 and the reader is referred to Figlus et al. (2021) [27], Brutsché et al. (2017, 2019) [7,28], and Young et al. (2020) [26] for more site characterization at SP, VB, and OD, respectively.

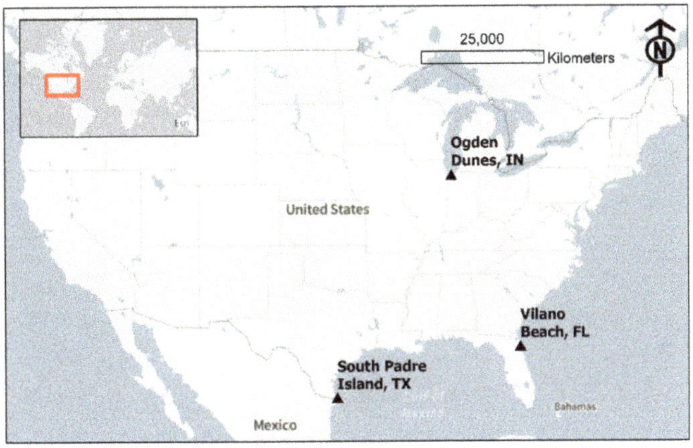

Figure 1. Site locations used in model set up.

2.1.2. Nearshore Profile Morphology

An idealized bathymetry for each site is created by fitting the underlying nearshore bathymetry to the equilibrium profile defined by Komar and McDougal (1994) as

$$h = B\left(1 - e^{-kx}\right), \tag{1}$$

where h is the depth, B is a parameter, k is the profile's decay coefficient, and x is the distance from the shoreline. The slope of the equilibrium profile,

$$\frac{dh}{dx} = kBe^{-kx}, \tag{2}$$

can be used to estimate B. Setting $x = 0$ m in Equation (2) yields the shoreline slope, $S_0 = kB$, which was estimated from grain size data [10] and Wave Information Studies (WIS) [29] wave height using the relationships proposed by McFall (2019) [30]. This reduces Equation (1) to a single parameter optimization problem, $h = \frac{S_0}{k}\left(1 - e^{-kx}\right)$, where k was determined by fitting Equation (1) to a set of shore-aligned, shore-normal bathymetric profiles.

The bathymetric profile data was collected by sampling shore-normal transects of the National Oceanographic and Atmospheric Administration's National Centers for Environmental Information (NCEI) digital terrain models for the study sites. This data was then aligned so that $z = MSL$ corresponded to $x = 0$ m. See Figure 2 for the sampled bathymetry data and resulting equilibrium profiles for each study site. It can be seen that, while some features of the profile data are not captured by the equilibrium profile, the general nearshore morphology is well represented and displays inter-regional variability. The assumption of a monotonic profile model neglects bar/trough features that are evident in the nearshore profile data and would naturally generate wave transformation. This assumption is justified in order to independently study wave transformation and sediment transport processes introduced by the nearshore nourishments. The equilibrium profiles were replicated in the alongshore to create alongshore uniform beaches for model input.

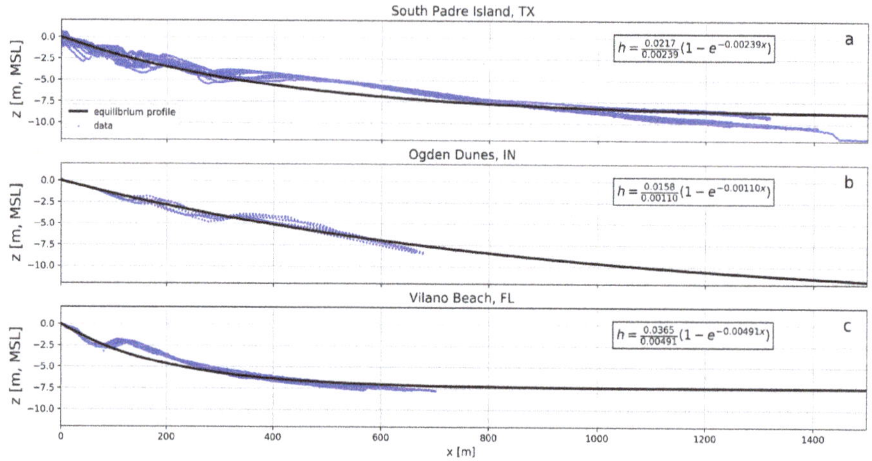

Figure 2. Nearshore bathymetry data (blue markers) and fitted equilibrium profile (solid black line) for the study sites. (**a**) Data and equilibrium profile for South Padre Island; (**b**) Ogden Dunes; and (**c**) Vilano Beach.

2.1.3. Nearshore Nourishment Shapes

Three nearshore nourishment shapes were investigated, a linear berm (LB), an undulated berm (UB), and a uniform distribution of discrete mounds (DM). The linear berm was alongshore uniform, except for the end caps that smoothly graded into the equilibrium profile. The undulated berm was similar to the linear berm, but the berm crest height (h_b) oscillated in the alongshore direction following a cosine function. Thus, at its highest point the UB crest was equal to the LB, but dips to half the LB crest height at its lowest point with a wave length of 160 m. The discrete mounds were constructed by uniformly distributing two-dimensional Gaussian functions within an area with crests spaced 25 m apart. Each nearshore nourishment shape was investigated centered at three different depths per equilibrium profile (i.e., region), 3 m, 5 m, and 7 m. The nourishment shapes and placement depths were chosen to represent nearshore nourishment construction practice observed in the field [6].

The combination of nourishment shape, depth, and region, plus one case per equilibrium profile without a nearshore nourishment to function as a control, resulted in a total of 30 numerical experiments. The nearshore placement volumes were held approximately constant at 100,000 m^3 to mitigate the influence of the variability on inter-comparisons. The geometric range of the nourishments (bed slope and curvature, crest height, etc.) introduced some variability in the absolute dimensions of the nearshore placements. Thus, the alongshore length and cross-shore width of the linear/undulated berms and the discrete mounds' footprints necessarily varied to a small degree between cases

The berm's cross-shore shape itself was derived from survey data of a nearshore berm constructed at Ft. Myers Beach, FL [23]. A Gaussian function,

$$f(x) = \alpha \exp\left[\frac{-(x-x_b)^2}{2\sigma^2}\right], \quad (3)$$

where x_b is the cross-shore distance to the berm crest, and α and σ are the scale and width parameters, respectively, used to approximate the berm as it intersected the underlying equilibrium profile. To generate congruent berm shapes across the range of cases, the berm's height above the bed (h_b) was held constant for each berm type, and the ratio of the Gaussian function's amplitude ($\alpha = h_b + \epsilon$) to its width (σ) was also held constant (see Figure 3). For each Gaussian curve generated, an algorithm incremented ϵ, then calculated α and σ, which continued until the seaward tail of the Gaussian mound was less than 1 mm from the equilibrium profile. The result was a smooth equilibrium profile exhibiting berms with similar geometric properties. See Figure 4 for a demonstration of the SP profile with the three LBs placed at depths of 3 m, 5 m, and 7 m. Note that the initial depth of the berm crest, F_b, is the same between profiles, but the berm's cross-sectional area may be somewhat smaller or larger depending on the profile's local slope and curvature. The discrete mound cases had smaller h_b in order to approximate a hopper dredge capacity of 360 ± 60 m^3.

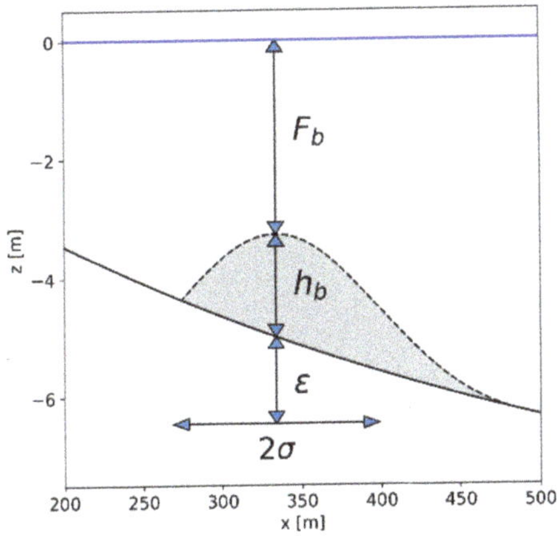

Figure 3. Cross-shore profile of idealized nearshore berm.

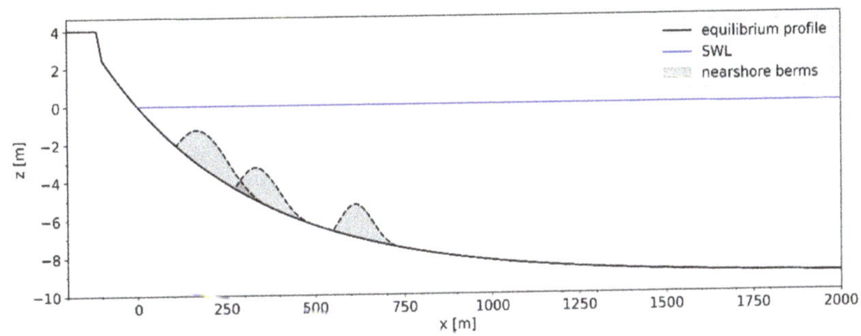

Figure 4. Cross-shore profile of SP linear berms.

2.1.4. Wave Climate

An incident wave field was used to drive the nearshore circulation and sediment transport. Since the goal was to provide realistic estimates of nearshore nourishment performance, real-world waves were required in contrast to schematized conditions. The WIS database provides long-term hindcast time-series of bulk wave parameters generated from a third-generation, phased-averaged wave model and was selected to create input wave boundary conditions. WIS employs WAVEWATCH III to simulate wave energy transformation along the U.S. coastline and outputs results at regularly spaced locations. The stations used were 73,020, 94,001, and 63,417 with nominal depths of 36 m, 17 m, and 19.5 m for the SP, OD, and VB scenarios, respectively. The wave parameters were shoaled to profiles' offshore boundary depths using Snell's law. Further, the wave angles were rotated from the geographic coordinate system of the WIS data to the idealized models' coordinate system

To obviate the need to select a particular time span to simulate, a method of generating random, realistic time-series of wave conditions from WIS data was adopted using Markov chain models [31]. The synthetic time-series retained the essential statistical properties of each site's sea state, i.e., the mean and variance of the significant wave height, peak period, and mean wave direction. Three-dimensional Markov chain models of these bulk parameters were generated from the 40 year hindcast records at each site. The fall and winter months were chosen to drive the model as they statistically exhibit more frequent and larger storm events. This may result in more sediment transport seaward than would occur if the summer/spring months were included in the simulations. Additionally, the historical projects at these sites were constructed in the summer, exposing the nourishment to fall and winter waves following construction. See Figures 5 and 6 for a comparison of the modeled and WIS hindcast wave height probability density functions (PDF) and an example realization of the VB Markov model, respectively.

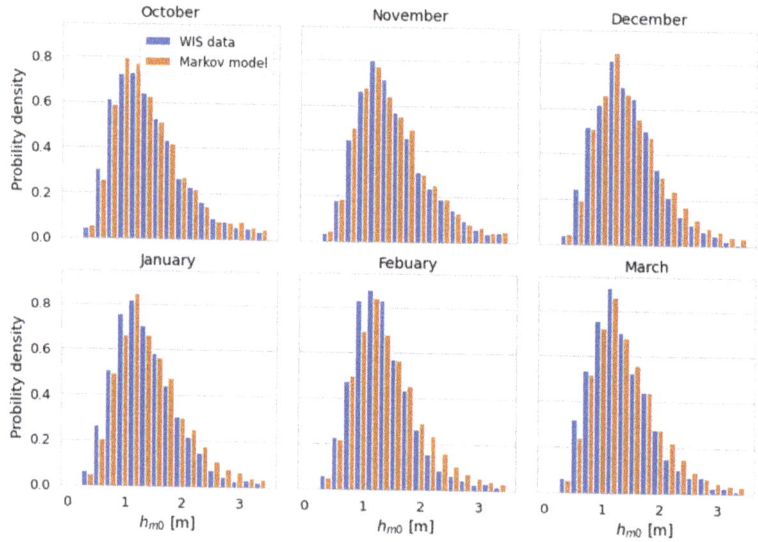

Figure 5. Monthly wave height PDFs of the WIS (blue) and Markov chain model results (orange) used as input to the VB scenario. Starting at the upper left, and from left to right, the panels show PDFs for October through March.

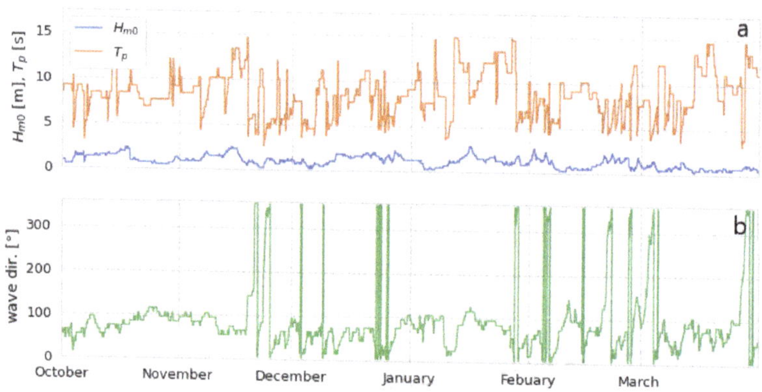

Figure 6. Realization of the Markov chain model for VB. (**a**) Zero-moment wave height (blue line) and peak period (orange line); (**b**) wave direction (green line).

The wave climates differed between the regions to a large extent (see Table 1). Wave heights and periods at OD were the smallest, where on average zero-moment wave heights

(H_{m0}) were less than 0.5 m. This is contrasted with the other two regions, which had mean wave heights above 1 m. While both SP and VB had large maximum wave heights, and comparable mean significant wave heights, their mean peak periods (T_p) were different.

Table 1. Statistics of regional wave climates, including the maximum zero-moment wave height, the mean (and standard deviation) zero-moment wave height, and the mean (and standard deviation) peak period.

Region	Max. H_{m0} (m)	Mean (std.) H_{m0} (m)	Mean (std.) T_p (s)
SP	4.16	1.11 (0.48)	6.3 (1.6)
VB	4.17	1.45 (0.63)	9.5 (2.6)
OD	3.22	0.47 (0.44)	3.8 (1.7)

2.2. Numerical Model

2.2.1. The Coastal Modeling System

The Coastal Modeling System (CMS) was employed to simulate the nearshore nourishment evolution and wave fields. The CMS consists of a combined suite of 2DH modules for flow, waves, and sediment transport with morphology change [32,33]. The CMS began as a class project in 1994 at Florida Institute of Technology and improved over several years to a full-fledged two-dimensional (2DH) circulation model named M2D [34,35]. The Coastal Inlets Research Program added many more capabilities to M2D, including sediment transport and simplified 2DH salinity, added an implicit solution scheme to the existing explicit offering, integrated a phase-averaged wave model, and eventually renamed the model from M2D to the Coastal Modeling System to represent its intended use in nearshore coastal regimes.

CMS-Flow is solved with numerical computations of the mass, momentum, and sediment transport equations on either a Cartesian or Quadtree grid and with either an implicit or explicit solution scheme using a finite-volume method. CMS-Wave is a steady-state, phase-averaged, spectral wave module that computes wave height, direction, and period, wave dissipation, and wave radiation stresses [36,37] and includes wave transformation processes such as refraction, diffraction, reflection, shoaling, and wave breaking. The link between the two modules is two-way. CMS-Wave calculated radiation stresses and wave dissipation values are passed to CMS-Flow to impose wave action on the currents. The wave-modified currents are then used to force sediment transport with one of several transport algorithms, as chosen prior to simulation start. Once CMS-Flow calculates the resulting hydrodynamics and sediment transport, the wave-modified currents and updated depths due to sediment transport are passed to CMS-Wave to compute the waves at the next time interval, as designated by a user-defined parameter, and therefore completes the two-way interaction between wave and flow processes.

2.2.2. Computational Domain

Two computational grids were created for the hydrodynamic/sediment transport/ morphology change model (CMS-Flow) and the phase-averaged wave model (CMS-Wave), which were used for all the simulations. The grids were used to couple the waves and hydrodynamics as described in Section 2.2.1, but all simulated results were output on the CMS-Flow grid. This was done to facilitate post-processing and ensure commensurability of the simulation results.

Overall, the computational domains extended 1800 m in the alongshore and 2180 m in the cross-shore (see Figure 7 for a diagram of the computational grids). An area of interest (AoI) was delineated within the domain. The AoI used a uniform grid spacing of 3 m. The AoI covered the extents of each nearshore nourishment across the regions, depths, and morphologies. The CMS-Flow grid utilized Quadtree refinement [37] to transition from a resolution of approximately 50 m at the offshore boundary to that of the AoI, while the CMS-Wave grid smoothly refined a Cartesian grid from a resolution of 100 m at the offshore to the AoI [38]. The minimum grid resolution was chosen to resolve the individual discrete mounds.

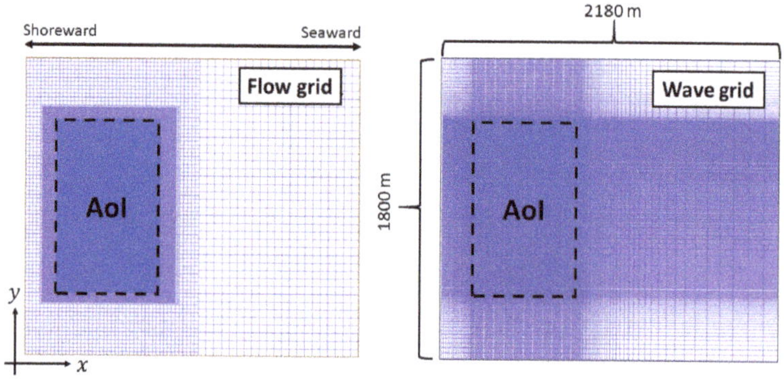

Figure 7. Computational grids used for the CMS-Flow and CMS-Wave models.

2.2.3. Model Set up and Inputs

The CMS-Flow model was driven with radiation stress gradients that were generated within CMS-Wave and interpolated onto its computation grid. Tides and wind stresses were neglected, and, thus, all water level displacements and currents were generated by the waves. A constant water level of zero meters was imposed on the open boundaries of the CMS-Flow model (i.e., the seaward and lateral boundaries). A moving shoreline forms the closed, landward boundary and is defined by a wetting/drying algorithm with a threshold depth of 5 cm.

CMS-Flow's implicit solver was used with a dynamic time-step with a maximum value of 900 s. The time-step was programmatically decreased by half if the solver did not converge within 28 iterations. The total simulation duration is 6 months, and a ramp was applied for the first hour of the simulation to smooth potential shocks.

A roller model was used to account for momentum stored in the breaking wave front before being transferred to water column, and the roller's contribution to wave mass flux was also included. A roughness coefficient, based on the local depth and a Manning's n value, was used to calculate the bottom shear stress. A spatially uniform Manning's n of 0.023 was applied. Sediment transport was calculated with the Lund-CIRP formula [39], using a single sediment fraction and a uniform grain size of 0.24 mm. Default sediment transport parameters were used for all the simulations. Bed updating occurred with every flow/sediment transport computation time-step.

2.3. *Analysis*

Three post-processing analyses were performed on the simulation results. One computed the wave energy dissipation introduced by the nourishment. The second analyzed the longevity of the nourishment in terms of sediment retention within the placement's original footprint. The last analyzed the alongshore and cross-shore migration of the nourishment. These analyses provide a quantitative summary of two typical objectives in the design and construction of nearshore nourishments: attenuate wave energy and renourish the nearshore profile.

2.3.1. Wave Energy Dissipation

The analysis considered the wave energy per unit area with the AoI as well as the wave energy flux through the landward, shore-parallel control surface of the AoI (see Figure 7). Wave energy per unit area was calculated using linear wave theory as,

$$E = \frac{1}{8}\rho g H_{m0}^2, \tag{4}$$

where E is the wave energy, ρ is the density of the water (1015 kg m^{-3}), g is the acceleration due to gravity, and H_{m0} is the zero-moment wave height. To analyze the effects of the nearshore nourishment, the wave energy within the AoI is normalized by the wave energy of the control case, E^*. To present the nourishment's cumulative effect, this quantity is averaged per grid cell over the first two months of the simulation, yielding

$$\overline{E} = \frac{1}{N}\sum_i^N \frac{E_i}{E_i^*}, \tag{5}$$

where $\overline{E}$ is the mean relative wave energy change and i indicates the output time-step. $\overline{E}$ varies throughout the AoI, where $\overline{E} > 1$ indicates wave energy amplification and $\overline{E} < 1$ indicates wave energy dissipation.

The wave energy flux per unit width (wave power) was calculated for each grid cell on the landward interface of the AoI as

$$P = Ec_g = \left[\frac{1}{8}\rho g H_{m0}^2\right]\left[n\frac{L}{T_p}\right], \tag{6}$$

where P is the wave power, c_g is the group velocity, L is the wave length, T_p is the peak period, and $n = \frac{1}{2}\left(1 + \frac{2kh}{\sinh 2kh}\right)$ is the group parameter, with $k = \frac{2\pi}{L}$ being the wave number and h the local depth. The local wave length was obtained by iteratively solving the linear dispersion relation using the local depth and peak period.

The mean wave power was calculated as

$$\overline{P} = \frac{1}{\Gamma}\int_0^\Gamma P dy, \tag{7}$$

where Γ is the length of the control surface. Finally, the relative energy dissipation introduced by the nourishment was calculated by normalizing the wave power with the nourishment by the mean wave power from the control case (i.e., without a nourishment),

$$\overline{\alpha} = \frac{\overline{P}}{\overline{P}^*}, \tag{8}$$

where $\overline{P}^*$ is the mean wave power from the control case. $\overline{\alpha}$ is a function of time and varies with the offshore wave height and instantaneous morphology of the nearshore profile.

2.3.2. Nourishment Longevity

The question of the placed sediment's fate is an import consideration when planning nearshore nourishments. The retention of sediment within the original placement's plan-view template or footprint is often an objective. To assess each scenario's performance vis-à-vis this objective, the amount of sediment volume within the original footprint was tracked through time. This provides a measure of the placement's longevity under the simulated wave conditions.

The excess (deficit) volume of the nearshore nourishment relative to the control was used to assess retained sediment volume and was calculated as,

$$V(t) = \int_A [z_b(t) - z_b^*(t)]dA, \tag{9}$$

where $V(t)$ is the sediment volume within the original nearshore placement's footprint throughout the simulation, t is the simulation time, A is the area delineated by the original placement, $z_b(t)$ is the simulated bed level of the case being analyzed, and $z_b^*(t)$ is the control's bed level. The excess sediment volume was normalized by the initial excess sediment volume (V_0) for inter-comparison. $V = 1$ implies no sediment transport from the nourishment, i.e., no nourishment deflation, and $V = 0$ implies total removal of the placed sediment.

2.3.3. Sediment Displacement

The sediment of the initial nearshore nourishment can be transported either onshore, offshore, or in the alongshore direction. This is relevant for project design to know if potential project sites are cross-shore or longshore transport dominated. The center of mass of the nourishment is tracked through time to assess this sediment's net displacement. Conceptually, this allows the analysis of the nourishment's movement in general. If more sediment moves onshore, as opposed to offshore, then the center of mass will shift onshore. The center of mass's movement can then be analyzed in terms of cross-shore and alongshore displacement, revealing the dominant transport direction.

The coordinates for the center of mass are calculated as,

$$(\overline{x}, \overline{y}) = \left(\frac{M_y}{M}, \frac{M_x}{M}\right) \tag{10}$$

where $\overline{x}$ is the cross-shore coordinate (positive onshore), $\overline{y}$ is the alongshore coordinate (positive to the left when looking at the shore from the sea), $M_y = \int xMdA$ is the moment about the alongshore axis, $M_x = \int yMdA$ is the moment about the cross-shore axis, and $M(x, y; t) = z_b(x, y; t) - z_b^*(x, y; 0) > 0$ is the accreted sediment volume over the initial equilibrium profile. Note that only positive differences are included in the calculation of volume.

3. Results

3.1. Waves

3.1.1. Planform Energy Dissipation

Surface waves undergo a number of transformations as they propagate from the continental shelf through the nearshore and eventually dissipate and/or reflect within the surfzone/swash zone. The most relevant processes for open coasts are shoaling, refraction, and wave breaking. The construction of a nearshore nourishment via external sediment sources will inevitably affect these processes. Figure 8 displays the mean relative wave energy change over the first 60 simulation days. The cases displayed are the LB, UB, and DM at every placement depth for the SP regional scenario.

In the LB case, it is expected that the waves will shoal and break further offshore, provided that the waves are large enough to "feel the bottom" before they reach the equilibirum profile's surfzone, i.e., where they break in the control. This is evidentally the case given the spatial distribution of $\overline{E}$ within Figure 8a–c, which shows an LB placed at depths of 3 m, 5 m, and 7 m, respectively. When the berm is placed at $h = 3$ m, there is an increase in shoaling and, thus, wave energy seaward of the berm (located at appoximately $x = 1800$ m). Landward of the berm, wave breaking dissipation occurs at a higher rate than in the control case leading to reduced wave energy. For the linear berms in 5 m and 7 m placement depths, minimal shoaling and breaking occur near the berms. In fact, the most obvious pattern is evident refraction-diffraction occuring landward of the berm, leading to wave energy amplification leeward of the berm.

Figure 8d–f shows the mean relative wave energy change for the UB at placement depths of 3 m, 5 m, and 7 m, respectively. The cross-shore distribution of wave energy change broadly corresponds to the LB cases, i.e., increased shoaling/breaking for the $h = 3$ m and some leeward amplification for the other placement depths. However, the undulations in the berm crest introduce an interesting alongshore pattern of wave energy amplification and dissipation. The local maxima in berm crests still caused wave breaking-induced energy dissipation, but the local minima allowed for shoaling. For the 5 m placement, the result is less amplification leeward of the berm, while the 7 m placement also shows slightly less amplification relative to the linear berm. However, the 3 m placement interestingly shows an increase in dissipation in the alongshore dimension at the landward interface of the AoI. The wave energy distribution introduced by the discrete mounds, shown in Figure 8g–i, is less pronounced. The area delineated by the mounds' placement shows slight amplification of wave energy, but a significant dissipation relative to the control leeward of the placement is not observed.

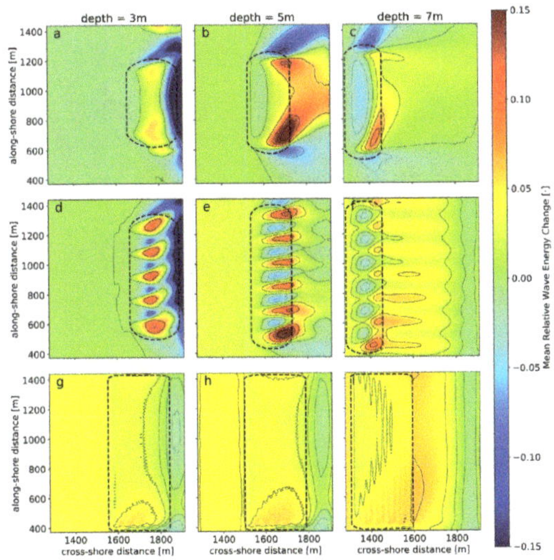

Figure 8. Mean wave energy attenuation averaged over the first two months of the South Padre Island scenario simulations. (**a–c**) LB at 3 m, 5 m, and 7 m depths, respectively. (**d–f**) UB at 3 m, 5 m, and 7 m depths, respectively. (**g–i**) DM at 3 m, 5 m, and 7 m depths, respectively. Dashed black lines indicate convex hull of nourishments initial footprint.

3.1.2. Aggregate Wave Energy Attenuation

Effects on wave transmission are typically important for the construction of shore protection features, and this is true of nearshore nourishments. To assess wave transmission through the nearshore nourishments, mean wave power at the landward shore-parallel interface of the AoI is compared to the control as described in Section 2.3.1. Since the methods employed the simulation of a realistic time-series, as opposed to schematized wave conditions, the morphology of the nourishment evolved thoughout the simulated period. Therefore, to assess wave transmission in aggregate, the relative mean wave power ($\overline{\alpha}$) is binned according to the offshore boundary condition wave height. In this manner, trends in wave fields can be elucidated without considering individual sea-states.

Figure 9 shows relative mean wave power distributions as a function of offshore wave height for the LB, UB, and DM nourishment configuations at the SP regional scenario, respectively. In Figure 9, the relative wave power for a linear berm at $h = 3$ m shows a clear trend of reducing wave energy flux ($\overline{\alpha} < 1.0$) through the AoI compared to the placements at 5 m and 7 m. The reduction in wave energy flux diminishes with smaller waves, which is expected as they are less affected by the berm. The placements at $h = 5$ m and $h = 7$ m do not appear to affect wave transmission at all as $\overline{\alpha} \approx 1.0$ for all the wave conditions, though there is slight amplification ($\overline{\alpha} > 1.0$) for the larger wave conditions. This may be due to saturation of the surfzone where the wave height at the AoI's landward interface is determined by the depth of the underlying profile at that location.

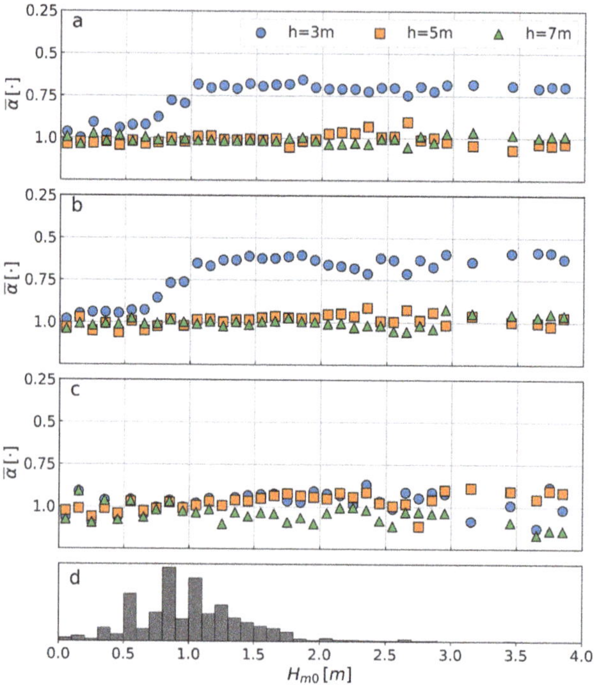

Figure 9. Relative wave power binned by offshore wave height at the South Padre Island site where the blue circles, orange squares, and green triangle markers represent the 3 m, 5 m, and 7 m placement depths, respectively. (**a**) Linear berm; (**b**) undulated berm; (**c**) discrete mounds; (**d**) PDF for offshore wave height.

The trend between placement depths for the UB (see Figure 9b) is similar to the linear berm for SP. However, the magnitude of the attenuation of the mean relative wave power is appreciably greater in the UB. The LB exhibits $\overline{\alpha} \approx 0.7$ for $H_{m0} = [1.0\text{ m}, 4.0\text{ m}]$, and the UB shows values closer to $\overline{\alpha} \approx 0.6$ for the same range of H_{m0}. While it is expected that berms placed in shallower water will attenuate more wave energy than those placed deeper, it is surprising that the UB reduces wave transmission to a larger degree than the LB. Complex free-surface dynamics around the undulations may play a role here but are inaccessible in a phase-averaged model. The deeper placements show similar distributions of mean relative wave power.

The discrete mounds shown in Figure 9c have a less pronounced impact on wave transmission compared to the linear and undulated berms. This is expected, considering the energy attenuation results shown in Figure 8. While some of the large enhancements of wave power introduced by the DM may not be convincing and some are likely outliers (i.e., $\overline{\alpha} \approx 1.4$ in the $H_{m0} = [3.0\text{ m}, 3.25\text{ m}]$ bin for $h = 7$ m), there is yet a noticeable trend of increased wave power ($\overline{\alpha} > 1.0$) for the $h = 7$ m placement. This trend holds for a majority of the wave height bins, potentially indicating that the small mounds induced wave shoaling.

The trends between the LB and UB for the OD regional scenario, shown in Figure 10a,b, respectively, are similar to that in the SP scenario. The shallower berms ($h = 3$ m) reduce wave transmission (approximately $0.8 > \overline{\alpha} > 0.6$) across all of the wave height bins, except the smallest waves ($H_{m0} < 0.5$ m). This is compared to the deeper placements, which, for the most part, do not affect wave transmission in the LB case. All shapes and depths show an outlier for waves less than 10 cm. The relative values shown can be particularly volatile when normalizing small values with small values. These small waves have a negligible effect on the nourishment morphology, but are shown for completeness. In the UB case, there is a notable amplification in wave transmission for waves with a wave height between 1 m and 0.5 m for the 7 m placement depth. Further, the trend between placement depths for the DM in the OD scenario, see Figure 10c, is also similar to that in the SP scenario (Figure 9c). The deepest placement ($h = 7$ m) exhibits some wave amplification for all the wave height bins ($1.2 > \overline{\alpha} > 1.0$), while the shallower placements have little effect on wave transmission.

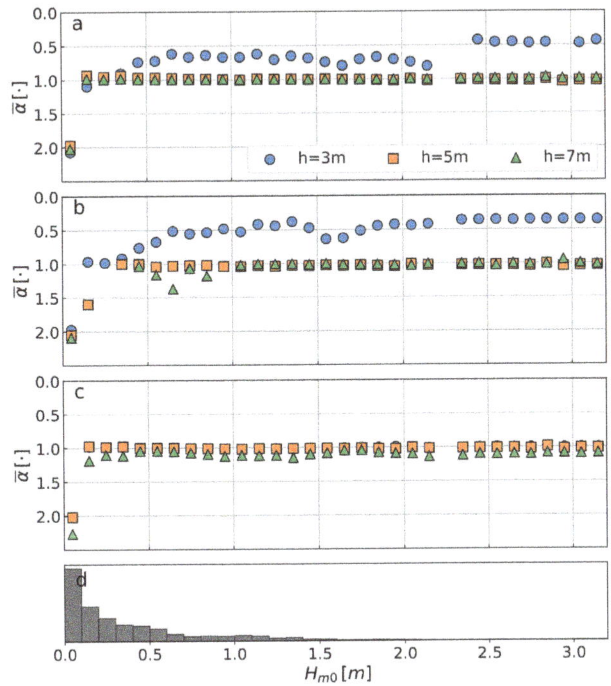

Figure 10. Relative integrated wave power binned by offshore wave height for the Ogden Dunes regional scenario where the blue circles, orange squares, and green triangle markers represent the 3 m, 5 m, and 7 m placement depths, respectively. (**a**) Linear berm; (**b**) undulated berm; (**c**) discrete mounds; (**d**) probability density function for offshore wave height.

The aggregate effects of the LB, UB, and DM on wave transmission for the VB regional scenario are presented in Figure 11. The general trends in the VB scenario differ from the SP and OD scenarios. Specifically, the UB does not clearly show more wave energy attenuation than the LB, as opposed to the SP and OD scenarios. Further, the difference in attenuation between the 3 m and 5 m placements at VB is not as pronounced as the other sites, which is potentially due to the longer period waves at VB. For both the LB and UB cases, the deepest placement ($h = 7$ m) leads to some wave energy amplification in medium wave heights (0.5 m $< H_{m0} < 1.75$ m). Also, for the LB, smaller waves ($H_{m0} < 0.75$ m) in the deepest placement depth are attenuated more than with the shallower placements, but attenuation for these wave bins is similar between placement depths for the UB.

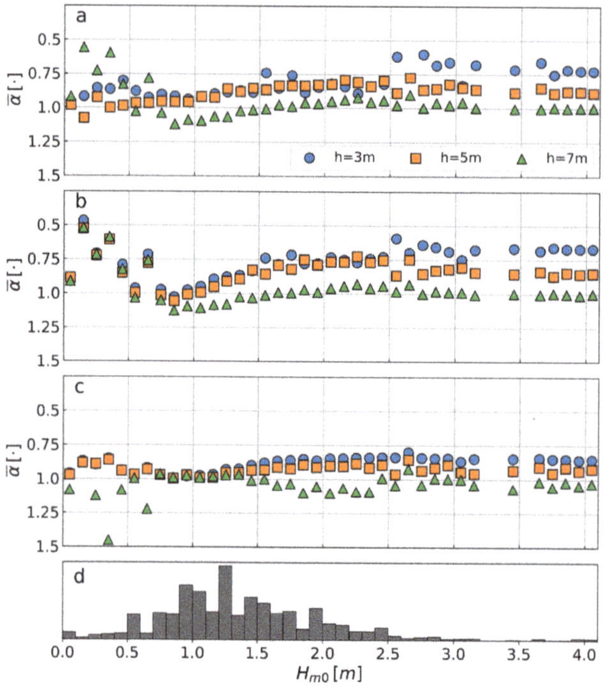

Figure 11. Relative integrated wave power binned by offshore wave height for Vilano Beach where the blue circles, orange squares, and green triangle markers represent the 3 m, 5 m, and 7 m placement depths, respectively. (**a**) Linear berm; (**b**) undulated berm; (**c**) discrete mounds; (**d**) probability density function for offshore wave height.

The wave height distributions (see panel d in Figures 9–11) indicate that the largest waves, which typically exhibit the highest energy attenuation value, are infrequent compared to the median wave height. However, both the infrequent, high-energy events and the more typical sea-states are attenuated to a greater degree by the shallower placements. The exception to this is the OD scenario, which follows a different wave height distribution that resembles the Gamma distribution, whereas the SP and VB scenarios more closely approximate a Rayleigh distribution. The OD site is located on the southern tip of Lake Michigan and is influenced by different wave generation mechanisms. The most frequent OD waves are less than 10 cm and show amplification under each nourishment shape and placement depth.

The LB and UB, when placed shallow enough, potentially attenuate 20% to 60% of the wave energy on average for wave heights larger than 0.5 m. On the other hand, the DM at any depth typically do not attenuate wave energy and, in some cases, cause wave amplification. This result is expected as the smaller height of the discrete mounds compared to the shore-parallel berms, due to distributing an equal amount of sediment over a larger extent, does not generate wave-breaking induced energy dissipation to the same degree. Presumably, the DM act to decrease the nearshore profile's depth over a larger distance, but not significant depth-limited breaking, and induce some shoaling that is not present in the control case leading to minor wave energy amplification.

3.2. Sediment Transport/Morphology Change

3.2.1. Nourishment Longevity

Nearshore nourishment projects are often constructed to nourish the beach profile for an intended period of time. Some are intended to deflate quickly to nourish adjacent beaches, and others are intended to retain sediment in the original placement for extended periods of time [5,40]. This is assessed here by tracking the excess (or deficit) sediment volume within the placement's footprint. Figures 12–14 show a time series of the percent volume retained through time, as well as the wave forcing, for the SP, OD, and VB regional scenarios, respectively.

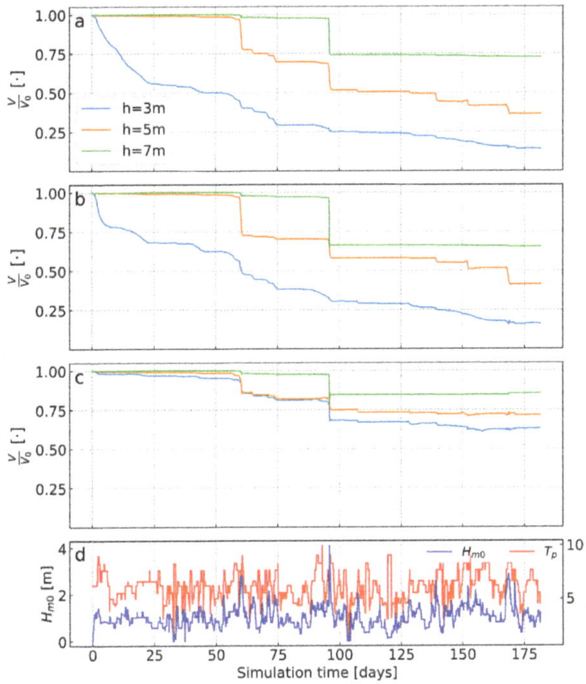

Figure 12. Longevity of nearshore nourishments at South Padre Island site in terms of percent placed volume remaining in original footprint through time where the blue, orange, and green colors represent the 3 m, 5 m, and 7 m placement depths, respectively. (**a**) Linear berm; (**b**) undulated berm; (**c**) discrete mounds; (**d**) time series of offshore zero-moment wave height and peak period.

The SP longevity curves (Figure 12) are clearly marked by episodic events, especially for the deeper placements ($h = 5$ m or $h = 7$ m). This is likely due to the limited potential of the smaller waves for sediment remobilization at the greater depths. The LB and, to a slightly lesser extent, the UB exhibit continuous sediment transport away from the original placement footprint for the shallowest placement ($h = 3$ m). The rate of nourishment deflation, i.e., reduction of placed sediment volume, for this placement depth appears approximately exponential for the LB, while the UB appears more linear after an initial adjustment. The DM placement at all depths more or less only responds to the two large events at approximately 60 and 95 days. Outside these two events, there is small, continuous deflation for the shallowest placement only.

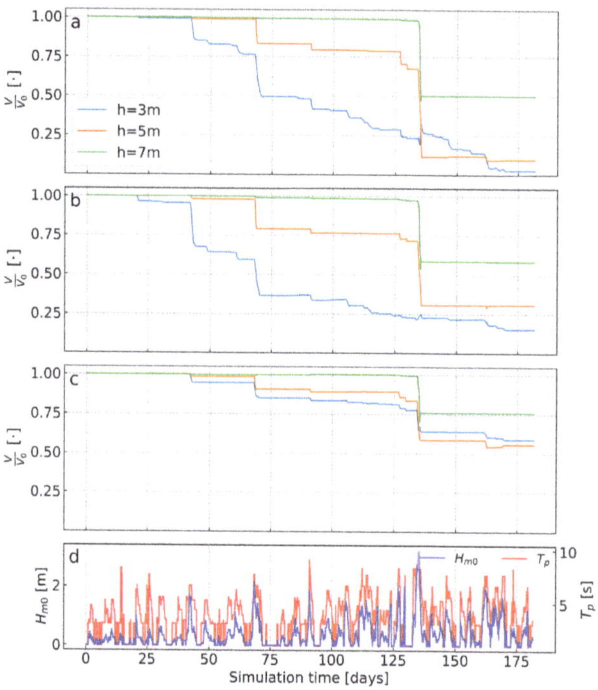

Figure 13. Longevity of nearshore nourishments at the Ogden Dunes site in terms of percent placed volume remaining in original footprint through time where the blue, orange, and green colors represent the 3 m, 5 m, and 7 m placement depths, respectively. (**a**) Linear berm; (**b**) undulated berm; (**c**) discrete mounds; (**d**) time series of offshore zero-moment wave height and peak period.

The longevity curves for the OD regional scenario appear more episodic than the SP curves, especially for the LB and UB placements at $h = 3$ m. As opposed to the SP scenario, there is little continuous nourishment deflation outside the events. The morphodynamics of the DM are also episodic. It is interesting that only one event (at approximately day 130) remobilized the deepest placement ($h = 7$ m), but it does so to a large degree, deflating almost 50%, 45%, and 25% of the sediment volume for the LB, UB, and DM placement shapes, respectively. This is also exhibited by the SP scenario for the large event at approximately day 95, except the relative quantity of berm deflation was not as large. These energetic events also deflate the deeper placements to a larger degree relative to the shallower placements. This suggest that large events are required to mobilize sediments placed deeper in the water column, as would be expected, and that, as they are more frequently reworked by waves, the shallower placements are more likely to be closer to equilibrium when large events occur.

The nourishment longevity at the energetic VB site (Figure 14) has some notably different features from the other two locations, but is comparable to the trends observed at SP. The LB and UB exhibit an initial period of adjustment at $h = 3$ m, followed by gradual evolution, similar to SP, while the deeper placements are markedly episodic. The longevity of the DMs at VB also shares characteristics with the other regional scenarios. The deepest placement is mostly unaffected by anything besides the largest high-energy events. The longevity of $h = 3$ m and $h = 5$ m placements track each other through time, but the shallower placement responds to forcing to a large degree, leading to a systematic difference. It should be noted that the 2015 nearshore berm constructed at VB in approximately 3 m became significantly deflated within 60 days and completely dispersed within 120 days [28], qualitatively replicating the current study's idealized results.

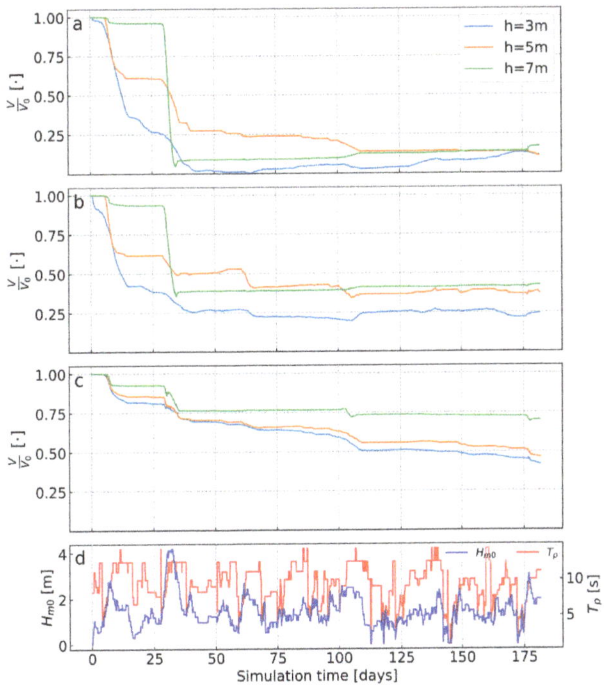

Figure 14. Longevity of nearshore nourishments at Vilano Beach site, in terms of percent placed volume remaining in original footprint through time where the blue, orange, and green colors represent the 3 m, 5 m, and 7 m placement depths, respectively. (**a**) Linear berm; (**b**) undulated berm; (**c**) discrete mounds; (**d**) time series of offshore zero-moment wave height and peak period.

A prolonged, high-energy event between day 5 and 18 significantly displaced sediment from the 3 m and 5 m LB and UB placements, but it did not mobilize sediment from the deeper placement. Significant deflation for the $h = 7$ m placement occurred approximately 10 days later (peak wave height at simulation day 30) where nearly 80% and 50% of the LB and UB berms sediment was transported away from the AoI, respectively. After this high-energy event, there is minor berm deflation for the LB or UB at any placement depth, in spite of additional high-energy events, suggesting that the profile has sufficiently adjusted towards equilibrium. The DMs' sediment volume at VB respond to the same events, but the deflation was reduced.

3.2.2. Cross-Shore/Alongshore Displacement

Nourishment deflation quantifies volumetric change rate and implies sediment transport away from the placement site, but does not elucidate the sediment's kinematics, which are important for design and management. Whether a nourishment migrates onshore, offshore, or alongshore figures substantially in the nearshore sediment budget and will influence shoreline morphodynamics. The evolution of cross-shore profiles is presented in Figures 15–17. The cross-shore profile is located in the middle of the domain. The shallow LB typically migrate onshore, while deeper placements show deflation of the nourishment, at least at the center profile. The shallow UB migrate on shore as well, but show more deflation than the LB. The DM appear to smooth out and contribute to the nearshore profile's elevation over a larger extent.

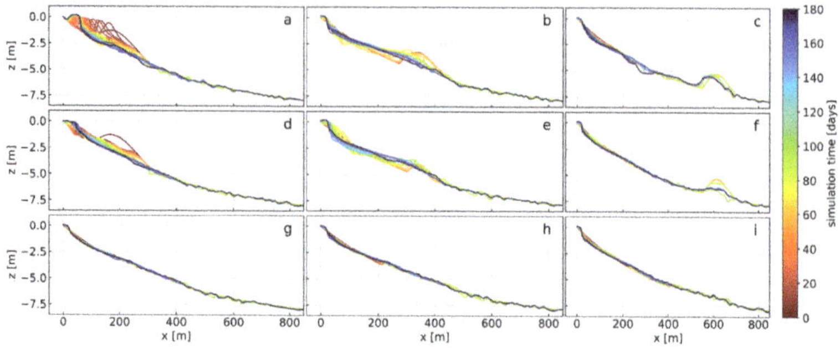

Figure 15. Central profile evolution for SP, where color indicates elapsed simulation time. (**a–c**) LB at 3 m, 5 m, and 7 m depths, respectively. (**d–f**) UB at 3 m, 5 m, and 7 m depths, respectively. (**g–i**) DM at 3 m, 5 m, and 7 m depths, respectively.

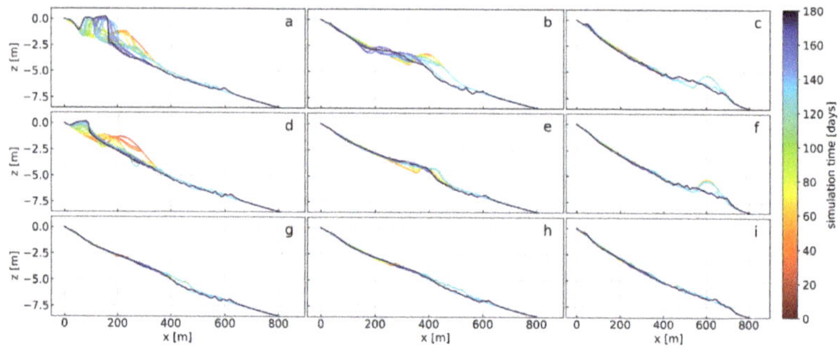

Figure 16. Central profile evolution for OD, where color indicates elapsed simulation time. (**a–c**) LB at 3 m, 5 m, and 7 m depths, respectively. (**d–f**) UB at 3 m, 5 m, and 7 m depths, respectively. (**g–i**) DM at 3 m, 5 m, and 7 m depths, respectively.

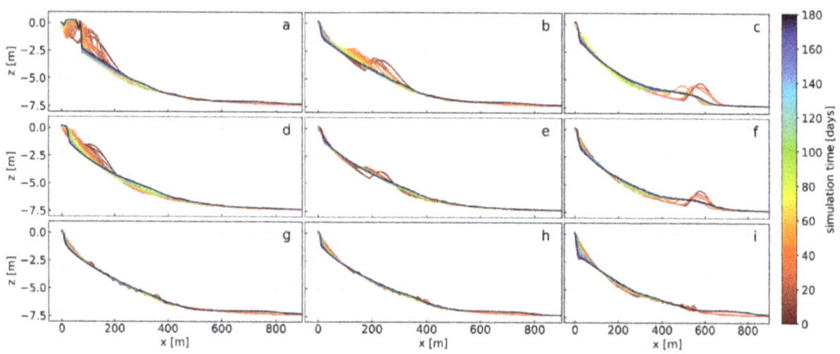

Figure 17. Central profile evolution for VB, where color indicates elapsed simulation time. (**a–c**) LB at 3 m, 5 m, and 7 m depths, respectively. (**d–f**) UB at 3 m, 5 m, and 7 m depths, respectively. (**g–i**) DM at 3 m, 5 m, and 7 m depths, respectively.

To better understand the sediment's bulk movement, the nourishment's center of mass is calculated through time and expressed as the cross-shore and alongshore displacements from its original position. Figures 18–20 show the different nourishment centers of mass for the SP, OD, and VB scenarios, respectively. The nourishment migration for the SP scenario in Figure 18 shows that there is considerable difference between the morphodynamics of the LB and UB placement types. While it was expected that the DM would display characteristics largely different from the other two nourishment types, the similarity between the LB and UB morphologies (alongshore linear features) was expected to lead to similar cross-shore/alongshore movement. This is clearly not the case, as the shallower placements ($h = 3$ m and $h = 5$ m) for the LB show large alongshore displacements, while the same placement depths for the UB do not. The cross-shore displacements for each nourishment shape are similar in character. There is either onshore migration or minimal movement outside high-energy events that transport sediment offshore. The tendency to migrate onshore seems to increase with increasing placement depth.

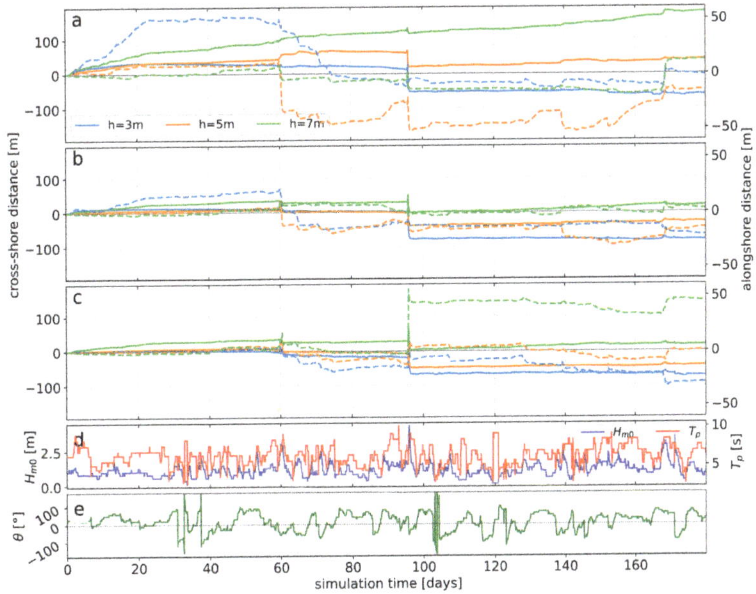

Figure 18. Time series of cross-shore (solid line) and alongshore (dashed lines) displacements of nearshore nourishment sediment for the South Padre Island scenario for 3 m (blue lines), 5 m (orange lines), and 7 m (green) placement depths, respectively. (**a**) Linear berm; (**b**) undulated berm; (**c**) discrete mounds; (**d**) offshore wave height and peak period; (**e**) wave direction relative to shore-normal.

The characteristic of nourishment migration at OD (Figure 19) shows similarity between the LB and UB, and to a lesser degree the DM, in contrast to the dissimilarity in the SP scenario. For the LB, UB, and DM, there is strict onshore movement with minimal offshore transport during high-energy events, which increases for the deeper placements ($h = 5$ m and $h = 7$ m). The alongshore movement increases with decreasing depth and is in the same direction for each nourishment shape and placement depth. Further, the patterns of cross-shore and alongshore movement for the 3 m placement depth, and to a lesser extent the 5 m and 7 m placements, is nearly identical between the LB and UB shapes.

Nearshore nourishment migration at the VB scenario (Figure 20) is distinctly marked by the high-energy event around day 30. For the LB, at the $h = 3$ m depth, the nearshore berm is displaced offshore more than 100 m in the course of a day. The pattern is the same for the $h = 5$ m placement but is reduced in magnitude, but the deepest placement migrates onshore during this event. These trends for the UB are different in that the event does not force offshore movement for the $h = 3$ m or $h = 5$ m placements but does generate some onshore migration for the 7 m placement. The dissimilarity between the UB and LB shapes at VB accords with the SP scenario, but contrasts with the conformity of responses found at the OD site.

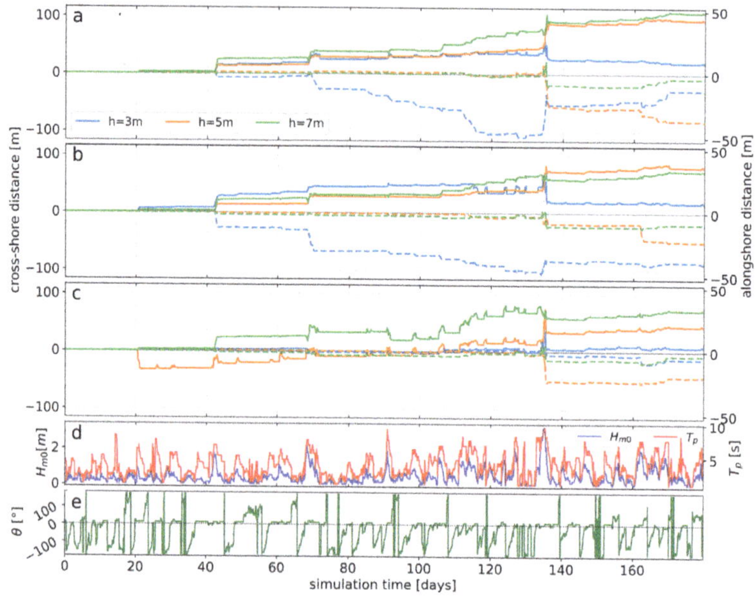

Figure 19. Time series of cross-shore (solid line) and alongshore (dashed lines) displacements of nearshore nourishment sediment for the Ogden Dunes scenario for 3 m (blue lines), 5 m (orange lines), and 7 m (green) placement depths, respectively. (**a**) Linear berm; (**b**) undulated berm; (**c**) discrete mounds; (**d**) offshore wave height and peak period; (**e**) wave direction relative to shore-normal.

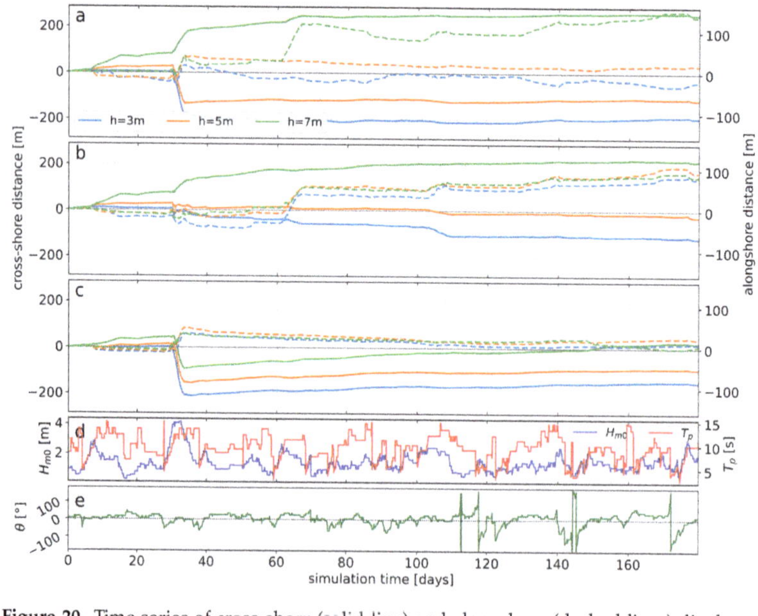

Figure 20. Time series of cross-shore (solid line) and alongshore (dashed lines) displacements of nearshore nourishment sediment for the Vilano Beach scenario for 3 m (blue lines), 5 m (orange lines), and 7 m (green) placement depths, respectively. (**a**) Linear berm; (**b**) undulated berm; (**c**) discrete mounds; (**d**) offshore wave height and peak period; (**e**) wave direction relative to shore-normal.

The DM are similarly displaced offshore by the high-energy event at day 30, except that the $h = 7$ m placement also moves offshore whereas the LB and UB migrate onshore. In each nourishment shape, there is initially negative alongshore displacement, followed by alongshore migration in the opposite direction after the same event. After that point, the alongshore patterns diverge, but the general trend is slow migration in the negative direction.

In general, the $h = 7$ m placements migrate onshore over the simulation period. The DM in the VB scenario appear to be an exception to this, but, in spite of the episodic offshore movement near day 35, the net displacement of the nearshore nourishment is onshore. The shallower placements ($h = 3$ m or $h = 5$ m) show considerably more variability between regions and nourishment shapes, outside the episodic nature of sediment transport. The net direction, i.e., onshore/offshore and $\pm$ alongshore, are typically the same for the LB and UB at a given region with comparable magnitudes of displacement. It is also noted that, although the OD nourishments deflate slowly, implying mild sediment transport rates, the direction of transport was dominantly onshore.

4. Discussion

In general, the results indicate several trends, including (1) shallower placements attenuate more wave energy; (2) shallower placements tend to evolve more continuously due to enhanced wave-driven sediment transport; (3) nearshore nourishment transport is typically onshore outside of periodic high-energy events, which generate offshore sediment transport.

4.1. Wave Attenuation

4.1.1. The Influence of Placement Depth

The clearest observable trend is that shallow placement depths are critical to appreciable wave energy attenuation. The central interquartile range of waves at the SP and VB regional scenarios exhibit greater mean wave energy attenuation for the shallower placements, thus indicating that appreciable energy attenuation is occurring for the majority of the waves. Consequently, both the infrequent, high-energy events and the more typical sea-states are attenuated to a greater degree by the shallower placements. Nearshore nourishments often produce shoreline morphodynamics, i.e., a landward salient, similar to submerged breakwaters [28]. The wave attenuation landward of the nourishments suggest that alongshore gradients in the alongshore component of the radiation stress generates convergent alongshore currents and, thus, accretion.

The amplified smaller waves at OD likely break near the shoreline and are not significantly influenced by the nearshore nourishment. Additionally, the large relative amplification for these small waves only indicates an increase of a few centimeters, which is within the range of numerical modeling accuracy. These smaller waves may play a marginal role in shoreline morphodynamics and wave heights > 0.3 m at this site, following the observable trend of increased wave energy attenuation with shallower placement. The observation that placement depth may produce wave amplification within the AoI is clearly the case for the DM at the SP and VB regional scenarios. When placed at $h = 7$ m there is nontrivial ($\overline{\alpha} > 10\%$) amplification for the larger wave heights. While this also occurs for the shallower placements, the trend is not as general. The causal mechanism is not clear, but it may be generated by refraction into the control interface of the AoI caused by the shallower placement area. This wave transformation process is evident in Figure 8, and the free-surface dynamics of AoI may be a future topic of investigation.

In practice, the placement depth should depend on the wave climate and be determined during the design phase of the nearshore nourishment strategy. For instance, only the $h = 3$ m placement attenuates wave energy for the SP and OD scenarios (see panel a and b, Figures 9 and 10, respectively), while there is comparable energy attenuation between the $h = 3$ m and $h = 5$ m placements for many wave height bins in the VB scenario (see panel a and b in Figure 11). In consideration of the longevity curves in Figures 12–14, where the deeper berms' deflation is driven by high-energy, episodic events, one could argue for a sophisticated approach to design where wave energy attenuation is optimized for a certain range of wave height, while longevity is optimized by placing the berm deep enough to avoid continuous evolution. This may be a zero-sum game, in that attenuating wave energy is a driver of berm deflation. However, the mechanisms of sediment transport (e.g., wave-averaged transport versus wave-breaking induced alongshore current transport) would vary depending on placement depth and influence berm deflation rates. Future in-depth analysis of the relevant sediment transport mechanisms may include phase resolving wave models.

4.1.2. The Influence of Nourishment Shape

A clear distinction between the alongshore bar-type berms, i.e., the linear and undulated berms, and the discrete mounds is evident in the wave attenuation results. While placement depth is critical for inducing wave-breaking energy dissipation, nourishment shape largely determines if wave energy attenuation occurs. For the SP and OD scenarios, appreciable wave attenuation only occurs for the LB and UB. The VB scenario is the exception, in that the DM attenuate wave energy at the 3 m and 5 m depths for the larger waves, however the most frequent waves do not exhibit much attenuation. Clearly, if shoreline protection through wave energy attenuation is a design objective, then mimicking an alongshore bar during berm construction should be pursued.

Comparing the influence of nourishment shape between the LB and UB wave attenuation analyses reveals an unexpected result. The UB clearly attenuates more wave energy at the higher wave heights and, to a lesser degree, at the middle wave heights. This is somewhat counterintuitive in that the undulations would presumably not generate as uniform wave-breaking energy dissipation within the incident wave field. Further, this is evidenced by the spatial distribution of mean relative wave energy in Figure 8. However, the UB clearly affects wave transmission through the AoI to a larger degree than the LB, particularly in the SP and OD scenarios. A plausible explanation is the difference in alongshore length between the LB and UB. The undulations in berm crest height in the UB necessitated an extension in the berm's alongshore dimension to maintain an approximately constant nearshore nourishment volume between placements. This is a positive result, considering most linear nearshore berm projects are constructed with undulations due to construction techniques such as periodic movement of the sediment discharge pipe [36]. Differences in the longevity of the 3 m UB and LB may also contribute to the different wave attenuation results. The shallow UB disperse at somewhat slower rates in the SP and VB cases, so there may have been a larger fraction of the berm impacting some of the waves for a longer duration.

This trend has important implications for nearshore nourishment design. If undulated berms consistently attenuate wave transmission to a larger degree, and do so by sheltering a larger stretch of shoreline, then berm design should optimize alongshore extent. The optimization needs to consider the undulation's maximum berm crest height required to induce wave breaking, as well as the alongshore wave length required to dissipate sufficient energy. Further numerical modeling may be able to explore a parameter space of undulated berm designs with the aim to investigate whether any easily constructible undulation dimensions clearly optimize wave energy dissipation.

4.2. *Sediment Transport and Nourishment Migration*

4.2.1. Nourishment Deflation Rates

The design lifespan of nearshore nourishments depends on the rate at which sediment is exported from the initial placement and, therefore, the rate at which the nourishment's volume decreases, i.e., the nourishment deflation rate. The deflation rate is a function of sediment transport that depends on the nearshore hydrodynamics and sedimentology. Here, we have controlled for these factors to explore the effects of nourishment shape and the cross-shore/alongshore distribution of nearshore nourishment material.

One evident trend in the longevity curves (see Figures 12–14) is the difference between episodic events and continuous background evolution. The DM do not exhibit appreciable background sediment transport and are instead dominated by episodic sediment transport. The longevity curves are punctuated by step-wise deflation corresponding to the high-energy events, where the wave height was large enough to mobilize the sediment. The movement of the DMs' centers of mass were similar between placement depths with generally the same alongshore/cross-shore trajectories per regional scenario. The deeper placements ($h = 7$ m) for LB and UB show a similar pattern and are only remobilized by the largest events within the simulations. Interestingly, the UB placements generally deflate less than the LB for the same events, e.g., for the event at day 130 at the SP scenario, the LB lost approximately 50% of volume while the UB lost approximately 40%. These results suggest that if retaining sediment on the profile is the primary objective, then deeper, more diffuse placements may optimize nourishment longevity.

4.2.2. Direction of Net Sediment Transport

In general, the direction of net sediment transport and the relative importance of the episodic deflation varies with nourishment shape and placement depth. However, the center of mass trajectories (see Figures 18–20) show that sediment is often transported onshore, especially for the deepest placements. During the high-energy storm events, a large amount of sediment is then transported offshore with a return to onshore transport that is inconsistent between nourishment types and placement depths. This result has been observed in other nearshore nourishment laboratory studies [13] and is expected, as it has long been recognized that storm events tend to move beach/beachface sediment offshore to form an offshore bar.

A few of the simulations exhibit net seaward transport, which is contrary to the expected behavior of onshore movement for nearshore nourishments [9]. However, this may be due to the number of storm events and the length of the simulation. Given the positive trend in the centroids' cross-shore position through time, it is reasonable to assume that net onshore movement would result after a longer duration. The steady onshore movement of some placements (e.g., the LB placed at 3 m at Vilano Beach in Figure 20a) outside the events suggests that the profiles are out of equilibrium with the wave climate or the bed's sediment composition, which would prefer to generate a steeper profile. This concept could be useful for the practice of beneficially placing nearshore nourishments. For example, if the antecedent conditions of the nearshore profile are out of equilibrium, then sediment could be placed deeper with the expectation that it will migrate onshore and nourish the profile. This may be advantageous for the beneficial use of dredge material, which incurs costs or lacks feasibility (due to dredge vessel limitations) when placing sediment in shallower water near the shoreline.

The sediment transport simulated within these idealized scenarios is driven by wind-waves alone. This simplification neglects the other drivers of nearshore sediment transport such as tides and winds. Tides modulate the cross-shore surfzone location and winds may generate significant nearshore circulation. Including these forcings would have changed the sediment transport calculations, but they are not the primary driver of nearshore circulation, which is free-surface waves. Since the objective of the study was to examine the influence of nourishment shape/depth on sediment transport and wave attenuation, these other processes were neglected and reserved for future work.

5. Conclusions

A numerical model (CMS) coupling phase-averaged wind-waves, nearshore hydrodynamics, and sediment transport/morphological change is used to simulate the evolution of nearshore nourishments and their effects on the wave field. The numerical experiments varied nearshore nourishment shape (linear, undulated, and discrete mounds morphologies), placement depth (i.e., 3 m, 5 m, and 7 m depth), and the wave climate/nearshore geomorphology (three coastal regions of the U.S.). A total of 30 idealized cases were simulated and analyzed for inter-comparison.

The analyses elucidate the nearshore nourishments' relative effect on wave energy attenuation, nourishment deflation rates, and cross-shore/alongshore nourishment migration patterns. The primary findings of the analyses were:

1. Shallower placements attenuate more energy than deeper placements for the linear and undulated berms.
2. The linear and undulated berms dissipate more energy than the discrete mounds, in spite of similar placement volumes.
3. The undulated berm dissipates more energy than the linear berm, which is presumably due to its greater alongshore length and may also be influenced by longer lifespans.
4. Longevity analysis shows that placement depth discriminates between continuous and episodic deflation. Shallower placements are subject to more continuous sediment transport, while the deeper placements respond primarily to high-energy events.
5. The trajectories of the modeled nourishments' centers of mass evidences onshore-directed transport of nourishment sediment, which is punctuated by offshore-directed sediment transport due to high-energy, episodic events.

Author Contributions: Conceptualization, C.L.J., B.C.M., D.R.K. and M.E.B.; methodology, C.L.J., B.C.M., D.R.K. and M.E.B.; software, C.L.J. and M.E.B.; formal analysis, C.L.J.; resources, M.E.B.; data curation, C.L.J.; writing—original draft preparation, C.L.J.; Investigation, C.L.J.; writing—review and editing, C.L.J., B.C.M., D.R.K. and M.E.B.; visualization, C.L.J.; supervision, B.C.M.; project administration, B.C.M.; funding acquisition, B.C.M. All authors have read and agreed to the published version of the manuscript.

Funding: This research was funded by the U.S. Army Corps of Engineers Coastal Inlet Research Program (CIRP) through the Inlet Geomorphic Evolution work unit.

Conflicts of Interest: The authors declare no conflict of interest.

References

1. Houston, J.R. The value of Florida beaches. *Shore Beach* **2013**, *81*, 4–11.
2. Elko, N.; Brutsché, K.; Robertson, Q.; Hartman, M.; Dong, Z. *USACE Navigation Sediment Placement: An RSM Program Database (1998–2018)*; CHETN-XX-21; US Army Corps of Engineers Engineer Research and Development Center: Vicksburg, MS, USA, 2021; under review.
3. Ahrens, J.P.; Hands, E.B. Velocity parameters for predicting cross-shore sediment migration. *J. Waterw. Port Coast. Ocean Eng.* **1998**, *124*, 16–20. [CrossRef]
4. Bain, R.L.; McFall, B.C.; Krafft, D.R.; Hudson, A. Evaluating transport formulations for application to nearshore berms. *J. Waterw. Port Coast. Ocean Eng.* **2021**, *147*, 04021031. [CrossRef]
5. Hands, E.B.; Allison, M.C. Mound migration in deeper water and methods of categorizing active and stable depths. In Proceedings of the Coastal Sediments '91, Seattle, WA, USA, 25–27 June 1991.
6. McFall, B.C.; Brutsché, K.E.; Priestas, A.M.; Krafft, D.R. Evaluation techniques for the beneficial use of dredged sediment placed in the nearshore. *J. Waterw. Port Coast. Ocean Eng.* **2021**, *147*, 04021016. [CrossRef]
7. Brutsché, K.E.; McFall, B.C.; Li, H.; McNinch, J.E.; Ousley, J.D.; Engle, J.A.; Maglio, C.K. Strategic nearshore placement of dredged sediment at Vilano Beach, Florida. *Shore Beach* **2017**, *85*, 77–84.
8. Huisman, B.J.; Walstra, D.J.R.; Radermacher, M.; de Schipper, M.A.; Ruessink, B.G. Observations and modelling of shoreface nourishment behavior. *J. Mar. Sci. Eng.* **2019**, *7*, 59. [CrossRef]
9. van Duin, M.J.P.; Wiersma, N.R.; Walstra, D.J.R.; Van Rijn, L.C.; Stive, M.J.F. Nourishing the shoreface: Observations and hindcasting of the Egmond case, The Netherlands. *Coast. Eng.* **2004**, *51*, 813–837. [CrossRef]
10. Brutsché, K.E.; McFall, B.C.; Bryant, D.B. *Literature Review of Nearshore Berms*; SR-19-2; US Army Corps of Engineers Engineer Research and Development Center: Vicksburg, MS, USA, 2019; 59p.
11. Bryant, D.B.; McFall, B.C. Transport of nearshore dredge material berms. In Proceedings of the 6th International Conference on Application of Physical Modelling in Coastal and Port Engineering and Science, Ottawa, ON, Canada, 10–13 May 2016.
12. Smith, E.R.; Mohr, M.C.; Chader, S.A. Laboratory experiments on beach change due to nearshore mound placement. *Coast. Eng.* **2017**, *121*, 119–128. [CrossRef]
13. Li, Y.; Zhang, C.; Cai, Y.; Xie, M.; Wang, Y. Wave Dissipation and Sediment Transport Patterns during Shoreface Nourishment towards Equilibrium. *J. Mar. Sci. Eng.* **2021**, *9*, 535. [CrossRef]

14. Grunnet, N.M.; Walstra, D.J.R.; Ruessink, B.G. Process-based modelling of a shoreface nourishment. *Coast. Eng.* **2004**, *51*, 581–607. [CrossRef]
15. Hoekstra, P.; Houwman, K.T.; Kroon, A.; Ruessink, B.G.; Roelvink, J.A.; Spanhoff, R. Morphological development of the Terschelling shoreface nourishment in response to hydrodynamic and sediment transport processes. In Proceedings of the 25th International Conference on Coastal Engineering, Orlando, FL, USA, 2–6 September 1996; pp. 2897–2910.
16. Kriauciuniene, J.; Gailiusis, B.; Rimaviciute, E. Modelling of shoreface nourishment in the Lithuanian nearshore of the Baltic Sea. *Geologija* **2006**, *53*, 28–37.
17. Kuang, C.; Han, X.; Zhang, J.; Zou, Q.; Dong, B. Morphodynamic Evolution of a Nourished Beach with Artificial Sandbars: Field Observations and Numerical Modeling. *J. Mar. Sci. Eng.* **2021**, *9*, 245. [CrossRef]
18. Larson, M.; Ebersole, B.A. *An Analytical Model to Predict the Response of Mounds Placed in the Offshore*; CHETN-II-42; US Army Corps of Engineers Engineer Research and Development Center: Vicksburg, MS, USA, 1999; 13p.
19. McFall, B.C.; Smith, S.J.; Pollock, C.E.; Rosati, J.; Brutsché, K.E. *Evaluating Sediment Mobility for Siting Nearshore Berms*; CHETN-IV-108; US Army Corps of Engineers Engineer Research and Development Center: Vicksburg, MS, USA, 2016; 11p.
20. Pollock, C.E.; Curtis, W.R.; Moritz, H.R. *Numerical Methods for Nearshore-Berm Evaluation, St. Johns County, Florida*; CHL-98-12; US Army Corps of Engineers Engineer Research and Development Center: Vicksburg, MS, USA, 1998; 131p.
21. Priestas, A.M.; McFall, B.C.; Brutsche, K.E. *Performance of Nearshore Berms from Dredged Sediments: Validation of the Sediment Mobility Tool*; TR-19-19; US Army Corps of Engineers Engineer Research and Development Center: Vicksburg, MS, USA, 2019; 63p.
22. Zhang, J.; Larson, M. A numerical model for offshore mound evolution. *J. Mar. Sci. Eng.* **2020**, *8*, 160. [CrossRef]
23. Brutsché, K.; Pollock, C.E. *Strategic Placement of Mixed Sediment in the Form of a Nearshore Berm along Fort Myers Beach, Florida*; TN-EWN-17-1; US Army Corps of Engineers Engineer Research and Development Center: Vicksburg, MS, USA, 2017; 3p.
24. Dolan, R.; Donoghue, C.; Stewart, D. Long-term Impacts of Tidal Inlet Bypassing on the Swash Zone Filter Feeder Emerita talpoida Oregon Inlet and Pea Island, North Carolina. *Shore Beach* **2006**, *74*, 23–27.
25. Dolan, R.; Stewart, D. A concept for reducing ecological impacts of beach nourishment and tidal inlet bypassing. *Shore Beach* **2006**, *74*, 28–31.
26. Young, D.L.; Brutsché, K.E.; Li, H.; McFall, B.C.; Maloney, E.C.; McClain, K.E.; Bucaro, D.F.; LeRoy, J.Z.; Dunker, J.J.; Johnson, K.K.; et al. *Analysis of Nearshore Placement of Sediments at Ogden Dunes, Indiana*; TR-20-4; US Army Corps of Engineers Engineer Research and Development Center: Vicksburg, MS, USA, 2020; 98p.
27. Figlus, J.; Song, Y.-K.; Maglio, C.K.; Friend, P.L.; Poleykett, J.; Engel, F.L.; Schoenblen, D.; Boburka, K. Particle Tracer Analysis for Submerged Berm Placement of Dredged Material Near South Padre Island, Texas. *WEDA J. Dredg.* **2021**, *19*, 14–30.
28. Brutsché, K.E.; McFall, B.C.; Li, H.; McNinch, J.E.; Ousley, J.D.; Engle, J.A.; Maglio, C.K. Strategic Placement of Dredge Material in Vilano Beach, Florida, USA. In Proceedings of the Coastal Sediments 2019: Proceedings of the 9th International Conference, St. Petersburg, FL, USA, 27–31 May 2019; pp. 2793–2804.
29. Bryant, M.A.; Hesser, T.J.; Jensen, R.E. *ERDC/CHL CHETN-I-91*; U.S. Army Engineer Research and Development Center: Vicksburg, MS, USA, 2016.
30. McFall, B.C. The Relationship between Beach Grain Size and Intertidal Beach Face Slope. *J. Coast. Res.* **2019**, *35*, 1080–1086. [CrossRef]
31. De Masi, G.; Bruschi, R.; Drago, M. *Synthetic Metocean Time Series Generation for Offshore Operability and Design Based on Multivariate Markov Model*; MTS/IEEE OCEANS; Discovering Sustainable Ocean Energy for a New World: Genova, Italy, 2015.
32. Militello, A.; Reed, C.W.; Zundel, A.K.; Kraus, N.C. *Two-Dimensional Depth-Averaged Circulation Model M2D: Version 2.0, Report 1, Technical Documentation and User's Guide*; TR-04-2; U.S. Army Engineer Research and Development Center: Vicksburg, MS, USA, 2004; 131p.
33. Reed, C.W.; Brown, M.E.; Sanchez, A.; Wu, W.; Buttolph, A.M. The Coastal Modeling System Flow Model (CMS-Flow): Past and Present. *J. Coast. Res.* **2011**, *59*, 1–6. [CrossRef]
34. Sanchez, A.; Wu, W.; Li, H.; Brown, M.; Reed, C.; Rosati, J.D.; Demirbilek, Z. *Coastal Modeling System: Mathematical Formulations and Numerical Methods*; TR-14-2; U.S. Army Engineer Research and Development Center: Vicksburg, MS, USA, 2014; 87p.
35. Sanchez, A.; Wu, W.; Beck, T.M. A depth-averaged 2-D model of flow and sediment transport in coastal waters. *Ocean Dyn.* **2016**, *66*, 1475–1495. [CrossRef]
36. Lin, L.; Demirbilek, Z.; Mase, H.; Zheng, J.; Yamada, F. *CMS-Wave: A Nearshore Spectral Wave Processes Model for Coastal Inlets and Navigation Projects*; TR-08-13; U.S. Army Engineer Research and Development Center: Vicksburg, MS, USA, 2008; 132p.
37. Wu, W.; Sánchez, A.; Zhang, M. An implicit 2-D shallow water flow model on an unstructured quadtree rectangular grid. *J. Coast. Res.* **2011**, *59*, 15–26. [CrossRef]
38. Lin, L.; Demirbilek, Z.; Mase, H. Recent capabilities of CMS-Wave: A coastal wave model for inlets and navigation projects. *J. Coast. Res.* **2011**, *59*, 7–14. [CrossRef]
39. Camenen, B.; Larson, M. A General Formula for Non-Cohesive Suspended Sediment Transport. *J. Coast. Res.* **2008**, *24*, 615–627. [CrossRef]
40. Brutsché, K.E.; Wang, P.; Beck, T.M.; Rosati, J.D.; Legault, K.R. Morphological evolution of a submerged artificial nearshore berm along a low-wave microtidal coast, Fort Myers Beach, west central Florida, USA. *Coast. Eng.* **2014**, *91*, 29–44. [CrossRef]

Journal of
Marine Science and Engineering

MDPI

Article

Subaqueous and Subaerial Beach Changes after Implementation of a Mega Nourishment in Front of a Sea Dike

Anna Kroon [1,2,*], Matthieu de Schipper [1], Sierd de Vries [1] and Stefan Aarninkhof [1]

[1] Department of Hydraulic Engineering, Faculty of Civil Engineering and Geosciences, Delft University of Technology, P.O. Box 5048, 2600 GA Delft, The Netherlands
[2] Svašek Hydraulics, Kratonkade 23, 3024 ES Rotterdam, The Netherlands
* Correspondence: j.kroon@tudelft.nl

Abstract: Sandy nourishments can provide additional sediment to the coastal system to maintain its recreational or safety function under rising sea levels. These nourishments can be implemented at sandy beach systems, but can also be used to reinforce gray coastal infrastructure (e.g., dams, dikes, seawalls). The Hondsbossche Dunes project is a combined shoreface, beach, and dune nourishment of 35 million m^3 sand. The nourishment was built to replace the flood protection function of an old sea-dike while creating additional space for nature and recreation. This paper presents the evolution of this newly created sandy beach system in the first 5 years after implementation based on bathymetric and topographic surveys, acquired every three to six months. A significant coastline curvature is created by the nourishment leading to erosion in the central 7 km bordered by zones with accretion. However, over the five-year period, net volume losses from the project area were less than 5% of the initial nourished sand volume. The man-made cross-shore beach profile rapidly mimics the characteristics of adjacent beaches. The slope of the surfzone is adjusted within two winters to a similar slope. The initially wide beaches (i.e., up to 225 m) are reduced to about 100 m-wide. Simultaneously, the dune volume has increased and the dune foot migrated seaward at the entire nourished site, regardless of whether the subaqueous profile gained or lost sediment. Our results show that the Hondsbossche Dunes nourishment, built with a natural slope and wide beach, created a positive sediment balance in the dune for a prolonged period after placement. As such, natural forces in the years after implementation provided a significant contribution to the growth in dune volume and related safety against flooding.

Keywords: nourishments; building with nature; coastline evolution; dune growth; Hondsbossche Dunes; coastal morphology

Citation: Kroon, A.; de Schipper, M.; de Vries, S.; Aarninkhof, S. Subaqueous and Subaerial Beach Changes after Implementation of a Mega Nourishment in Front of a Sea Dike. *J. Mar. Sci. Eng.* **2022**, *10*, 1152. https://doi.org/10.3390/jmse10081152

Academic Editor: Carlos Daniel Borges Coelho

Received: 14 July 2022
Accepted: 16 August 2022
Published: 20 August 2022

1. Introduction

Acceleration of sea level rise will require an increasing capacity to adapt to adverse changes e.g., [1,2] due to coastal erosion or reduced safety against flooding. At eroding coastlines, sandy (beach) nourishments have been used to mitigate the loss of sediment [3,4] to preserve recreational or safety functions. Nourishments can be implemented at beach systems, but can also be used to adapt gray coastal infrastructure (e.g., dams, dikes, seawalls) to new coastal management views, e.g., [5–7]. Coastal management views in which soft, nature-inclusive, and adaptive measures are preferred over more traditional hard protection measures [8,9], often referred to as Building with Nature, Engineering with Nature, or Living Shorelines.

In the Netherlands, nourishments are a key part of the coastal zone management. Nourishment strategy has become more large-scale and proactive in the last decades, including the use of mega-nourishments [10,11]. These mega-nourishments are sandy interventions in the coastal zone where large amounts of sediment (i.e., >500 m^3/m alongshore for beach nourishment or >1000 m^3/m alongshore for shoreface nourishment) are implemented. When mega nourishments are added to a beach, both cross-shore profile and alongshore shoreline curvature are strongly altered. This brings the coastal system out of equilibrium compared with its long-term average topography [12].

The Hondsbossche Dunes mega-nourishment project is a man-made sandy beach system constructed of 35 million m^3 sand. It is placed in front of a sea dike that was considered a weak link in the Dutch sea defense (Figure 1). The new sandy coastal defense consists of a shoreface, beach, and dune, aimed to increase safety against flooding while creating space for nature and recreation. As such, this project transferred (part of) the safety function from gray infrastructure to a soft sandy defense. The nourishment project significantly altered the coastal system in both along- and cross-shore directions. After placement, the newly created beach was on average 1.5–2 times wider, the subaqueous slope 1.5–2.5 times steeper, and the coastline curvature about 4 times larger than the adjacent coastal sections. The addition of large sediment volumes and the new cross-shore profile in a region that was protected by a dike for decades are bound to invoke a strong coastline response.

Figure 1. (**a**) Hondsbossche sea dike before placement of the nourishment (courtesy of Rijkswaterstaat, https://beeldbank.rws.nl/, accessed on 6 September 2021) and (**b**) the Hondsbossche Dunes Nourishment just after placement in April 2015 (courtesy of Aannemerscombinatie Zwakke Schakel–Van Oord-Boskalis). The orange line marks the crest of the original sea dike.

The planform adaptation of nourishments on the timescales of years is generally assumed to be governed by wave-driven alongshore sediment transport gradients [3,13]. Planform adaptation herein is symmetric in the alongshore, resulting in sediment accumulation in both adjacent beaches. Ludka et al. [14] show that this spreading can also be asymmetric, with wave direction correlating to the displacement of the center of mass of the nourishment.

Cross-shore profile equilibration after nourishment implementation occurs typically in the order of weeks to years [3]. Steep post-nourishment profiles adapt initially fast with sediment from the subaerial beach moving downslope [3,12,15]. This first adaptation in the cross-shore can be strongly impacted by high-energy events [16]. Over time, the profile equilibration slows down as the profile approaches a new dynamic equilibrium shape and steepness.

The beach width of the nourished beach is an important profile indicator from an economic and recreational perspective, e.g., [17–19]. After implementation of a beach nourishment, the beach is often at its widest and the reduction in beach width thereafter can be attributed to either horizontal movement of the waterline position, changes in dune foot position, or both. The variations in dune foot position originate from dune erosion during storms and deposition of aeolian sand transport. Aeolian transport can bring sediment from the intertidal zone and beach to the dunes at different timescales [20,21], with many processes influencing the magnitude of wind-driven sediment transport [22]. The resulting beach width is closely related to the shape of the dynamic equilibrium profile, as described by [23,24]. As such, the remaining dry beach width after nourishment can depend significantly on the size of nourished grains compared with native [3,24].

Large beach widths at mega-nourishments could lead to an increased dune volume over time [25] and create accommodation space for embryo-dune development [26]. Yet, [27] report that at least half of the subaerial sediment deposits at a dune landward of a mega-nourishment may originate from the intertidal zone. Intertidal beach geomorphology may therefore be as important as the total beach width. Moreover, complex subaerial beach configurations (e.g., variability in dune front orientation) can cause (locally) increased or decreased deposition [28]. Spatial differences in wave energy have also been related to differences in dune volume increase [29,30]. High-energy events with wave run-up and storm surge can furthermore result in sediment deposition near the dune foot [31].

Expectations on the evolution of coastal state indicators, such as beach width (as proxy for recreation) and subaerial sediment volume (as proxy for coastal safety), are important in engineering practice and creating stakeholder support of (mega-)nourishments. Our aim is to document the subaqueous and subaerial beach changes at the Hondsbossche Dunes nourishment as a nature-based solution for integrated coastal development. Therefore, we investigate how beach width and profile volumes coevolve in the first five years after implementation of the Hondsbossche Dunes mega nourishment. We analyze the change in profile volume, beach width, and profile steepness at 250 m spaced transects measured on a quarterly to yearly basis and compare this evolution with that of the adjacent coast.

The next section will describe the Hondsbossche Dunes nourishment in more detail. In Section 3, the data and method are elaborated on, followed by the results of the data analysis in Section 4. In Section 5, the results are discussed and, in Section 6, the conclusions can be found.

2. Case Study

2.1. Coastal Setting

The Hondsbossche Dunes nourishment is situated at the Northern North-Holland (NNH) coast in the Netherlands. The NNH coast is a sandy, wave-dominated coast bounded by the Marsdiep tidal inlet in the north and the breakwaters of IJmuiden harbour in the south (Figure 2b).

Prior to 2015, the NNH the sandy coastline was interrupted in the middle by the Hondsbossche and Pettemer sea defense. The sea dike had protected the low-lying hinterland since 1887. This location protruded seaward with respect to the surrounding coastline, making it an erosional hotspot and a location where human interventions date back several centuries [32]. Both this sea dike and the adjacent sandy coastal sections contained regular-spaced shore parallel beach groins (Figure 1). The nearshore bathymetry south of the sea dike (km 24–55, Figure 2c) is characterized as a cyclic multiple (2–3) bar system, with offshore migration in a cyclic period of approximately 15 years [33]. The section north of the original sea dike (km 8-23, Figure 2c) is characterized by a single nearshore bar without cyclic behavior. In this northern section, a shoreface shoal (the Pettemer Polder, km 20, Figure 2c) is also present at a depth of about 10 m.

The native sand has a gradation between 250 and 300 µm around the waterline [33]. The material found in the dune ranges between 220 and 280 µm [34]. At deeper water (i.e., −5 to −10 m + NAP), the material is, in general, finer [35], ranging between 170 and 200 µm.

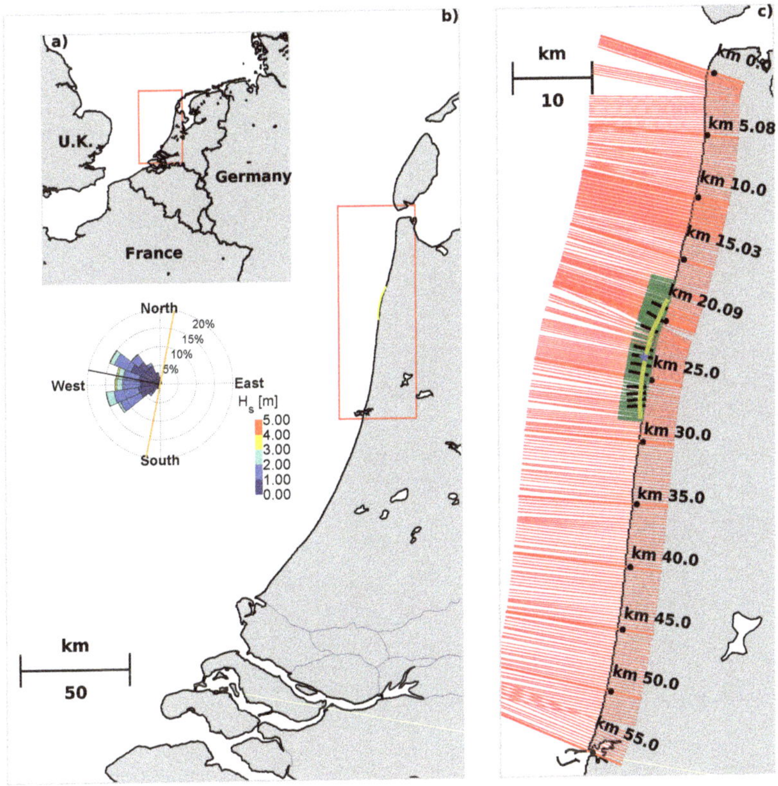

Figure 2. Location of case study site and an overview of the survey transects from available datasets. (**a**) Location of Dutch coastline within Europe. (**b**) Location of Hondsbossche Dunes nourishment (yellow line) at the Dutch coast including local nearshore wave climate at HD at −10 m + NAP depth based on a 20-year hindcast time-series derived by Kroon et al. [36]. (**c**) Location of transects from Jarkus (red), contractor survey transects (green), and 11 constructor high-temporal-resolution transects (black); the blue star indicates the location of the nearshore wave climate.

The NNH coast is exposed to a semidiurnal tide with a range of about 1.6 m. Mean low and mean high water are at ±0.8 m + NAP [37] (NAP is the Dutch reference level, roughly equal to mean sea level). Wind waves are mainly approaching from a southwesterly and northwesterly direction with longer period waves arriving mostly from the north [33]. The annual mean wave height H_s is 1.0 m, coinciding with wave periods typically of 4.3 s, at a depth of −10 m + NAP in the central part of the project site (location indicated with blue star in Figure 2c). More extreme wave heights with a 1/20 y recurrence, at the same location, have a height H_s = 4.7 m and period of $T_{m-1,0}$ = 8 s. The full wave rose based on a 20-year time-series of hindcast waves [36] is presented in Figure 2b. The spatial variation in the offshore wave climate along the Holland coast is small.

The net alongshore sand transport is estimated to increase over the NNH coast from 250,000 south to 550,000 m^3/m/y north [38]. According to this estimate, the gradient over the project area is about 100,000–250,000 m^3/y. Since the 1980s, almost 70 nourishments have been placed along the NNH coast within the framework of the Dutch coastal maintenance program [10]. Before 2015, the nourishment volume was, on average, 1.55 million m^3/y at the NNH coast and 250,000 m^3/y at the project area, Table 1. In the post-construction period of the Hondsbossche dunes project evaluated here, five nourishments have been implemented along the NNH coast: a combined beach and shoreface nourishment was placed between

12.13 and 14.21 km in 2017 with a combined volume of 720 m^3/m; this was repeated in 2019 for the shoreface with a volume of 530 m^3/m. A shoreface nourishment with a volume of 280 m^3/m was placed between km 31 and 40 in 2015 and repeated in 2019. Two beach nourishments with a volume of 220 m^3/m were placed between km 32 and 34 and between km 37 and 39 in 2015 [39,40].

Table 1. Total nourishment volumes [$\times 10^6$ m^3/y] at the NH coast [41].

Section	North	HD	South	Total
Period\KM	5–18.8	18.8–28	28–50	5–50
1986–2015	0.65	0.25	0.65	1.55
2015–2020	0.55	7	1.4	8.95

2.2. Hondsbossche Dunes Nourishment

The Hondsbossche and Pettemer sea dike was marked as a weak link in the Dutch sea defense with safety against flooding being below the desired level. Instead of dike reinforcement (e.g., heightening, widening), it was decided to increase safety against flooding by placing a sandy beach and (multirow) dune system in front of the dike. To achieve this purpose, a nourishment was designed that could meet the desired safety standard. This resulted in a nourishment of 35 million m^3 sand, with an average nourished volume of over 4000 m^3/m that exceeds locally 5000 m^3/m. The Hondsbossche Dunes nourishment is placed over a length of approximately 9 km coastline and covers almost 800 ha of subaqueous and 350 ha of subaerial domain.

The subaqueous profile was designed as a Dean [24] profile fitted to observed profiles of the adjacent coast. Although in practice placed steeper, the placed subaqueous volume was sufficient to accommodate adjustment to a profile slope close to this expected dynamic equilibrium slope. This design philosophy was adopted to prevent large subaerial readjustments, albeit with uncertainty regarding the applicability of a Dean [24] profile for a nourishment of this magnitude. The subaqueous profile is supplemented with a spatially varying buffer layer to compensate for expected alongshore losses, resulting in a design with wide beaches. The subaerial design is primarily based on a safety assessment with Duros+ [42] and an additional surcharge to account for the curved coastline [43]. Furthermore, the design included a dune valley, lagoon, and lookout dune to meet ecological and recreational demands (Figure 1b).

Placement of the nourishment took almost a year, and finished in April 2015. The construction works were phased from south to north and from shoreface to dune. The nourished sand was dredged 10–15 km offshore of the project site. The nourishment is constructed with a sediment grain size varying between 225 and 350 μm to match the native sediment size as best as possible. After placement of the sand, the dune was immediately planted with marram grass and dune foot fences were installed to prevent nuisance of sand transport over to the landward side of the dike. In February 2018, an additional nourishment was placed in the south of the Hondsbossche Dunes project area between km 25 and 26.5 to increase the local beach width for recreation. This extra nourishment had a volume of approximately 1 million m^3 and was placed on the beach.

3. Methodology

3.1. Morphological Datasets

Beach width and profile volume response are examined using monthly to quarterly surveys acquired in the first five years after construction. The available data originate from three different sources (Figure 3).

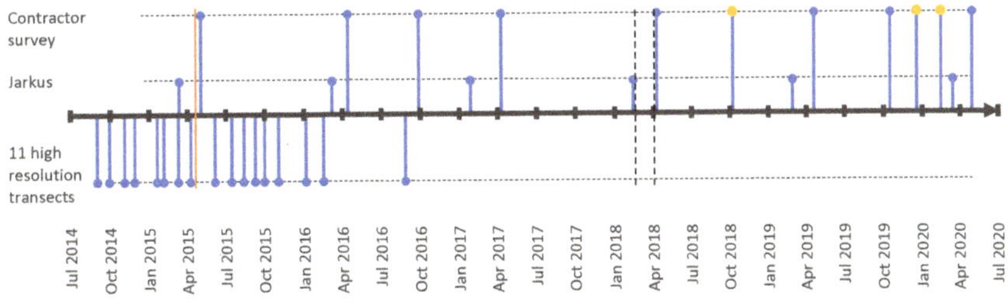

Figure 3. Timeline of the survey data. Three datasets are used: 1. JARKUS surveys covering the full coastal cell; 2. Contractor surveys covering the 9 km project site; 3. Contractor transect surveys of 11 profiles with high temporal resolution shortly after the construction period. Red vertical line indicates the formal end of construction; black vertical lines bound the additional 2018 beach nourishment period. Yellow dots indicate subaerial surveys; blue dots indicate surveys that include both subaerial and subaqueous data.

The first is the JARKUS dataset from Rijkswaterstaat covering the entire Dutch coast with transects spaced approximately 250 m-apart (red lines, Figure 2c) and acquired on a yearly basis [44]. These surveys display the response of the entire coastal cell, ranging from the port of IJmuiden to the Marsdiep tidal inlet. The second dataset is acquired more frequently (approximately four subaerial and two subaqueous surveys per year) by the contractor and maps the response of the 9-km project site. These surveys consist of cross-shore transects spaced 250 m-apart. Originally, these were surveyed perpendicular to the newly created coastline and, therefore, deviated slightly in position and orientation from the JARKUS transects. To enable better comparisons, the transect locations were revised in summer 2017 to match the JARKUS lines (green lines in Figure 2c). The third dataset contains 11 transects approximately 1 km-apart on which monthly to two-monthly measurements were taken by the contractor for a period of 1.5 years since the end of construction of each transect (black lines in Figure 2c). For several southern transects, data are already available from the end of 2014 while construction at the northern transects was only finished in April 2015. These 11 transects follow the orientation of the postconstruction coastline. The surveys obtained by the contractor (dataset 2 and 3) use single beam echosounder for the subaqueous data and the subaerial data are gathered mostly with LIDAR scans with occasionally walking GPS-RTK measurements.

The surveys that cover the entire nourished site in both the subaerial and subaqueous domain are gridded with a resolution of 2 m to examine the temporal evolution of the total volume change.

3.2. Data Reduction

To analyze the subaqueous and subaerial changes, the complex bathymetrical datasets are reduced to indicators that can be tracked in space and time and correlated to each other. First, these indicators are derived in the transect system of each dataset. Next, for overlapping or close positioned transects, these indicators are combined.

3.2.1. Sediment Volumes at Different Elevations in the Profile

The sediment budget of the nourishment and the adjacent coast are analyzed for various vertical slices representing the different subsections of the profile (dunes, beach, shoreface). The total profile-integrated volume is defined as the volume above −10 m + NAP, the level beyond which no significant bed level change is visible ($\sigma_{\Delta z} < 20$ cm) and seaward of the landward boundary. The lower shoreface volume is the lowest subsection analyzed, defined as the volume slice between −10 m + NAP and −4.8 m + NAP. The −4.8 m + NAP level represents the level above which 90% of the waves in the local long-term wave climate break, and is used to delineate between shoreface and surfzone subsections. The morphologically most active cross-shore section, the beach and surfzone, is bounded by −4.8 m + NAP and

3 m + NAP. Finally, the dune volume subsection is analyzed from the bed levels above 3 m + NAP and seaward of the landward boundary, Figure 4. The landward boundary is chosen sufficiently far into the dune such that there is no significant sediment transport over it and its position does not affect the computed volume changes. Not all measurements extend until the landward boundary at each time step, for each transect. If not, these measurements are discarded for the total and dune volume change.

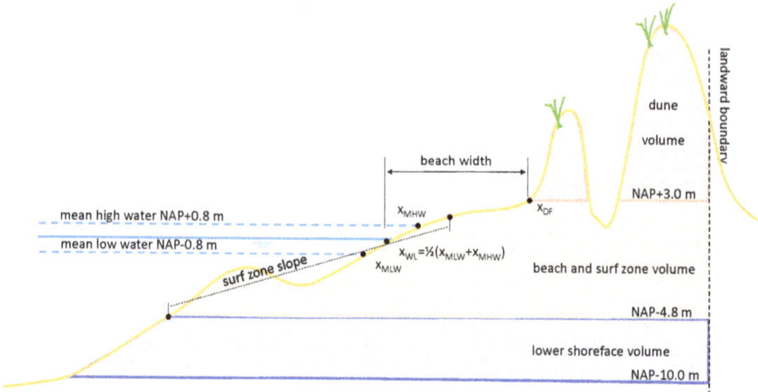

Figure 4. Schematic cross-shore profile to illustrate the various profile parameters extracted from the survey data. NAP is the Dutch reference level, roughly equal to mean sea level.

3.2.2. Beach Width, Shoreline Position, and Dune Foot

The beach width is defined as the distance between the waterline position and the dune foot position, Figure 4. The waterline position is taken from the average of the cross-shore positions of the bed at the MHW and MLW levels, similar to the definition used by de Vries et al. [45]. The dune foot elevation is taken constantly at a level of 3 m + NAP, and dune foot position follows from the intersect with the cross-shore profile with the 3 m + NAP. There are many alternative definitions of dune foot, e.g., using a maximum slope change criterion [46]. The volume change analysis in this paper is only marginally affected by this definition and, for simplicity, a fixed horizontal level is assumed. Similarly, beach width changes on the time scales considered are hardly affected by the definition; however, absolute values of beach width can be more sensitive.

3.2.3. Surfzone Slope

The surfzone slope is determined to examine cross-shore adaptation in the period directly following the construction. The slope is obtained from a least squares linear fit through the vertical elevations of the beach and surfzone bounded by the closest crossings with −4.8 and 1.5 m + NAP levels (Figure 4), similar to the approach of de Vries et al. [45]. The upper level of +1.5 m NAP is chosen well-above the high water level to increase the robustness of the determined slope to inaccuracies around the interface of the subaqueous and subaerial measurements. Nevertheless, cross-shore slope values are sensitive to sandbars moving in and out of the evaluated elevation points. For the temporal evolution of the surfzone slope, the surveys before the formal end of construction are also included to map the response rapidly after the nourishment works.

3.2.4. Coastline Curvature

Coastline curvature is defined as the gradient in the orientation of the shoreline curve. This shoreline curve is obtained by connecting the cross-shore shoreline positions (as defined in Section 3.2.2) in the different transects. Before determination of the gradient, the shoreline curve is filtered with a uniform filter with an alongshore length of 1 km and averaged over the five-year evaluation period to remove small fluctuations. Finally, the gradient is computed using a second-order accurate central difference scheme.

4. Results

This section presents the data analysis of the Hondsbossche Dunes nourishment project. First, an overview of the morphological development in the five years after implementation of the nourishment is given. Second, the volumetric evolution over time is discussed with focus on different vertical zones (dunes, beach, surfzone, shoreface) and in the context of the larger coastal cell. Next, profile adjustment is presented and the subaerial evolution of the nourishment is assessed in more detail. Finally, the relationship between sediment budgets and concurrent changes in beach width and planform coastline curvature are examined.

4.1. General Morphodynamic Response of the Nourished Beach in Five Years

The bathymetric change in the five years after the nourishment works is illustrated using gridded plan view topographies (Figure 5a,b) and profiles of selected transects (Figure 6). The placement of the nourishment in front of the sea dike created a curved coastal section, protruding seaward with respect to adjacent beaches (Figure 5a). The central part of this new beach system eroded (7 km out of the 9 km) in the first years after implementation of the project. This erosion is focused around the waterline and in the surfzone (Figure 5c). During these years, both adjacent coastline sections experienced accretion. This accretion is present in the lower shoreface at the coastline sections where erosion transitions to accretion (km 20 and km 26) and extends in the cross-shore to the surfzone in sections further north and south. The coastline section with, predominantly, accretion in the north is larger in magnitude and intensity compared to the accretive section in the south.

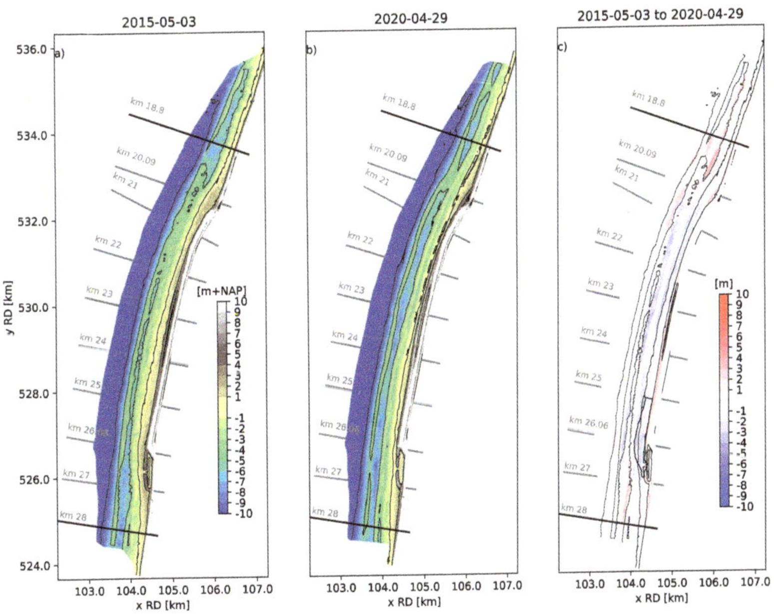

Figure 5. Bed elevation data shortly after construction in May 2015 (panel (**a**)) and after five years in April 2020 (panel (**b**)). Colors show the bed elevation in meters with respect to the NAP datum (approximately MSL). Panel (**c**)) represents the erosion (blue colors) and sedimentation (red colors) between the two surveys five years apart. The black dotted contour lines are based on the bed elevation of May 2015 (in panel (**a**)/(**c**)) or 2020 (in panel (**b**)) and indicate the −10, −4.8, 0, and 3 m + NAP isobaths. The black solid lines delineate the nourishment boundaries (km 18.8–28). The location of the additional 2018 beach nourishment is indicated with the black polygon.

The cross-shore profiles at the nourished site show the development of a double barred system with an outer bar approximately 500 m from the waterline with a crest height around −4 m + NAP and a more dynamic inner bar around 200 m from the waterline with a crest around −2 m + NAP (Figure 6). Bed level variations extend until a depth of approximately −10 m + NAP (Figure 5c). Dune growth can be observed along the entire nourished site, either in the form of embryo dune establishment on the nourished beach, or dune face progradation and heightening of the first dune row (Figures 5c and 6).

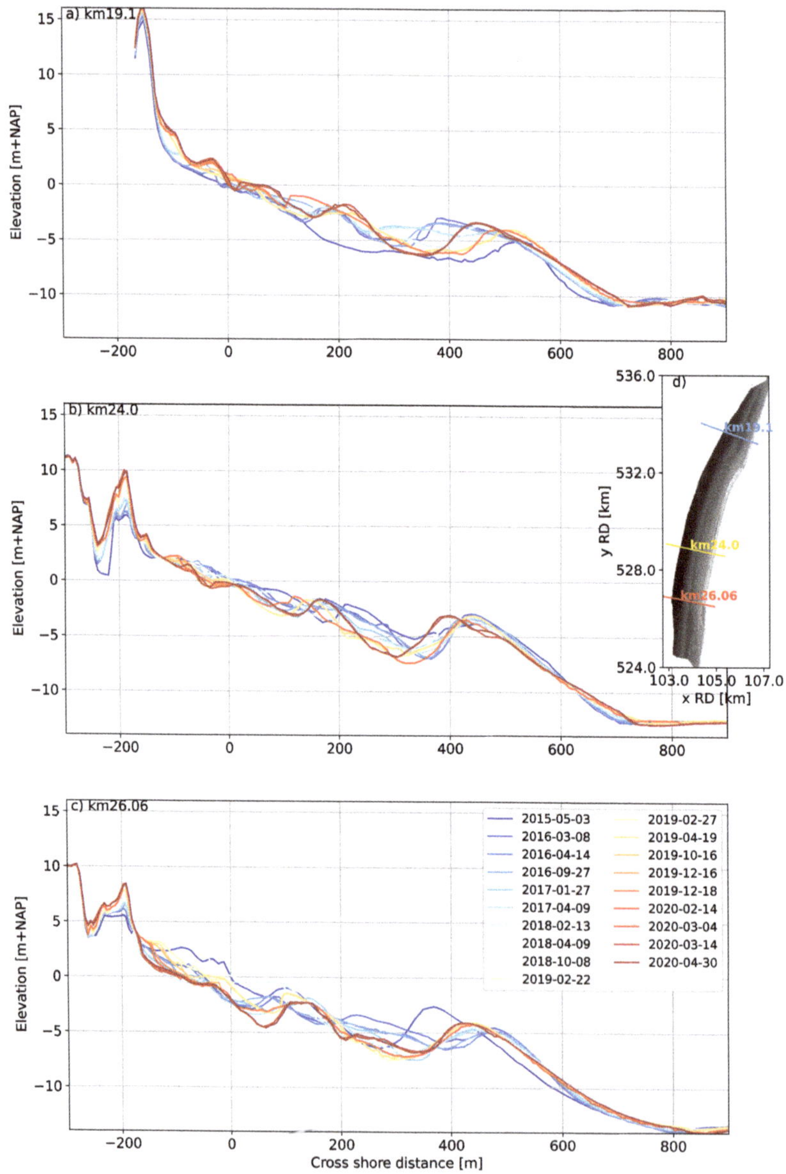

Figure 6. Bed levels on three cross-shore transects. (**a**) Northern transect at km 19.1. (**b**) Central transect at km 24.0. (**c**) Southern transect at km 26.06. Colors show the different surveys in time. (**d**) the location of the transects in plan view. 0 m + NAP is approximately MSL.

4.2. Volumetric Changes

Although large volumes of sediment are displaced, the majority of the added sediment in the project area can be retraced over the five-year period. A volumetric budget of the entire 9 km placement area shows a loss of 1.6×10^6 m^3 by 2018 (Figure 7, black symbols), which is less than 5% of the 35×10^6 m^3 initially added to the region. The 2018 nourishment of 1×10^6 m^3 resulted in a positive jump in volumes, but a small negative trend remained and, by 2020, (after 5 years) a loss of 1.4×10^6 m^3 is observed. Small fluctuations in volumes can be observed, which are likely due to survey inaccuracies and gridding of the transect data.

The 9 km-long sediment budget area is subdivided in the alongshore into a central section and two lateral sections to examine lateral diffusion of the nourishment. The central part of the nourishment shows a negative trend (Figure 7, orange symbols) with exception of the moment of placement of the 2018 nourishment. After five years, the 7 km-long central area has lost a volume of $\sim 3 \times 10^6$ m^3. The erosion from the center of the nourishment in the first year is $\sim 1.3 \times 10^6$ m^3, about twice as high as the following second and third years (Figure 7, orange symbols). After the third year, the volume in the central section increases with the additional 2018 nourishment. After this intervention, the central volume decreases further in the fourth and fifth years.

Opposed to the erosive central part, both lateral sides show a gradual accumulation of sediment with small fluctuations. The gained volume in the northern (Figure 7, blue symbols) sediment budget section is about three times larger than in the south (Figure 7, green symbols), while the coastal section is 1.5 times longer in the alongshore direction. This asymmetry is in agreement with the northward direction of the alongshore transport in this region [38].

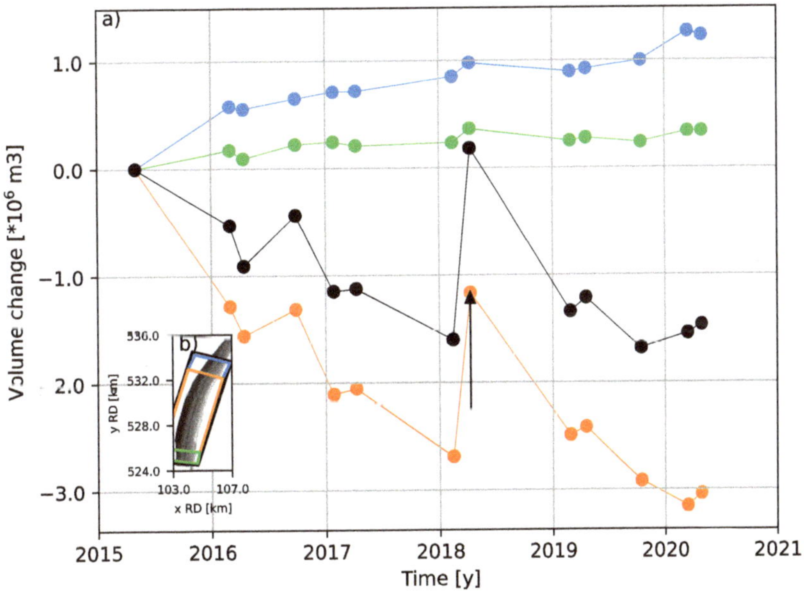

Figure 7. Volume change above −10 m + NAP since April 2015 in three alongshore sections. (**a**) The center of the project (km 20.23 to 27) (orange), the adjacent beach north (km 18.8 to 20.23) (blue), the adjacent beach south (km 27 to 28) (green), and the total volume change for all sections combined (black). The black arrow indicates the moment and size of the additional 2018 nourishment of $\pm 1 \cdot 10^6$ m^3. (**b**) Top view of the nourishment and locations of the volume polygons.

To obtain insight in the alongshore and cross-shore variations in time, the volumetric changes are further examined using seven transects. The profile-integrated volume changes of the transects located near the edges of the project area show accretion up to 1000 m^3/m alongshore (Figure 8b, red and blue symbols), where transects in the center of the project area show a loss in volume varying between 300 and 700 m^3/m after 5 years (60 to 140 $m^3/m/y$). Especially, the first two surveys after construction show large changes, followed by several surveys with more moderate adaptation (Figure 8b).

Regardless of whether profiles are eroding or accreting (based on the profile-integrated volume), dune volumes mostly increase from survey to survey (Figure 8c). The dune volume growth shows a significant variation among transects, between 40 to 300 m^3/m after 5 years (i.e., 8 to 60 $m^3/m/y$). The dune volume growth is largest for a profile at the center of the nourishment (km 24, yellow symbols in Figure 8c). Here, volume gain in the dune is a significant part of the cross-shore sediment balance and of the same order of magnitude as volume losses below 3 m + NAP. At transects with a more moderate dune volume increase (e.g., km 21.23, 22.63, and 26.06) the dune volume increase is about five times smaller than the profile-integrated volume decrease.

Likewise, the volume change in the lower shoreface (−10 to −4.8 m + NAP) can locally be a significant contribution to the profile-integrated volume change (Figure 8d). Especially for the central transects (km 22.63 to 26.06), the magnitude of the lower shoreface losses is ~50% of the profile-integrated sediment loss.

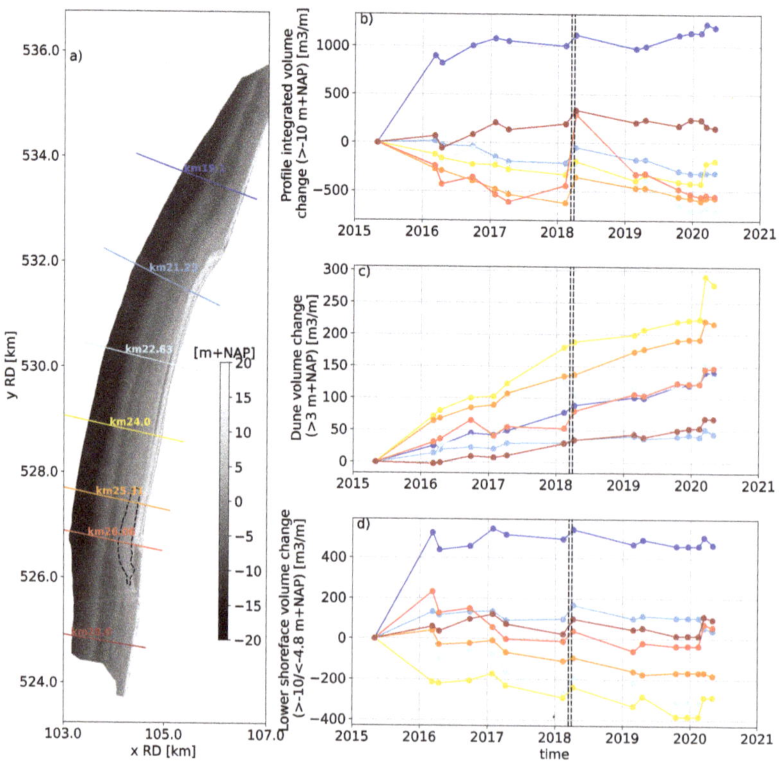

Figure 8. Volume change over time for transects at km 19.1, 21.23, 22.63, 24, 25.31, 26.06, and 28. (**a**) Location of transects (colored lines) and 2018 nourishment (dashed contour), (**b**) Profile-integrated volume change (>−10 m + NAP), (**c**) Dune volume change (>3 m + NAP), and (**d**) lower shoreface volume change (between −10 and −4.8 m + NAP). The black dotted vertical lines indicate the period of the additional 2018 beach nourishment between km 25 and 27.

4.3. Changes in Perspective of the Morphodynamics of the Coastal Cell

The coastline curvature at the nourishment is stronger than that of the coastal cell (Figure 9a). The original coastline slowly rotates 10° from 285 to 275° N over approximately 20 km (∼0.5°/km) to the south of the nourishment project, and 20° from 295 to 275° N over 15 km to the north of the nourished site (∼1.3°/km), while at the center of the nourishment, the coastline orientation changes 25° from 275 to 300° N in less then 7 km (∼4°/km) (Figure 9b).

Expansion of the sediment budget analysis to a larger 45 km-long coastal stretch shows that average annual volume changes per cross-shore profile vary in the alongshore direction between +250 and −250 m^3/m/y over the period of 2015–2020 (Figure 9c, blue line), with large fluctuations between years (Figure 9c, blue shading), particularly for locations where nourishments have been implemented. The central beach of the Hondsbossche Dunes stands out as an erosional zone, with large volume losses of 100 m^3/m/y (Figure 9c,d). This erosional zone is consistent with the area where the coastline is strongly curved (Figure 9b, km 20 to 27). Volume variations at adjacent coastal sections are, in general, of a smaller order of magnitude. Repeated nourishments (km 12.13–14.21 and 31–40, gray blocks in Figure 9c,d,f) at the adjacent coast are visible as sections with a positive volume balance averaged over the 5 years. Sections with a positive volume balance are also visible next to the Hondsbossche Dunes nourishment and are related to spreading of the earlier-placed nourishment, feeding the adjacent coast (Figure 9c, km 17–20 and 27–28).

Subdivision in the different elevations shows that, at the nourished site, profile-integrated volumetric changes (Figure 9c) are dominated by variations in the beach and surfzone elevations (Figure 9d). Beach and surfzone changes are, respectively, 5 and 3 times larger than the average volume changes in the dunes or lower shoreface (Figure 9e,f), but this ratio varies strongly alongshore. For instance, at the center of the nourishment (between km 24 and 25), a peak in deposition in the dune is observed, in combination with a reduction in beach and surfzone volumes resulting in a ratio of almost 1.

4.4. Cross-Shore Adaptation

The steepness of the cross-shore profile over time is examined and compared with values of the surrounding coast. The datasets contain several transects that are initially steeper than the adjacent coast. These steeper slopes adapt to a similar magnitude as the adjacent coast in the first two winters, Figure 10b. This is not observed for all transects, as for some transects, monitoring started several months after construction had finished. Moreover, not all transects were constructed with the same initial steepness. For transects where the first surveys indicated a profile steepness similar or lower than the adjacent coast, no large change in slope is observed, Figure 10c. The southern transects are initially very steep, two to four times steeper as the adjacent coast, but readjust and flatten over the first winter period (2014/2015) to a steepness about the maximum of that of the adjacent coast. In the next year, the profiles remain quite constant to become even flatter in the second winter. Placement of the 2018 nourishment steepens the local transect in the south, yet, also this is temporary and the profile steepness returns to its previous range. Construction at the more northern transects was still ongoing in the winter of 2014/2015. Nevertheless, these transects show a similar pattern with steepness values outside the range of the adjacent coast that decrease after the winter of 2015/2016.

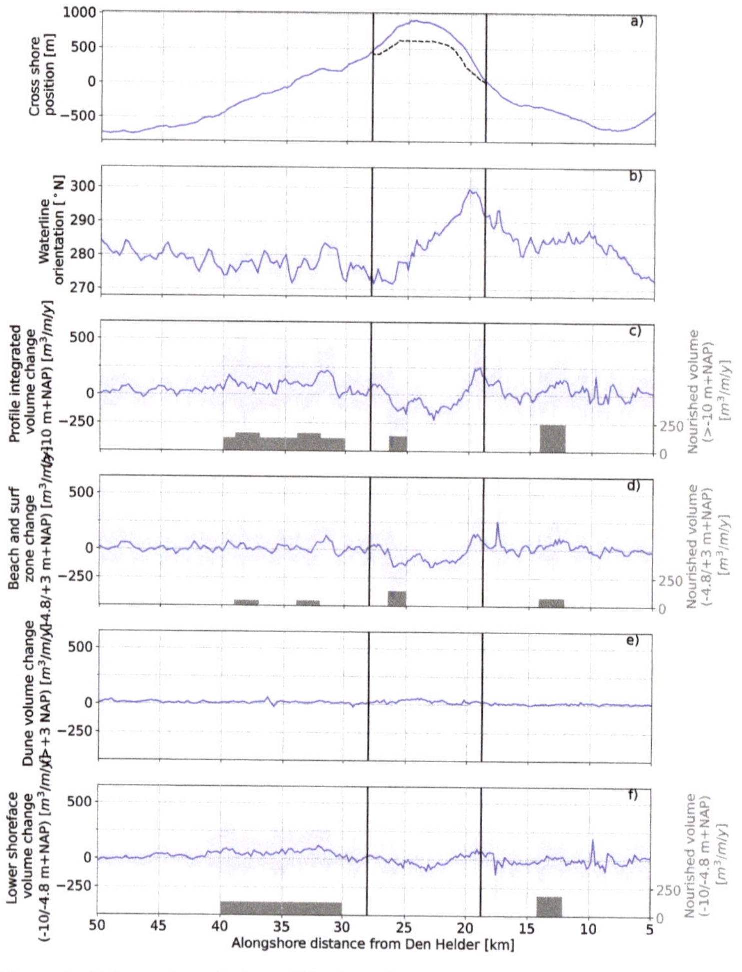

Figure 9. Volumetric evolution of North-Holland coast with the nourished section in between the black vertical lines (JARKUS dataset only). (**a**) Cross-shore position of the waterline (relative to the waterline position at km 0.2), the position the prenourishment waterline in black stripes. (**b**) Orientation of the waterline. (**c**) Yearly profile-integrated volume change (>−10 m + NAP). (**d**) Yearly volume change in the beach and surfzone (−4.8 m + NAP and 3 m + NAP). (**e**) Yearly dune volume change (above 3 m + NAP). (**f**) Yearly volume change on the shoreface (−10 m + NAP to −4.8 m + NAP). Blue shadings gives the $\pm\sigma$ interval and solid blue lines the average over the period April 2015–April 2020. The gray blocks indicate the nourishment volumes placed within the coastal cell in the period April 2015–April 2020.

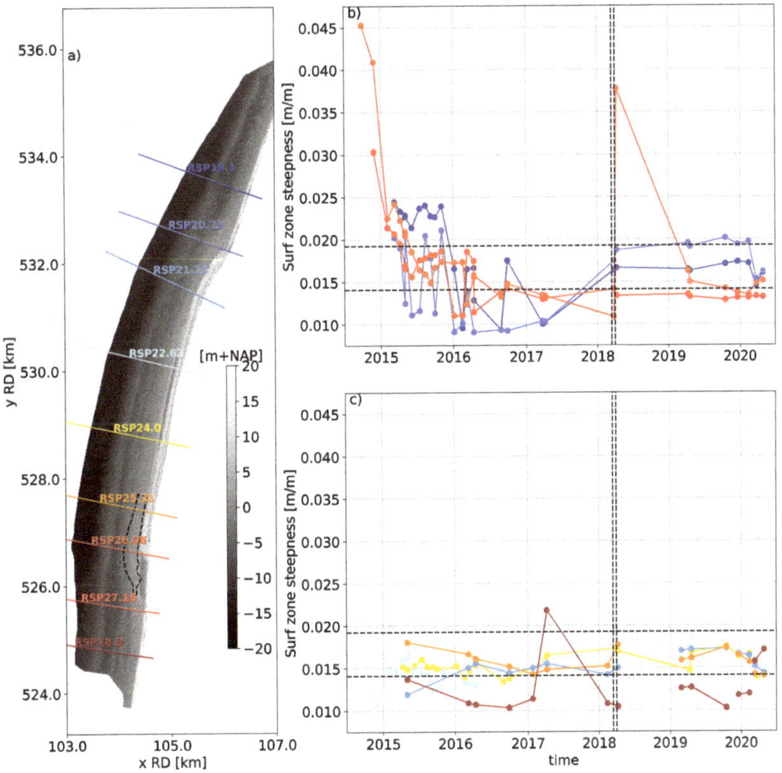

Figure 10. Surf and intertidal zone slope (−4.8/1 m + NAP) adaptation in time. (**a**) Location of transects and 2018 nourishment; (**b**) slope at initially steeper profiles at km 19.1, 20.23, 26.06, and 27.16; (**c**) slope at initially smoother profiles at km 21.23, 22.63, 24.0, 25.31, and 28.0. The black dotted horizontal lines indicate the $\mu \pm \sigma$ profile steepness of the adjacent coast. The black dotted vertical lines indicate the period of additional beach nourishment between km 25 and 27.

4.5. Subaerial Evolution

Subaerial volume changes are relatively small compared with subaqueous zones but can be important for the assessment of safety against flooding or recreation potential. In this section, we examine the evolution of the subaerial beach through the spatial and temporal patterns in dune foot position and volume, and beach width.

The average dune volume increase over 5 years at the nourished beach is about 30 $m^3/m/y$ (Figure 11c) and peaks at almost 60 $m^3/m/y$ at the center of the nourished section (km 24.0), showing there is a significant alongshore variation of deposition in the dune. In comparison, during the same period, dune growth at the adjacent coast is 15 $m^3/m/y$. Dune growth can occur through lateral expansion of the dune (i.e., seaward change in dune foot position), heightening of the dune, or both. At the Hondsbossche Dunes, both elements are visible. Dune growth in volume is, however, not always connected to lateral expansion (Figure 11c,d). In the nourished section, the dune foot position changes on average 4 m/y seaward (Figure 11d), twice as high as at the adjacent coast, at which the dune foot position changes 2 m/y seaward. Two nourished sections in the larger coastal section (km 12.13–14.21 and 31–40) stand out as zones with a large seaward movement of the dune foot. An exceptionally large dune foot change is found at the northern transition zone of the nourishment (±km 20), where the local dune foot has progressed with over 50 m. This is the result of the local infill of a discontinuity in the first dune row. This discontinuity resulted from a stepback in the constructed dune at the transition to the adjacent coast. Although the dune foot migrates forward by 50 m at this location, this is not reflected in larger dune volume changes (Figure 11c, ±km 20).

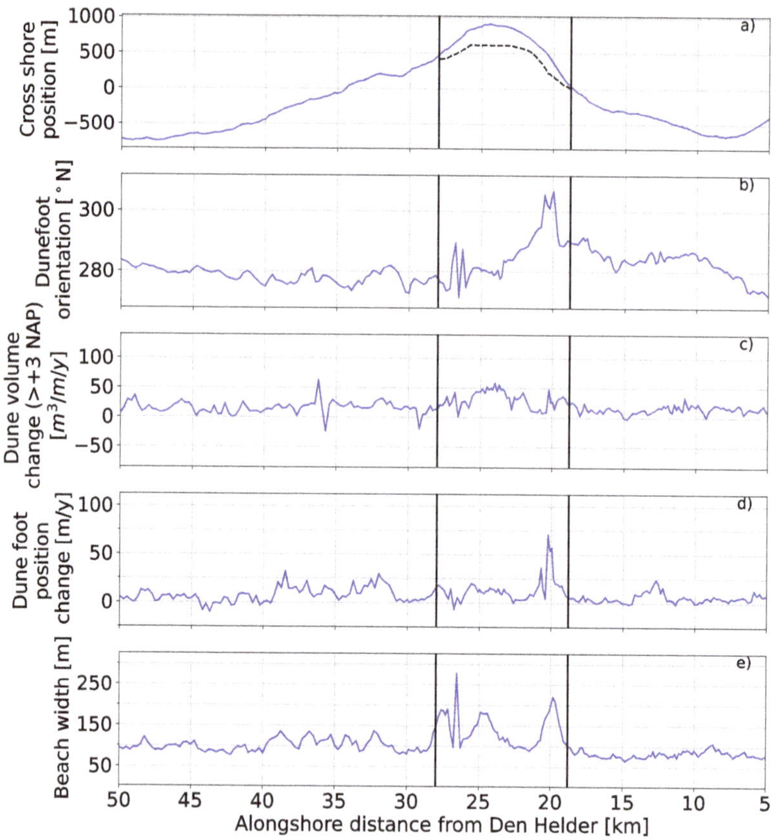

Figure 11. Dune evolution of North-Holland coast with the nourished section in between black lines (JARKUS dataset only). (**a**) Cross-shore position of the waterline (relative to the waterline position at km 0.2); the position of the prenourishment waterline is in black stripes. (**b**) Orientation of the dune foot. (**c**) Dune volume change, above 3 m + NAP. (**d**) Dune foot position change. (**e**) Beach width.

The nourished beach at the Hondsbossche Dunes is on average 180 m-wide directly after construction. This is about twice the width of the surrounding beaches where the average beach to the south is 83 m and to the north is 101 m-wide (Figure 11e). The beach width adapts rapidly over the years as a consequence of both fluctuations in the land–water interface, and seaward migration of the dune foot (Figure 12c,d). The contribution of dune foot migration to the changes in the beach width is substantial at the nourished site—on average, 20% of the total beach width change and, locally, more than 35%.

At the eroding part of the nourished site, between km 20.23 and 27, the beach width decreases by about 80 m (Figure 12b, round markers). The majority of the reduction occurs in the first two years, after which the beach width adaptation slows down and converges to beach width values similar to the adjacent coastal sections. Transects 25.31 and 26.06 are an exception as the beach width is disturbed by the placement of the additional nourishment in 2018. The beach width reduction at the eroding transects is primarily caused by a landward change in waterline position (60 m of the 80 m reduction in beach width). This change in waterline position is initially fast but slows down over time (Figure 12c). From the landward end, beach width is reduced by a more constantly prograding dune foot of, on average, 20 m seaward.

The beach width in the accretive lateral sections of the nourishment displays no clear trend (Figure 12b, triangular markers). At these transects, the seaward migration of the waterline (20–50 m) is similar to the seaward migration of the dune foot (30–40 m) (Figure 12c,d).

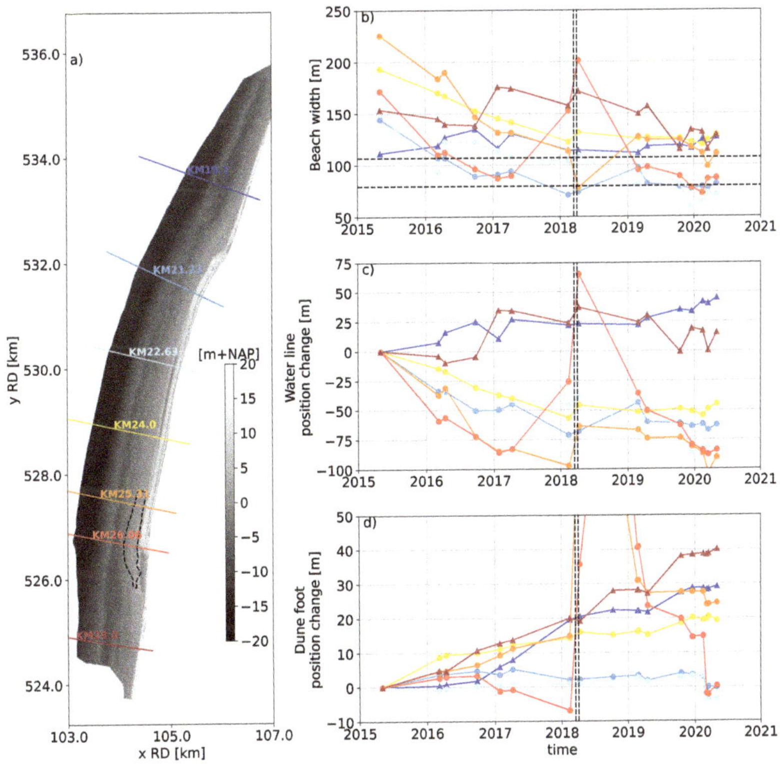

Figure 12. Beach width in time for transects 19.1, 21.23, 22.63, 24, 25.31, 26.06, and 28. (**a**) Location of transects. (**b**) Beach width. (**c**) Waterline position change. (**d**) Dune foot position change. The black dotted horizontal lines indicate the $\mu \pm \sigma$ beach width at the adjacent coast. The black dotted vertical lines indicate the period of the additional 2018 beach nourishment. Eroding and accretive transects are marked with circle and triangle symbols, respectively.

4.6. Profile Volumes and Beach Width Changes as Function of Coastline Curvature

The observations of volume change and beach width change at the Hondsbossche Dunes are correlated to evaluate their dependence. This is performed using the JARKUS dataset only, because it is the most consistent dataset to cover both the subaqueous and subaerial domains for the entire evaluation period. For the nourished section, the beach width change ΔW_b and total profile-integrated volume changes ΔV are well-correlated ($r^2 = 0.75$, Figure 13a). Changes in waterline position Δx_{sl} correlate even better with the volume change ($r^2 = 0.84$) as noise introduced by dune foot position changes is excluded from the relation (Figure 13c). The coastline curvature is an important driver of nourishment adjustment, e.g., [3,13]. For the Hondsbossche Dunes, both volume change and beach width change indeed correlate well with curvature ($r^2 = 0.47$ and $r^2 = 0.54$, respectively, Figure 13b,d), confirming the importance of alongshore wave-driven sediment transport for both indicators.

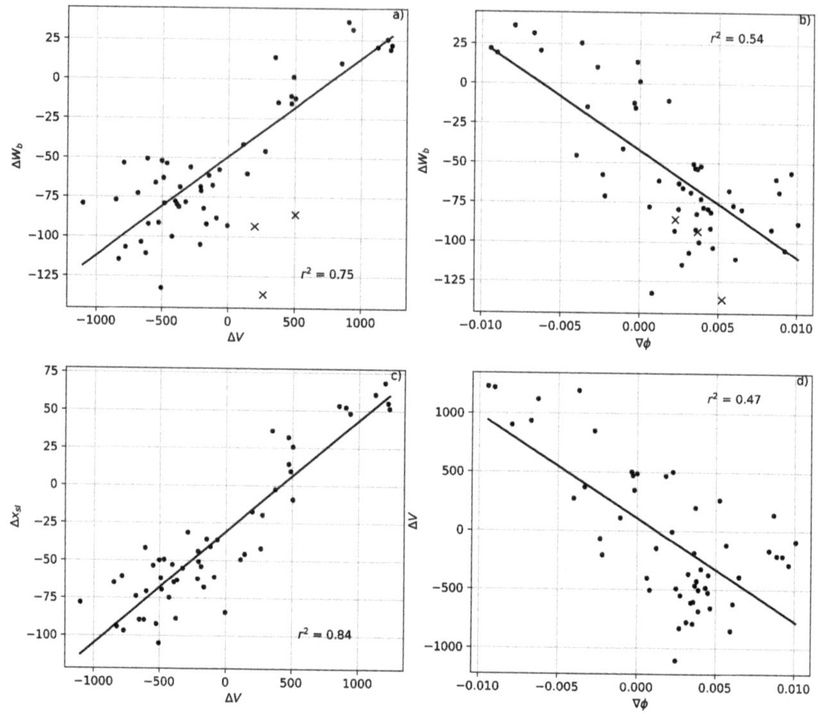

Figure 13. Correlations of profile volume change, beach width, coastline position, and coastline gradient. (**a**) Correlation of profile volume change and beach width change, (**b**) coastline gradient and beach width change, (**c**) coastline position change and volume change, and (**d**) coastline gradient and volume change. Data points are for the project area only and based on JARKUS dataset, for the period between April 2015 and 2020. Observations marked with a cross are treated as outliers in the determination of the correlation coefficients. Beach width changes at these locations are dominated by alongshore infilling of a local stepback in the dunefront at the edge of the original sea dike (see Figure 11a,d km 20–20.5).

The least squared linear regression line relating beach width and volume change intersects the $\Delta V = 0$ m^3 line at negative values, estimating around 50 m reduction in beach width for profiles with no net volume change (Figure 13a). Similarly, the regression line relating volume change and waterline position crosses $\Delta x_{sl} = -30$ m at $\Delta V = 0$ m^3 (Figure 13b). These offsets are considerable compared to the range of beach width changes observed ($\pm$100 m). Suggesting that, next to profile volume changes, cross-shore redistribution of volume is an important contribution to beach width change in the first five years.

5. Discussion

5.1. Beach Width and Dune Growth

The initially wide beach at the Hondsbossche Dunes project significantly reduced over the five years investigated. The landward trend in waterline position is found to be the main contribution to the beach width changes (in the order of several meters per year) and about two to four times the magnitude of the seaward shift in the horizontal dune foot position (Figure 8). This dominance of waterline position changes over fluctuations in dune foot are similar to observations at a nearby natural beach [47]. In general, fluctuations and trends in waterline position occur on a range of scales (i.e., seasonal and storm scales to interannual scales, e.g., [47–50]).

The observed net landward migration of the waterline after implementation of a nourishment may be caused by alongshore gradients in sediment transport as well as redistribution of the (nourished) sand downslope in the cross-shore direction. The cross-shore adaptation at smaller nourishments is reported to occur primarily in the first months or year after implementation [51–53], with high-energy wave events playing an important role [16,54]. Volume and beach width changes at the Hondsbossche Dunes in the last years are also smaller than the first year (Figure 12b), which is in line with results of these earlier studies where cross-shore redistribution is initially strongest. Further, the rapid adjustment of the cross shore slope in the first two winters (Figure 10) is in line with reported reductions in slope adjustment at other nourished sites [16,55]. For the large nourishment investigated here, the cross-shore equilibration of the profile is not limited to the subaqueous profile and sediment moving downslope, as sketched by Dean [3], Elko and Wang [16]. Dune growth is a substantial part of the sediment budget and postnourishment equilibration (Figure 14).

Alongshore sediment transport gradients impact waterline position through both the planform adaptation of the nourishment and the pre-existing background erosion rate [56,57]. We observe a total shoreline retreat of up to 100 m for transects with large net volume losses (Figure 13c), about 70 m more than locations with minimal volume losses (Figure 13c, points near $\Delta V = 0$). This underlines that alongshore effects are critical to understand changes in terms of beach width at this nourishment.

Furthermore, the analysis reveals contrasting trends of the waterline migration (driven by marine processes) and dunefoot migration (dominated by aeolian processes), resulting in a reduction in the beach width from both land and seaward sides over the past years. This is indicative of nourished beaches during the first period with no or minimal dune erosion, and similar to findings of Bezzi et al. [58]. At these wide, nourished beaches, the dune foot may move seaward (Figure 15b) due to aeolian transport, but as the beach erodes in the following years the dunes become more prone to erosion [45,59,60]. More severe and frequent dune erosion events will likely shift the dune foot landward and can bring the waterline seaward. Several transects have reached beach widths similar to the adjacent coast (within the $\mu \pm \sigma$ range of 89 to 107 m) after five years (Figure 12a). At these locations, some initial retreat in dune foot position (Figure 12c) and an increase in beach width suggests stabilization after the last winter (Figure 12a). These observations are early signs that future beach widths at the nourished site may fluctuate within a similar range as the adjacent coast.

Figure 14. Schematic view of observed cross-shore behavior at the Hondsbossche Dunes nourishment including an increase in dune volume and equilibration of beach width, modified from [16].

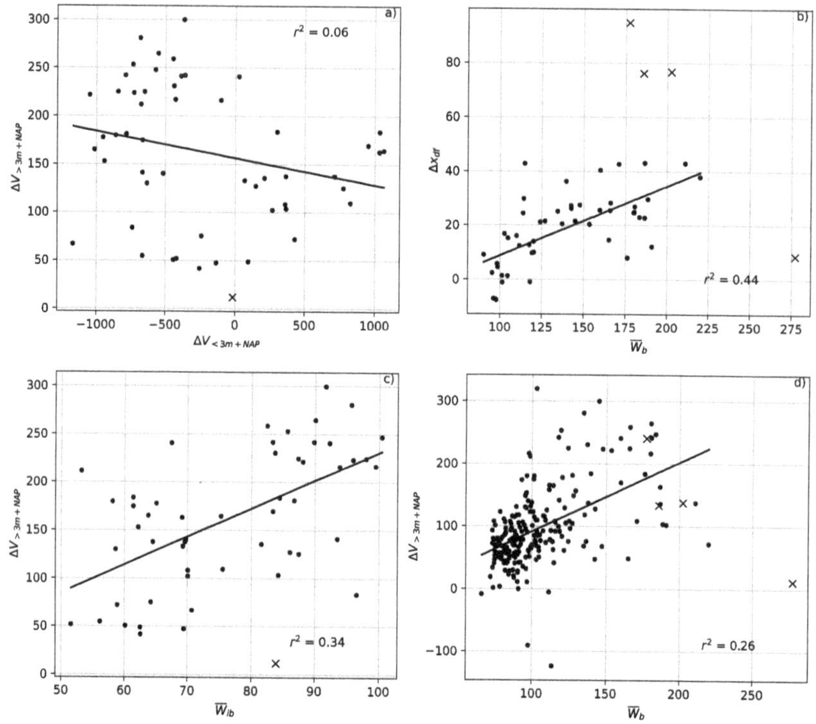

Figure 15. Correlations of beach width and volume changes. (**a**) Correlation of volume change below 3 m + NAP and volume change above 3 m + NAP; (**b**) average beach width and dune foot position change; (**c**) average intertidal beach width and volume change above 3 m + NAP at the Hondsbossche Dunes; and (**d**) average beach width and volume change above 3 m + NAP at the NNH coast for the period between April 2015 and 2020. All values are based on the JARKUS dataset only. Observations marked with a cross are treated as outliers in the determination of the correlation coefficient. At these locations, dune foot changes are dominated by an alongshore infilling of a local stepback in the dune front at the edge of the original sea dike (Figure 11a,d km 20–20.5) or beach width and dune volume cannot be determined appropriately according to our definition due to a local opening in the dune in front of the lagoon (Figure 11e km 26.5).

In the cross-shore direction, an increase in dune volume is observed regardless of the net trend in the profile volume below ($r^2 = 0.06$, Figure 15a). While dune volume gains are up to five times smaller than total volume change (50 m^3/y compared to 250 m^3/y, Figure 9), sediment accumulation in the dune can locally be a significant component in the cross-shore sediment balance and contribute to the safety against flooding. The observed average dune volume increase over the nourished site is 30 $m^3/m/y$ (Figure 11), which is similar to the upper limit of measured dune growth at the Dutch coast [45]. In contrast to earlier observations of larger dune growth in the first year after implementation of a nourishment e.g., [27,61], we observe no significant difference in dune volume gains between the subsequent years. From an engineering perspective, the observed average growth of 150 m^3/m in the postnourishment period is significant and an important contribution to safety against flooding. To put the observed dune growth in perspective of the Dutch coast, the average dune erosion is estimated to be around 20–100 m^3/m during a 1/30 year event [62] and 100–800 m^3/m (100–250 m^3/m at the NNH coast) in case of an extreme event (1/10,000 y) [63].

The dune volume increase shows significant variability in the alongshore direction and peaks in the center of the nourishment around km 24 (Figure 11). Dune development can be sensitive to beach width, beach sediment budget, shoreline orientation, vegetation and sediment, or surface properties amongst others, e.g., [29,64–66]. For the Hondsbossche Dunes, we observe that dune foot migration is correlated to beach width ($r^2 = 0.44$, Figure 15b), where larger beach widths result in more seaward migration of the dunefoot. A similar correlation between beach width and changes in dune volume cannot be found at the nourishment. However, Figure 15d shows that the behavior of the nourishment compared to the adjacent coast by including the entire coastal cell into the evaluation does result in a correlation. Although small, the larger dune growth at the Hondsbossche Dunes does show a positive correlation with the locally large beach width ($r^2 = 0.26$, Figure 15d). The variation in dune growth at the nourishment itself is correlated with the width (W_{ib}) of the intertidal beach instead ($r^2 = 0.34$, Figure 15c). This is in line with previous observations that large parts of aeolian sediment deposits originate from the low-lying beach that is regularly reworked by waves [27].

5.2. Predicting Beach Width and Volume Change

Predictions of coastal dynamics are complex due to the large range of phenomena involved—e.g., [67]—and remain a challenge despite the advances made in the last decade [68]. Simple curve-fitting methods can provide good results to describe aggregated parameters of nourished beaches, such as remaining volume in the project area [69].

For a detailed view of the shoreline behavior, one-line type coastline models—e.g., [13,70–73]—can provide information on postnourishment development of the beach in a computationally efficient manner [4]. These models simulate the redistribution of sediment alongshore based on alongshore gradients in wave action, coastline gradient, or sediment availability. At the Hondsbossche Dunes, volume changes are indeed correlated with shoreline curvature ($r^2 = 0.47$, Figure 13d), confirming one of the key underlying assumptions of one-line models. One of the important parameters in these one-line models is selecting an active profile height, which is often estimated from an upper limit on the beach (e.g., the dunefoot or berm height at 3 m above MSL) and a depth of closure. At the Hondsbossche Dunes, the inner and outer closure depth are approximately 8 to 12 m below MSL using Hallermeier's formulas [74,75]. This roughly coincides with the observations of the lowest point of observed bed level changes at ± -10 m + NAP (Figure 6). The Hondsbossche Dunes analysis also gives an additional option to estimate the active profile height based on the the linear regression line between profile volume change and shoreline position (Figure 13c). The reciprocal of the slope of the regression line can be interpreted as a representative active profile height of 14 m (i.e., 1/0.07). The similarity in values indicates that Hallermeier's formulas could have been used to estimate the closure depth in the nourishment's design phase.

The obtained correlations between volume change, coastline curvature, and beach width indicate that one-line models can be used to predict the response of nourishments on the scale of the Hondsbossche Dunes. Nevertheless, several limitations of one-line models can be illustrated with the data of the Hondbossche Dunes. First, the correlation between volume change and coastline curvature for several subsections of the nourishment is significantly higher ($r^2 > 0.6$) than the averaged value ($r^2 = 0.47$) for the project site, suggesting that the relative importance of coastline gradient as a driver of volume changes varies alongshore. Alongshore variations in offshore wave conditions, grain size d_{50}, or lower shoreface slope may potentially explain part of the remaining variance. Not all these parameters are or can be captured by a one-line model. Secondly, one-line models typically do not include cross-shore redistribution of sediment. Considering the significance of the cross-shore equilibration on initial waterline position changes, it is important to calibrate a one-line model on net volume changes rather than waterline position changes for nourished beaches. Calibrating on waterline positions may result in an overestimation of shoreline retreat since the initial cross-shore adjustment could falsely be extrapolated to a longer time-scale. Furthermore, the subaerial response and dune foot migration is often not resolved separately by one-line models, while this is essential to estimate the beach width development and dune growth.

For the aforementioned nourishment aspects, planform process-based modeling techniques may potentially be used, such as in previous mega-nourishment design studies, e.g., [36,76,77]. To predict the beach width response including alongshore variability and dunefoot migration, as observed at the Hondsbossche Dunes site, an explicit representation of the subaerial beach may be necessary. The recently developed coupled hydrodynamic and aeolian processed-based models [59,78,79] show potential to predict this behavior in the near future.

5.3. Implications for Design

In the Hondsbossche Dunes project, a beach dune system was created in front of an old sea dike. The man-made sandy system was aimed to increase safety against flooding while adding additional value. The design was inspired by Building with Nature principles, amongst others aiming to use natural forces to support our human objectives [9].

Our analysis suggests that after an initial adjustment period, the man-made beach system obtains comparable characteristics to the adjacent coast in terms of beach width and surfzone slope, despite the large quantities of added sand. Its exposed location in front of the sea dike and inherent coastal curvature will likely result in continued redistribution of sediment to the adjacent coast. The adjacent coast will benefit from the spreading of this sediment similar to the feeding effect of the sand engine [12,55]. The sediment supply to the adjacent coast could be considered an amenity of reinforcement of gray infrastructure with a seaward sandy extension.

For creating and maintaining safety against flooding with a coastal nourishment, sediment volume above storm surge level is critical. Adding sediment at higher elevations in the profile during construction is often costly due to the machinery shaping the profile. Rapid downslope transport of this sand during erosion events is undesired and inefficient. Our results show that the design of the Hondsbossche Dunes nourishment, with a natural slope and wide beach, proved to be successful in creating a positive sediment balance in the dune for a prolonged period after placement. With a dune volume increase at the nourished site three times higher than the surrounding beach sections, natural forces are indeed partaking in the building of strong flood defenses.

6. Conclusions

Sandy nourishments have been applied as coastal engineering interventions for decades. Lately, projects using millions of m^3 of added sand to replace gray coastal infrastructure (e.g., dams, dikes, seawalls) have been initiated. This paper presents the morphological development of the Hondsbossche Dunes (the Netherlands), a nourishment of 35 million m^3 in front of an old sea-dike. The Hondsbossche Dunes is a unique area of newly created beach and dunes, aimed to increase safety against flooding while creating space for nature and recreation. Nearly twenty topographic surveys in the first 5 years after placement are used to examine the redistribution of sediment volume in both the along- and cross-shore directions and the adaptation of the postnourishment profiles with initially wide beaches.

In the first years after implementation, large local volume changes up to 1000 m^3 per meter alongshore were observed. Yet, net volume losses in the 9 km coastal section were less than 5%, indicating that reworking was mostly local. The central part of the nourished site stands out as an erosive zone, with large (60 to 140 $m^3/m/y$) erosion. This erosion is predominantly in the subaqueous part of the profile and coincides with a shoreline retreat of about 80 m. Lateral coastal sections on the other hand show large accretion and a waterline migration of around 30 m-seaward. The man-made cross-shore beach profile rapidly mimics the adjacent beaches, as the surfzone slope is adjusted within two winters to a similar slope.

The seaward sandy extension of the sea dike creates a significant coastline curvature. The observed net profile volume change is, at several sections of the nourishment, strongly correlated ($r^2 > 0.6$) with this planform curvature. In our observations, the local change in waterline position, ΔX_s, correlates well with the volume change in the full profile ΔV_p ($r^2 = 0.84$). Finding this strong correlation suggests that the use of one-line models is appropriate when predicting the volume and coastline change of mega-nourishments.

An important observation is that the subaerial and subaqueous parts of profiles display contrasting behavior in the first years after placement. The dune volume increases in the first years after implementation with 30 $m^3/m/y$; for many profiles, this net gain in volume is found regardless of the erosive trend in the lower part of the profile. The magnitude of the dune volume increase at the nourished site is three times higher than at the adjacent coast. This implies that the nourishment is bringing additional sediment volume above surge level, which is key to coastal safety.

As the dune foot migrates in the seaward direction and the shoreline moves landward, the beach width is reduced from two sides. The initially wide beaches (i.e., up to 225 m) are transformed in five years to about 100 m-wide, similar to adjacent beaches. The similarity in beach widths near the end of the five years investigated suggests that upcoming storm events may be able to erode sediment from the dunes, potentially reducing the excessive net growth of dune volume in the near future. Our results demonstrate that several years may be needed for the sandy cross-shore profile to reach characteristics similar to the nearby coast, after reinforcement of gray infrastucture. Natural forces can provide a significant additional contribution to the building of dunes during these years, further increasing the safety against flooding.

Author Contributions: Conceptualization, A.K., M.d.S., S.d.V. and S.A.; methodology, A.K. and M.d.S.; software, A.K.; validation, A.K., M.d.S., S.d.V. and S.A.; formal analysis, A.K.; resources, M.d.S. and S.A.; data curation, A.K.; writing—original draft preparation, A.K.; writing—review and editing, A.K., M.d.S., S.d.V. and S.A.; visualization, A.K.; supervision, M.d.S.; project administration, M.d.S.; funding acquisition, S.A. All authors have read and agreed to the published version of the manuscript.

Funding: This research was funded by TKI Deltatechnology, Rijkswaterstaat and Hoogheemraadschap Hollands Noorderkwartier. Van Oord/Boskalis and Svašek Hydraulics provided in-kind funding to the study.

Data Availability Statement: The JARKUS dataset used in this research is publicly available at https://opendap.deltares.nl/thredds/catalog/opendap/rijkswaterstaat/jarkus/profiles/catalog.html, accessed on 9 February 2021. Restrictions apply to the availability of the contractor datasets. These data were obtained from Van Oord and Boskalis and are available from the authors with the permission of Van Oord/Boskalis only.

Acknowledgments: The authors want to thank TKI Deltatechnology, the Dutch government Rijkswaterstaat, the water board Hoogheemraadschap Hollands Noorderkwartier, Svašek Hydraulics, and Msc students Linde de Jongh and Tom Pak for their support of this research. Finally, a special thanks is for the contractors Van Oord and Boskalis, who made their survey data from the Hondsbossche Dunes project available for this research.

Conflicts of Interest: The funders had no role in the design of the study; analyses, or interpretation of data; in the writing of the manuscript, or in the decision to publish the results. Van Oord, Boskalis, and Rijkswaterstaat collected the data used in this study.

References

1. Nicholls, R.J.; Cazenave, A. Sea-level rise and its impact on coastal zones. *Science* **2010**, *328*, 1517–1520. [CrossRef]
2. Hinkel, J.; Aerts, J.C.; Brown, S.; Jiménez, J.A.; Lincke, D.; Nicholls, R.J.; Scussolini, P.; Sanchez-Arcilla, A.; Vafeidis, A.; Addo, K.A. The ability of societies to adapt to twenty-first-century sea-level rise. *Nat. Clim. Chang.* **2018**, *8*, 570–578. [CrossRef]
3. Dean, R.G. *Beach Nourishment: Theory and Practice*; World Scientific: Singapore, 2002.
4. de Schipper, M.A.; Ludka, B.C.; Raubenheimer, B.; Luijendijk, A.P.; Schlacher, T.A. Beach nourishment has complex implications for the future of sandy shores. *Nat. Rev. Earth Environ.* **2021**, *2*, 70–84. [CrossRef]

5. Ton, A.; Vuik, V.; Aarninkhof, S. Longshore sediment transports by large-scale lake circulations at low-energy, non-tidal beaches: A field and model study. *Coast. Eng.* 2022, *submitted.*
6. Perk, L.; van Rijn, L.; Koudstaal, K.; Fordeyn, J. A rational method for the design of sand dike/dune systems at sheltered sites; Wadden Sea Coast of Texel, The Netherlands. *J. Mar. Sci. Eng.* **2019**, *7*, 324. [CrossRef]
7. Vuik, V.; Jonkman, S.N.; Borsje, B.W.; Suzuki, T. Nature-based flood protection: The efficiency of vegetated foreshores for reducing wave loads on coastal dikes. *Coast. Eng.* **2016**, *116*, 42–56. [CrossRef]
8. Bridges, T.S.; Bourne, E.M.; King, J.K.; Kuzmitski, H.K.; Moynihan, E.B.; Suedel, B.C. *Engineering with Nature: An Atlas*; U.S. Army Engineer Research and Development Center: Vicksburg, MI, USA, 2018. [CrossRef]
9. de Vriend, H.J.; van Koningsveld, M.; Aarninkhof, S.G.J.; de Vries, M.B.; Baptist, M.J. Sustainable hydraulic engineering through building with nature. *J. Hydro-Environ. Res.* **2015**, *9*, 159–171. [CrossRef]
10. Brand, E.; Ramaekers, G.; Lodder, Q. Dutch experience with sand nourishments for dynamic coastline conservation—An operational overview. *Ocean. Coast. Manag.* **2022**, *217*, 106008. [CrossRef]
11. Stive, M.; de Schipper, M.; Luijendijk, A.; Aarninkhof, S.; van Gelder-Maas, C.; van Thiel De Vries, J.; de Vries, S.; Henriquez, M.; Marx, S.; Ranasinghe, R. A New Alternative to Saving Our Beaches from Sea-Level Rise: The Sand Engine. *J. Coast. Res* **2013**, *29*, 1001–1008. [CrossRef]
12. de Schipper, M.A.; de Vries, S.; Ruessink, G.; de Zeeuw, R.C.; Rutten, J.; van Gelder-Maas, C.; Stive, M.J.F. Initial spreading of a mega feeder nourishment: Observations of the Sand Engine pilot project. *Coast. Eng.* **2016**, *111*, 23–38. [CrossRef]
13. Pelnard-Considère, R. Essai de théorie de l'évolution des formes de rivage en plages de sable et de galets. *Journées de l'hydraulique* **1957**, *4*, 289–298.
14. Ludka, B.C.; Guza, R.T.; O'Reilly, W.C. Nourishment evolution and impacts at four southern California beaches: A sand volume analysis. *Coast. Eng.* **2018**, *136*, 96–105. [CrossRef]
15. Liu, G.; Cai, F.; Qi, H.; Zhu, J.; Lei, G.; Cao, H.; Zheng, J. A method to nourished beach stability assessment: The case of China. *Ocean. Coast. Manag.* **2019**, *177*, 166–178. [CrossRef]
16. Elko, N.A.; Wang, P. Immediate profile and planform evolution of a beach nourishment project with hurricane influences. *Coast. Eng.* **2007**, *54*, 49–66. [CrossRef]
17. Chang, J.I.; Yoon, S. The economic benefit of coastal erosion control in Korea. *J. Coast. Res.* **2016**, *1*, 1317–1321. doi: 10.2112/SI75-264.1. [CrossRef]
18. Liu, Z.; Cui, B.; He, Q. Shifting paradigms in coastal restoration: Six decades' lessons from China. *Sci. Total. Environ.* **2016**, *566–567*, 205–214. [CrossRef]
19. Gopalakrishnan, S.; Smith, M.D.; Slott, J.M.; Murray, A.B. The value of disappearing beaches: A hedonic pricing model with endogenous beach width. *J. Environ. Econ. Manag.* **2011**, *61*, 297–310. [CrossRef]
20. de Vries, S.; Arens, S.; De Schipper, M.; Ranasinghe, R. Aeolian sediment transport on a beach with a varying sediment supply. *Aeolian Res.* **2014**, *15*, 235–244. [CrossRef]
21. Walker, I.J.; Davidson-Arnott, R.G.; Bauer, B.O.; Hesp, P.A.; Delgado-Fernandez, I.; Ollerhead, J.; Smyth, T.A. Scale-dependent perspectives on the geomorphology and evolution of beach-dune systems. *Earth-Sci. Rev.* **2017**, *171*, 220–253. [CrossRef]
22. Bauer, B.; Davidson-Arnott, R.; Hesp, P.; Namikas, S.; Ollerhead, J.; Walker, I. Aeolian sediment transport on a beach: Surface moisture, wind fetch, and mean transport. *Geomorphology* **2009**, *105*, 106–116. [CrossRef]
23. Bruun, P. *Coast Erosion and the Development of Beach Profiles*; Beach Erosion Board: Washington, DC, USA, 1954.
24. Dean, R.G. Equilibrium beach profiles: Characteristics and applications. *J. Coast. Res.* **1991**, *7*, 53–84.
25. Galiforni Silva, F.; Wijnberg, K.M.; de Groot, A.V.; Hulscher, S.J. The effects of beach width variability on coastal dune development at decadal scales. *Geomorphology* **2019**, *329*, 58–69. [CrossRef]
26. van Puijenbroek, M.E.; Limpens, J.; de Groot, A.V.; Riksen, M.J.; Gleichman, M.; Slim, P.A.; van Dobben, H.F.; Berendse, F. Embryo dune development drivers: Beach morphology, growing season precipitation, and storms. *Earth Surf. Process. Landf.* **2017**, *42*, 1733–1744. [CrossRef]
27. Hoonhout, B.; de Vries, S. Aeolian sediment supply at a mega nourishment. *Coast. Eng.* **2017**, *123*, 11–20. [CrossRef]
28. Smyth, T.A.; Hesp, P.A. Aeolian dynamics of beach scraped ridge and dyke structures. *Coast. Eng.* **2015**, *99*, 38–45. [CrossRef]
29. Moulton, M.A.; Hesp, P.A.; Miot da Silva, G.; Keane, R.; Fernandez, G.B. Surfzone-beach-dune interactions along a variable low wave energy dissipative beach. *Mar. Geol.* **2021**, *435*, 106438. [CrossRef]
30. Thornton, E.B.; MacMahan, J.; Sallenger, A.H. Rip currents, mega-cusps, and eroding dunes. *Mar. Geol.* **2007**, *240*, 151–167. [CrossRef]
31. Cohn, N.; Ruggiero, P.; de Vries, S.; Kaminsky, G.M. New insights on coastal foredune growth: The relative contributions of marine and aeolian processes. *Geophys. Res. Lett.* **2018**, *45*, 4965–4973. [CrossRef]
32. Wijnberg, K.M.; Terwindt, J.H.J. Extracting decadal morphological behaviour from high-resolution, long-term bathymetric surveys along the Holland coast using eigenfunction analysis. *Mar. Geol.* **1995**, *126*, 301–330. [CrossRef]
33. Wijnberg, K.M. Environmental controls on decadal morphologic behaviour of the Holland coast. *Mar. Geol.* **2002**, *189*, 227–247. [CrossRef]
34. Technische Adviescommissie voor de Waterkeringen. *Leidraad voor de Beoordeling van Duinen als Waterkering*; Technische Adviescommissie voor de Waterkeringen: The Hague, The Netherlands, 1984.

35. van Alphen, J. *De Morfologie en Lithologie van de Brandingszone Tussen Terheijde en Egmond aan Zee*; Technical Report NZ-N-87.28; Rijkswaterstaat: Utrecht, The Netherlands, 1987.
36. Kroon, A.; van Leeuwen, B.; Walstra, D.J.; Loman, G. Dealing with Uncertainties in Long-term Predictions of a Coastal Nourishment. In *Coastal Management: Changing Coast, Changing Climate, Changing Minds*; ICE Publishing: London, UK, 2016; pp. 9–18. [CrossRef]
37. Rijkswaterstaat. *Kenmerkende Waarden, Getijgebied 2011.0*; RWS Centrale Informatievoorziening: Delft, The Netherlands, 2013.
38. van Rijn, L.C. Sediment transport and budget along the Dutch coastal zone. *Coast. Eng.* **1997**, *32*, 61–90. [CrossRef]
39. Rijkswaterstaat. *Kustlijnkaartenboek 2019*; Ministerie van Infrastructuur en Waterstaat, Rijkswaterstaat: Utrecht, The Netherlands, 2018.
40. Rijkswaterstaat. *Kustlijnkaartenboek 2021*; Ministerie van Infrastructuur en Waterstaat, Rijkswaterstaat: Utrecht, The Netherlands, 2020.
41. Rijkswaterstaat. Nourishments along the Dutch Coast Since 1952. Available online: https://opendap.deltares.nl/thredds/catalog/opendap/rijkswaterstaat/suppleties/catalog.html (accessed on 12 April 2021).
42. van Gent, M.R.; van Thiel de Vries, J.S.; Coeveld, E.M.; de Vroeg, J.H.; van de Graaff, J. Large-scale dune erosion tests to study the influence of wave periods. *Coast. Eng.* **2008**, *55*, 1041–1051. [CrossRef]
43. Den Heijer, C.K.; Baart, F.; van Koningsveld, M. Assessment of dune failure along the Dutch coast using a fully probabilistic approach. *Geomorphology* **2012**, *143–144*, 95–103. [CrossRef]
44. Rijkswaterstaat. Annual Cross-Shore Transect Bathymetry Measurements along the Dutch Coast Since 1965. Available online: https://opendap.deltares.nl/thredds/catalog/opendap/rijkswaterstaat/jarkus/profiles/catalog.html (accessed on 9 February 2021).
45. de Vries, S.; Southgate, H.N.; Kanning, W.; Ranasinghe, R. Dune behavior and aeolian transport on decadal timescales. *Coast. Eng.* **2012**, *67*, 41–53. [CrossRef]
46. Smith, A.; Houser, C.; Lehner, J.; George, E.; Lunardi, B. Crowd-sourced identification of the beach-dune interface. *Geomorphology* **2020**, *367*, 107321. [CrossRef]
47. Quartel, S.; Kroon, A.; Ruessink, B.G. Seasonal accretion and erosion patterns of a microtidal sandy beach. *Mar. Geol.* **2008**, *250*, 19–33. [CrossRef]
48. Ludka, B.C.; Gallien, T.W.; Crosby, S.C.; Guza, R.T. Mid-El Niño erosion at nourished and unnourished Southern California beaches. *Geophys. Res. Lett.* **2016**, *43*, 4510–4516. [CrossRef]
49. Yates, M.L.; Guza, R.T.; O'Reilly, W.C. Equilibrium shoreline response: Observations and modeling. *J. Geophys. Res. Ocean.* **2009**, *114*. [CrossRef]
50. Griggs, G.B.; Patsch, K. Natural changes and human impacts on the sand budgets and beach widths of the Zuma and Santa Monica littoral cells, Southern California. *Shore Beach* **2018**, *86*, 1–14.
51. Park, J.Y.; Gayes, P.T.; Wells, J.T. Monitoring Beach Renourishment along the Sediment-Starved Shoreline of Grand Strand, South Carolina. *J. Coast. Res.* **2009**, *25*, 336–349. [CrossRef]
52. Marinho, B.; Coelho, C.; Larson, M.; Hanson, H. Short- and long-term responses of nourishments: Barra-Vagueira coastal stretch, Portugal. *J. Coast. Conserv.* **2018**, *22*, 475–489. [CrossRef]
53. Larson, M.; Hanson, H.; Kraus, N.C.; Newe, J. Short- and long- term responses of beach fills determined by eof analysis. *J. Waterw. Port Coasta Ocean Eng.* **1999**, *125*, 285–293. [CrossRef]
54. Browder, A.E.; Dean, R.G. Monitoring and comparison to predictive models of the Perdido Key beach nourishment project, Florida, USA. *Coast. Eng.* **2000**, *39*, 173–191. [CrossRef]
55. Roest, B.; de Vries, S.; de Schipper, M.; Aarninkhof, S. Observed changes of a mega feeder nourishment in a coastal cell: Five years of sand engine morphodynamics. *J. Mar. Sci. Eng.* **2021**, *9*, 37. [CrossRef]
56. Verhagen, H.J. Analysis of beach nourishment schemes. *J. Coast. Res.* **1996**, *12*, 179–185.
57. Dean, R.G.; Yoo, C. Beach-Nourishment Performance Predictions. *J. Waterw. Port Coast. Ocean. Eng.* **1992**, *118*, 567–586. [CrossRef]
58. Bezzi, A.; Fontolan, G.; Nordstrom, K.F.; Carrer, D.; Jackson, N.L. Beach nourishment and foredune restoration: Practices and constraints along the venetian Shoreline, Italy. *J. Coast. Res.* **2009**, *1*, 287–291.
59. Cohn, N.; Ruggiero, P.; García-Medina, G.; Anderson, D.; Serafin, K.A.; Biel, R. Environmental and morphologic controls on wave-induced dune response. *Geomorphology* **2019**, *329*, 108–128. [CrossRef]
60. Itzkin, M.; Moore, L.J.; Ruggiero, P.; Hacker, S.D.; Biel, R.G. The relative influence of dune aspect ratio and beach width on dune erosion as a function of storm duration and surge level. *Earth Surf. Dyn.* **2021**, *9*, 1223–1237. [CrossRef]
61. van der Wal, D. Beach-Dune Interactions in Nourishment Areas along the Dutch Coast. *J. Coast. Res.* **2004**, *20*, 317–325. [CrossRef]
62. Athanasiou, P.; van Dongeren, A.; Giardino, A.; Vousdoukas, M.; Antolinez, J.A.; Ranasinghe, R. A Clustering Approach for Predicting Dune Morphodynamic Response to Storms Using Typological Coastal Profiles: A Case Study at the Dutch Coast. *Front. Mar. Sci.* **2021**, *8*, 747754. [CrossRef]
63. den Heijer, C. The Role of Bathymetry, Wave Obliquity and Coastal Curvature in Dune Erosion Prediction. Ph.D. Thesis, Delft University of Technology, Delft, The Netherlands, 2013.
64. Silva, F.G.; Wijnberg, K.M.; de Groot, A.V.; Hulscher, S.J.M.H. On the importance of tidal inlet processes for coastal dune development. *Coast. Dyn. Proc.* **2017**, *104*, 106–110. [CrossRef]
65. Wijnberg, K.; Poppema, D.; Mulder, J.; van Bergen, J.; Campmans, G.; Galiforni-Silva, F.; Hulscher, S.; Pourteimouri, P. Beach-dune modelling in support of Building with Nature for an integrated spatial design of urbanized sandy shores. *Res. Urban. Ser.* **2021**, *7*, 241–259. [CrossRef]

66. de Vries, S.; van Thiel de Vries, J.S.; van Rijn, L.C.; Arens, S.M.; Ranasinghe, R. Aeolian sediment transport in supply limited situations. *Aeolian Res.* **2014**, *12*, 75–85. [CrossRef]
67. Thieler, E.R.; Pilkey, O.H.; Young, R.S.; Bush, D.M.; Chai, F. The use of mathematical models to predict beach behavior for U.S. coastal engineering: A critical review. *J. Coast. Res.* **2000**, *16*, 48–70.
68. Ranasinghe, R. On the need for a new generation of coastal change models for the 21st century. *Sci. Rep.* **2020**, *10*, 2010. [CrossRef] [PubMed]
69. Weathers, H.; Voulgaris, G. Evaluation of beach nourishment evolution models using data from two South Carolina, USA beaches: Folly Beach and Hunting Island. *J. Coast. Res.* **2013**, *69*, 84–98. [CrossRef]
70. Hanson, H. Genesis-A Generalized Shoreline Change Numerical Model. *J. Coast. Res.* **1988**, *5*, 1–27.
71. Dabees, M.; Kamphuis, J.W. ONELINE, a numerical model for shoreline change. *Proc. Int. Conf. Coast. Eng.* **1998**, 1, 2668–2681.
72. Hurst, M.D.; Barkwith, A.; Ellis, M.A.; Thomas, C.W.; Murray, A.B. Exploring the sensitivities of crenulate bay shorelines to wave climates using a new vector-based one-line model. *J. Geophys. Res. Earth Surf.* **2015**, *120*, 2586–2608. [CrossRef]
73. Roelvink, D.; Huisman, B.; Elghandour, A.; Ghonim, M.; Reyns, J. Efficient Modeling of Complex Sandy Coastal Evolution at Monthly to Century Time Scales. *Front. Mar. Sci.* **2020**, *7*, 535. [CrossRef]
74. Hallermeier, R.J. A profile zonation for seasonal sand beaches from wave climate. *Coast. Eng.* **1981**, *4*, 253–277. [CrossRef]
75. Hallermeier, R.J. Sand Transport Limits in Coastal Structure Design. In *Coastal Structures'83*; American Society of Civil Engineers: Reston, VA, USA, 1983; pp. 703–716.
76. Mulder, J.P.; Tonnon, P.K. "Sand Engine": Background and Design of a Mega-Nourishment Pilot in the Netherlands. *Coast. Eng. Proc.* **2011**, *1*, 35. [CrossRef]
77. Bodde, W.; McCall, R.; Jansen, M.; van den Berg, A.; Roelvink, J. Long-term morphological modelling: Combining storm impact and daily conditions in an integrated modeling framework. In Proceedings of the Coastal Dynamics 2017, Helsingør, Denmark, 12–16 June 2017; pp. 1809–1820.
78. Luijendijk, A.; Velhorst, R.; Hoonhout, B.; de Vries, S.; Ranasinghe, R. Integrated Modelling of the Morphological Evolution of the Sand Engine Mega-Nourishment. *Proc. Coast. Dyn. 2017* **2017**, 1874–1885.
79. Itzkin, M.; Moore, L.J.; Ruggiero, P.; Hovenga, P.A.; Hacker, S.D. Combining process-based and data-driven approaches to forecast beach and dune change. *Environ. Model. Softw.* **2022**, *153*, 105404. [CrossRef]

Journal of
Marine Science and Engineering

MDPI

Article

Morphological Development and Behaviour of a Shoreface Nourishment in the Portuguese Western Coast

Celso Aleixo Pinto [1,*], Rui Taborda [2], César Andrade [2], Paulo Baptista [3], Paulo Alves Silva [4], Diogo Mendes [5] and Joaquim Pais-Barbosa [4]

[1] Coastal Monitoring and Risk Division, Portuguese Environment Agency (Agência Portuguesa do Ambiente), 2610-124 Amadora, Portugal

[2] IDL—Instituto D. Luiz, Departamento de Geologia, Faculdade de Ciências, Universidade de Lisboa, 1749-016 Lisboa, Portugal; rui.taborda@fc.ul.pt (R.T.); candrade@fc.ul.pt (C.A.)

[3] Departamento de Geociências e Centro de Estudos do Ambiente e do Mar (CESAM), Campus de Santiago, Universidade de Aveiro, 3810-193 Aveiro, Portugal; renato.baganha@ua.pt

[4] Departamento de Física e Centro de Estudos do Ambiente e do Mar (CESAM), Campus de Santiago, Universidade de Aveiro, 3810-193 Aveiro, Portugal; psilva@ua.pt (P.A.S.); joaquim.paisbarbosa@ua.pt (J.P.-B.)

[5] HAEDES, Casais do Arrocho, 2025-452 Azóia de Cima, Portugal; diogo.mendes@haedes.eu

* Correspondence: celso.pinto@apambiente.pt

Abstract: Current coastal protection strategy in Portugal defines beach and shoreface nourishment as a valid measure to mitigate coastal erosion in some erosional hot-spots, being considered as an adaptation measure under the present climate change scenario, including the impacts of sea level rise. However, scant objective data on shoreface nourishments are available to evaluate performance of this type of intervention in mitigating beach erosion and managing coast risk. We present the first monitoring results of a $\approx 2.4 \times 10^6$ m^3 shoreface nourishment on the Aveiro coast (Costa Nova—Ílhavo), the largest until now in Portugal, focusing on its morphological development, impacts on adjacent beaches due to alongshore spreading and cross-shore redistribution, and contribution to the sediment budget of the nourished sediment cell. The analyses are based on high-resolution coastal monitoring data, provided by the Portuguese COaStal MOnitoring Program (COSMO). A Multiple Monitoring Cell (MMC) approach was used to evaluate local and feeder efficiency of the nourishment, sediment budget exchanges within both the placement and wider survey domains (≈ 1 km^2 and 12 km^2, respectively). Results show rapid (ca. 6 months) morphological change over the placement area, with a decrease of about 40% of the initial volume. Fast onshore sediment redistribution explains part of this change, placed sand having merged with the pre-existing bar system increased the volume of the shallower nearshore. Longshore transport is reflected by increasing the robustness of the bar downdrift of the placement area and also explains the negative sediment budget (0.75×10^6 m^3) of the survey domain, which corresponds to losses through its southern boundary. Sediment spreading also induced accretion of the subaerial section of Costa Nova beaches in front of the placement area, reversing their long-term erosive trend. In contrast, this trend persisted at downdrift beaches. This suggests that the time lag of the subaerial beach response to this intervention increases with the distance to the placement area, and reversal of the erosive trend will only be noticeable in the following years. This study provides new insights on the time scales of beach response to high-magnitude shoreface interventions in high-energy wave-dominated sandy coasts, which will support decision making regarding similar operations designed to manage erosional hot-spots elsewhere.

Keywords: wave dominated coast; coastal erosion; shoreface nourishment; cross-shore and longshore processes; coastal protection strategy

Citation: Pinto, C.A.; Taborda, R.; Andrade, C.; Baptista, P.; Silva, P.A.; Mendes, D.; Pais-Barbosa, J. Morphological Development and Behaviour of a Shoreface Nourishment in the Portuguese Western Coast. *J. Mar. Sci. Eng.* **2022**, *10*, 146. https://doi.org/10.3390/jmse10020146

Academic Editor: Zhen-Gang Ji

Received: 16 December 2021
Accepted: 19 January 2022
Published: 22 January 2022

1. Introduction

Coastal areas are inherently dynamic, driven by meteorological, oceanographic, geological, and anthropogenic factors [1]. Sandy beaches occupy more than one-third of the world's coastline, with one-quarter under erosion [2], and this trend may be aggravated due to climate change effects, namely, sea level rise [3–5] and changes in frequency and duration of storminess [6,7]. Historical and ongoing development of the coastal area has induced coastal squeeze, leading to the loss of scenic values, recreational beach areas and habitats, and increasing hazards to people, built environments, and infrastructures, especially in densely populated and developed coastlines [8]. Management of these areas is particularly challenging, and can only be adequately addressed if grounded upon comprehensive knowledge of coastal changes and driving processes based on the existence of coastal monitoring data.

Beach nourishment (also referred to as sand replenishment or beach fill) comprises the addition of good quality sand to increase the width or volume of a specific beach or coastal stretch [9]. It is a coastal management technique used in risk reduction and adaptation to climate change worldwide [10]. Beach nourishment has been undertaken in emergency contexts, as a local and short-term remedy solution (e.g., in the aftermath of storm-induced erosion [11]) and as a regional and long-term management strategy to counteract erosive trends and reduce coastal vulnerability [10–14].

The nourishment type depends, among other factors, on coastal management objectives and on sand availability. Sediment is either placed above mean sea level (over the beach berm or foredune), or in the inner (upper) shoreface. While the former is typically called beach nourishment, the latter is often designated as shoreface nourishment. Borrow sediments are typically obtained from maintenance dredging of inlets or navigation channels associated with nearby harbours/fishing ports/recreational marinas. In addition, the outer (lower) shoreface may also provide excellent borrow areas [14–18].

In mainland Portugal, beach nourishment started in 1950, with up to 170 interventions concluded until 2020 (data updated from [14]). Altogether, they comprised placement of approximately 42 M m^3, over the inner shoreface (thus landward of the depth of closure—DoC) and over the beach/berm and foredune. Only 8% of these interventions may be considered of high magnitude (i.e., volume > 1×10^6 m^3 [14]). To date, nourishments carried out have essentially resorted to sediment dredged from inlets and channels associated with port activities (commercial/fishing/recreation) (about 90% of all operations and of total volume involved). Dredged material was used to minimize erosion caused by port infrastructures, such as sand starvation downdrift of jetties, and enhanced accumulation updrift of jetties and in inlets and channels.

Extensive monitoring of sand redistribution following beach nourishments is crucial to assess project performance and impacts, in addition to improving understanding of the associated coastal dynamics [10,13]. In opposition to subaerial beach nourishment, behaviour of shoreface nourishment is still not well-understood [13]. This contrasts with the increasing importance of the latter worldwide, mostly due to its cost-effectiveness [13]. For example, 43% of nourishment operations carried out in mainland Portugal were of the shoreface type, and comprised 50% of the total placement volume [14].

So far, most data associated with shoreface nourishments have been acquired in a limited suite of environmental settings (e.g., The Netherlands [10,12,13] and USA [19]) with poor representation of high-energy wave dominated coasts, where tides and storm surge are secondary drivers.

The objective of this work is to provide new insights on the morphological evolution, behaviour, and efficiency of a high-magnitude volume shoreface nourishment, undertaken in a high-energy wave-dominated coast, downdrift of a stabilized tidal inlet (Aveiro, Portugal).

The study relies upon high resolution monitoring following the largest shoreface nourishment undertaken in Portugal, encompassing the downdrift domain potentially affected by short-term sediment redistribution. This work extends and complements previous study on the same field area [20], but it differs by introducing a sediment budget approach that allowed the understanding of post-nourishment sediment dispersion. In addition, we investigate the time scale required for the beach to acquire a condition of equilibrium replicating the equilibrium beach profile of Bruun's, following the imposition of a significant sand volume.

2. Study Area

2.1. Coastal Setting

The study area comprises a low-lying beach–dune system that extends 20 km south of Aveiro inlet until Praia de Mira (MI) (north), including Barra (BA), Costa Nova (CS), Vagueira (VG1; VG2), Labrego (LB), Duna Alta (DA), Areão (AE), and Poço da Cruz (CZ) beaches (Figure 1). This coastal stretch has an NNE–SSW orientation, being included in coastal cell 1 (sub-cell 1b), according to an established classification [21]. Coastal defence works in the area include two jetties at Aveiro harbour entrance, nine groynes (E1 to E9), and three rock armour revetments (two in Costa Nova and one at Vagueira beach).

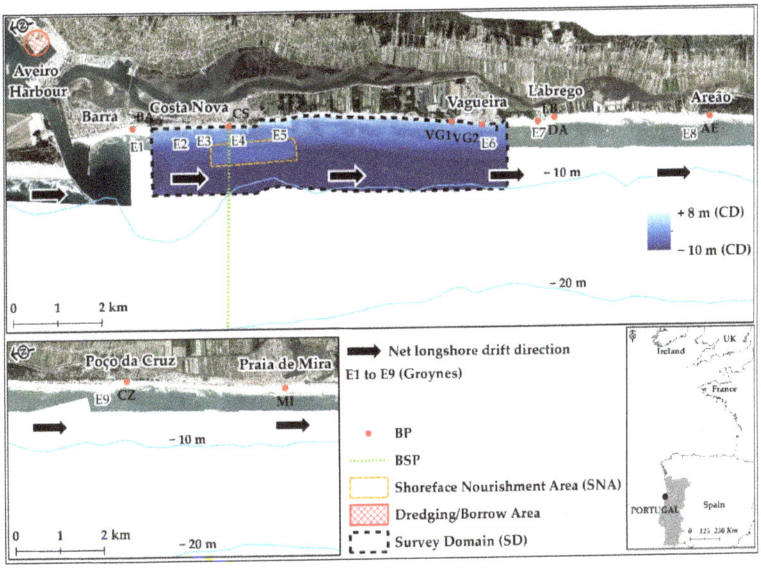

Figure 1. Location of study area with detailed position of the dredging/borrow area and shoreface nourishment area (SNA). Beach profiles (BP), Beach shoreface profiles (BSP), and topo-bathymetric surveys (TBS) in the survey domain (SD) and southward until Praia de Mira.

According to [22], the averaged offshore significant wave height (Hs) is = 2.36 m, with a monthly averaged value of 1.77 m during summer that increases to 3.04 m during winter. Mean wave peak-period (Tp) is 10.7 s, ranging between 8.7 s and 12.3 s during summer and winter, respectively. Mean wave direction is predominantly from NW (71%), with a mean value of 310.5° and monthly averaged value ranging from 298° to 324°. The number of storm events (Hs > 4.5 m) per year is, on average, 15 per year, with a mean value of Hs_{Max} = 5.6 m and maximum values higher than 10 m. Tides are semi-diurnal, ranging from 1.2 m to 3.6 m during neap and spring tides respectively.

Longshore sediment transport is about 1×10^6 m^3/year directed towards south [22–24], resulting from a southward and northward fluxes of circa 1.5×10^6 m^3 and 0.5×10^6 m^3, respectively. However, mean annual values show an irregular and noncyclical pattern, with yearly averages ranging from 0.11×10^6 m^3/year to 2.24×10^6 m^3/year between 1953 and 2010 [24]. The authors in [22,25] also identified this annual inter-variability in longshore sediment transport rates, reporting values of 0.6×10^6 m^3/year to 3.2×10^6 m^3/year (1952 to 2010) and 0.16×10^6 m^3/year to 1.52×10^6 m^3/year (2000 to 2019), respectively.

Beach sediments consist of well and very well sorted medium sand (median grain size between 0.4 mm and 0.6 mm) [26], whereas shoreface sand is finer (medium to fine sand, median grain size between 0.35 mm and 0.21 mm). Sands of the beach and shoreface are essentially quartzic with minor contributions of carbonate bioclasts (shell fragments of bivalves).

2.2. *Shoreline Evolution*

Shoreline evolution south of the Aveiro inlet, between Barra and north of Mira beach varies according to the considered time scale (i.e., long, medium or short-term). The authors in [27] mention retreat rates of −15 m/year between 1947 and 1954 in Barra-Costa Nova, and −5.2 m/year and −3.0 m/year in Vagueira from 1954 to 1990, while [28] reported a maximum retreat of 400 m of the waterline during 1948–2005 in Costa Nova-Vagueira. The authors in [29] obtained retreat rates between 1958 and 2018 of −1.0 m/year and −4.4 m/year in Barra-Costa Nova and Costa Nova-Vagueira, respectively. To the south until Mira beach, [30] obtained retreat rates ranging from −1.9 m/year to −4.5 m/year for the period 1958–2010. The above-mentioned results, obtained by different authors and methods, clearly show the existence of long-term erosion in this coastal sector. Causes are mostly related to: (1) the negative sediment budget that is verified towards south of the Aveiro inlet due to the reduction in sediment supply from the north (mainly Douro River) [21,31]; and (2) the interruption in longshore sediment transport caused by the north jetty of the Aveiro harbour, which blocked most of the available sand coming from the north [23].

More recently there is a decrease in erosion rates. Between 2010 and 2018, [29] obtained retreat rates of −0.40 m/year and −0.02 m/year in Barra-Costa Nova and Costa Nova-Vagueira sectors, respectively. Further south and until Mira, [29] obtained erosion rates ranging from −0.01 m/year to −0.57 m/year for the period 2013–2018. The most recent coastline (i.e., dune foot) comparison, from 2018 to 2020, carried out within the scope of this work using ortophotos and Digital Elevation Models (DEM) provided by the Portuguese COaStal MOnitoring Program—COSMO (https://cosmo.apambiente.pt, accessed date on 15 November 2021), shows a positive rate of 0.4 m/year between Barra-Vagueira, and 1.4 m/year from this point to the north of Praia de Mira, with 70% of the coastline being characterized as stable or accreting, according to the classification proposed by [2] (Figure 2).

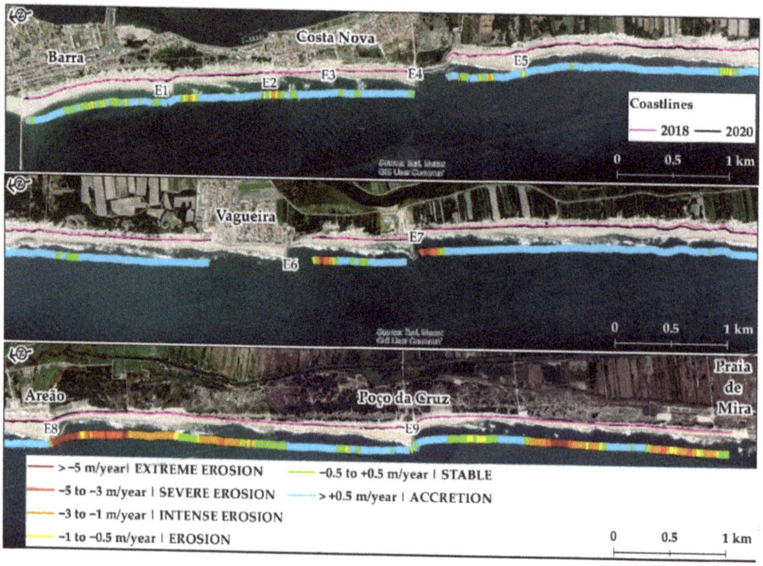

Figure 2. Coastline evolution from 2018 to 2020.

The causes for the reduction of coastal erosion rates are most certainly related to previous nourishments performed in this coastal cell [14,32], which promoted localized replacement of the existing negative sediment budget.

2.3. Previous Nourishments

Since 1965, a total of 26 beach fills have been carried out between Barra beach and Mira beach, 20% on the dry beach/dune and 80% in the shoreface, comprising the deposition of 15.3 M m^3 (36% of the total amount deposited in Portugal from 1950 to 2020), of which almost 50% was deposited in the last 10 years (data updated from [14]). This demonstrates the growing awareness regarding the beneficial use of sediments dredged by Port of Aveiro to counteract coastal erosion, and is a result of an established policy of integrated sediment management involving different stakeholders (i.e., Port and Environmental Authorities).

3. Methods

3.1. Shoreface Nourishment

Shoreface nourishment intervention comprised the dumping of $\approx$ 2.375 M m^3 of sediment in May to early September 2020. Sand was dumped in front of Costa Nova beach (Figures 1 and 2a), between 4 and 8 m below Chart Datum (CD—chart datum lies 2.0 m below mean sea level—MSL). The sand mound acquired a broadly trapezoidal shape with a flat summit at −4 m CD, and extended over 0.95 km^2 (length = 1.9 km; width = 0.5 km). Sediments were dredged from a nearby borrow area ($\approx$6.5 km) and consist of dredge spoil stored within the Aveiro harbour (Figure 3).

Figure 3. (**a**) Dumping area and (**b**) dredging area.

Borrow materials are essentially made of quartz and consist of moderately well sorted, coarse and medium sand (mean grain size of 0.92 mm) [33], the mean size spanning over a wider size-spectrum than the beach. Coarser sand (about 1/3 of the samples analysed) is somewhat less well sorted than medium sand. All samples analysed (n = 36) for textural characterization of borrow materials yielded less than 1% mud (particles < 63 µm) contents. However, about 20% of the samples revealed non-negligible (higher than 5%) amounts of particles finer than 0.250 mm, in one case reaching up to 14% of the whole sample. Assuming that the sampling scheme correctly represents the total volume of borrow materials, and that fine and very fine sand particles are especially prone to be readily removed from the dump site to the offshore, we estimate that up to 8% of the sand volume dumped in the nourishment area could have been lost during (or shortly after) the nourishment operations, slightly reducing the sediment volume in the dump area.

An additional volume of $\approx$320,000 m^3 ($\approx$295,000 m^3 after the above-mentioned textural correction) of sand was deposited between September 2020 and March 2021, adjacent to the seaward slope of the main dump, outside the pre-defined SNA of the main intervention, but within the SD. Finally, more 212,000 m^3 were deposited between March 2021 and Aug 2021, at the location of the main dump, but that were excluded from sediment budget calculations within the SD, given that last Topo-Bathymetric Surveys (TBS) dates from March 2021.

3.2. Data Collection

Three hourly deep water wave parameters (Hs—significant wave height of combined wind waves and swell; Tp—peak wave period; —mean wave direction) covering the period from June 2018 to September 2021 were obtained for a point located broadly 60 km WNW of Aveiro harbour, at 9.5° W; 40.5° N from the Climate Data Store (https://cds.climate.copernicus.eu/cdsapp#!/dataset/reanalysis-era5-single-levels?tab=form; accessed on 3 January 2022).

Topographic and hydrographic data used for this study were provided by the COSMO Program [34] containing three types of datasets: Beach Profiles (BP), Beach-Shoreface Profiles (BSP), and Topo-Bathymetric Surveys (TBS), performed during the monitoring period, from July 2018 to September 2021 between Praia da Barra and Praia de Mira beaches (Figure 1). An additional Multi-Beam Hydrographic Survey (MBHS) was made by the Aveiro Port Administration in early September 2020, immediately after the main dump, covering only the deposition area (survey boundaries indicated in Figure 1).

BP consist of cross-shore transects at several coastal locations (Figure 1) that were repeatedly surveyed using a GPS/RTK between a fixed reference onshore and extending at least to the +1 m (CD) contour line. Horizontal resolution was better than 1 m and vertical accuracy better than 0.05 m.

BSP incorporate and extend BP into the shoreface down −20 m (CD) contour line, using a jet sky equipped with a GPS/RTK coupled with a single beam echo sounder, with planimetric and vertical accuracy similar to BP.

TBS results from the combination of topographic and hydrographic data. Topography was acquired using aerial photogrammetry techniques over stereoscopic imagery captured by a high-resolution camera mounted on a fixed wing UAV, equipped with GPS/RTK, supported by several ground control points. RMS error in planimetry and altimetry were better than 0.05 m. Hydrography extends into the beach submarine domain down to ca. −10 m (CD) and data were obtained using a single-beam (transect spacing of 30 m in the depth range of +1 to −3 m CD) and multi-beam echo sounders, the latter with vertical accuracy of 0.05 m, the same as the MBHS.

3.3. Data Processing and Analysis

Wave data were used to characterize deep water wave regime over the monitoring period and to estimate potential longshore drift in the survey domain. Wave parameters at breaking were computed using Airy wave theory and Snell's Law. Longshore drift estimates were obtained by the energy flux method, using the CERC formula [35]. In agreement with the findings of [36], which reported an overestimation of 7.85 × regarding the CERC formula when spectral effects are not considered in computations of longshore drift, the formula was parametrized with a smaller empirical factor (k'= 0.39/7.85). Bathymetric data were processed with QIMERA (multi-beam) and HYPACK software (single-beam) by ATLANTICLAND (a consortium member of the COSMO Program). The obtained XYZ point cloud was interpolated to produce DEM with 0.03 m pixel resolution. For the topographic surveys it was used the AGISOFT software, with a processed point cloud of 100 points/m^2, generating a DEM with a 0.01 m resolution (data processed by GEOGLOBAL, a consortium member of the COSMO Program).

In order to analyse the morphological development and behaviour of the shoreface nourishment, including spreading and diffusion processes over the survey domain (SD), a volumetric, cut-fill, and bed level changes analysis was performed over the TBS using ArcGIS and Globalmapper softwares. To analyse sediment budget exchanges in terms of cross-shore/longshore processes and evaluate feeder efficiency of the nourishment, the SD was divided in different areas, here designated as Multi-Monitoring Cells. Shoreface and subaerial beach response through time in relation to the nourishment, was based on the comparison of selected contour lines extracted from the TBS.

Shoreface and subaerial beach variability (i.e., volume and width) and its response to the nourishment was analysed through the available BP and BSP. For the BP, volume was calculated above + 1 m (CD) up to a fixed height, where no variations occur; in turn, beach width variation was estimated measuring the horizontal displacement of the +2 m (CD). Shoreface morphological variability was addressed using BSP and evaluated through the assessment of horizontal displacement of selected contour lines down to the closure depth. To highlight the potential effect of the shoreface nourishment (SN), width and volume values obtained from BP and BSP were normalized by subtraction from the pre-nourishment time-averaged values, July 2018 to July 2020 and August 2018 to July 2020, respectively.

The depth of closure, seaward of which no significant morphological changes occur, according to the original definition proposed by [37] was estimated from statistics of bed-level variability, as proposed by [36]. The value of 0.15 m was set as the threshold on the standard deviation (s) of the bed level. This is slightly higher than the vertical accuracy reported for the Jet Ski Single-Beam acquisition platform (0.10 m) and allows for some additional uncertainty due to wave conditions and water depth as suggested by [38].

A Multiple Monitoring Cell (MMC) approach was used to evaluate local and feeder efficiency of the SN and to assess post-nourishment pathways of sediment dispersion within survey domain (SD). This was applied to the SD by splitting this area into twenty rectangular cells, organized in four cross-shore rows and five longshore columns.

To quantify shoreline evolution, namely, comparison of contour lines of TBS), the Digital Shoreline Analyses System [39] was used, an extension in ArcGIS (ESRI) software that allowed for net shoreline movement and retreat rates calculations.

4. Results

4.1. *Wave Forcing and Longshore Drift in the Monitoring Period*

Figure 4 shows the time distribution of deep-water wave parameters over the monitoring period, highlighting the seasonality of wave forcing. Mean wave conditions (Hs = 2.23 m; Tp = 10.9 s) are in close agreement with long-term estimates indicated in Section 2.1. High-energy events essentially cluster in winter/autumn months with high directional variability in drift currents. In contrast, during summer/spring, wave height and period are lower and drift currents are more consistently directed southward.

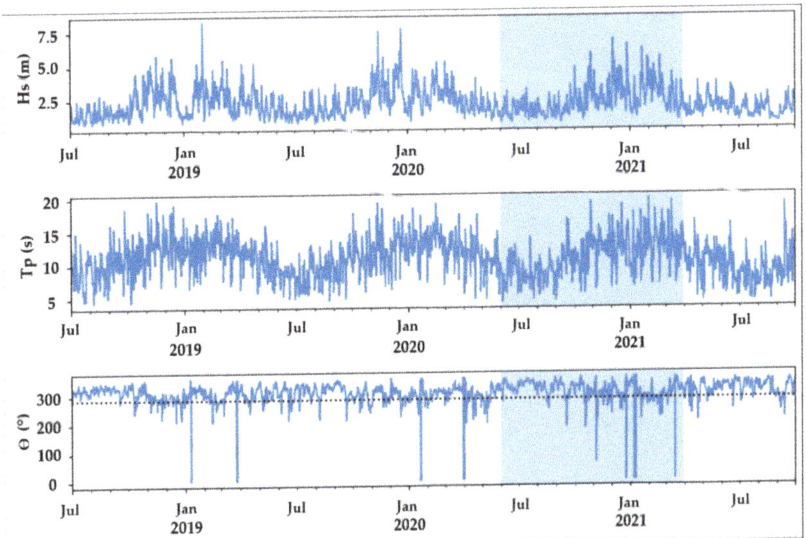

Figure 4. Time-series of deep-water parameters (Hs, Tp,) during the monitoring period (July 2018 to September 2021). The highlighted blue region corresponds to the period covered by the pre- and post-nourishment TBS (June 2020 to March 2021). The dotted line indicates shore normal direction.

During the monitoring period, monthly longshore drift was almost exclusively directed towards south (Figure 5), with a grand total of 3.15×10^6 m^3. This corresponds to a yearly average of 0.97×10^6 m^3/year, which is remarkably similar to longer term estimates of mean annual longshore drift reported by [22–24].

In the period covered by the pre- and post-nourishment TBS (June 2020 to March 2021) (highlighted blue region in Figure 5) the only reversal on longshore drift direction is observed in February 2021 in correspondence with persistency of westerly waves. Notwithstanding this singularity, the net drift was directed southward with a magnitude of 0.87×10^6 m^3, strongly influenced by high-energy winter waves (Figures 4 and 5).

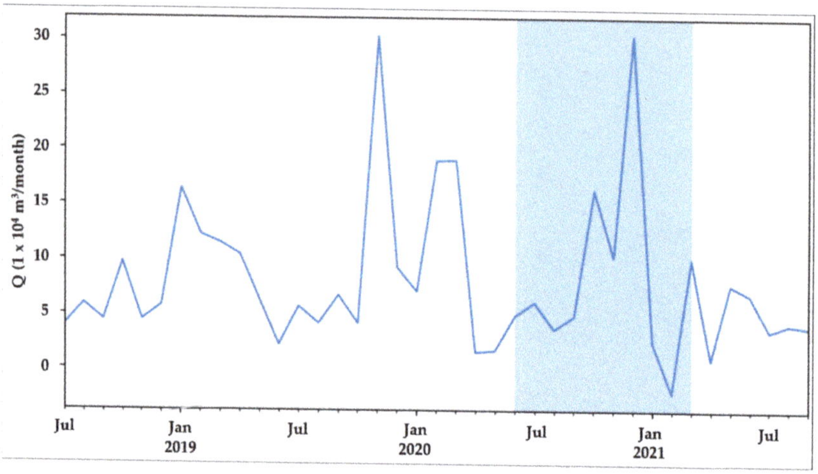

Figure 5. Monthly longshore drift during the monitoring period (positive values indicate southward drift) The highlighted blue region corresponds to the period covered by the pre- and post-nourishment TBS (June 2020 to March 2021).

4.2. Beach Profiles (BP)

Results (Figure 6) show the variability of subaerial beach volume and width related to short term (seasonal) wave forcing. Significant changes following the nourishment were only observed in (CS) (the profile closest to, and aligned with the SNA), with a consistent increase in width and volume of the subaerial beach over time. In this case, magnitude of changes clearly exceeds the range of seasonal variability over the monitoring period. All the remaining profiles, both updrift and downdrift, do not appear to have been substantially influenced by the nourishment.

4.3. Beach Shoreface Profiles (BSP)

Figure 7 illustrates changes in cross-shore profiles extending from the foredune until -10 m (CD) over the monitoring period at the location where the shoreface nourishment was dumped, complementing data on subaerial beach changes mentioned above.

Depth of closure (DoC) used in this study was estimated from profile convergence ($s < 0.15$ m) and is—9.3 m (CD) (Figure 7). This depth is lower than previous wave-based empirical estimates of DoC for the same area, which can vary from -8.5 m CD to -17 m CD [28,40,41]. Discrepancies are mostly related to the different approaches (empirical versus morphological) and time scale of the analyses [42].

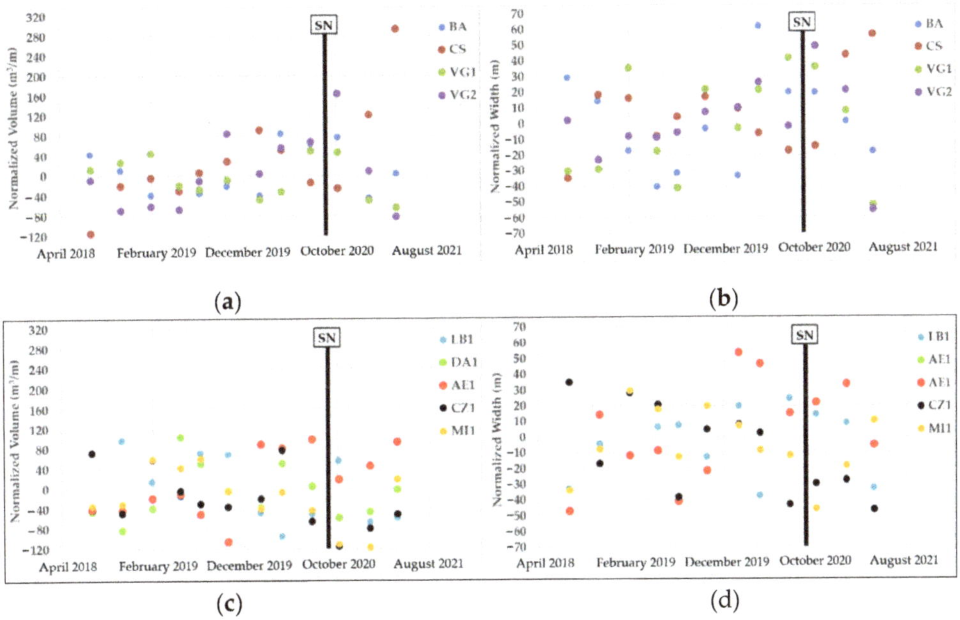

Figure 6. Normalized beach volume and width variation along the monitoring period (**a**) Normalized volume in the SD; (**b**): Normalized width in the SD; (**c**): Normalized volume downdrift of SD; (**d**) Normalized width downdrift of SD.

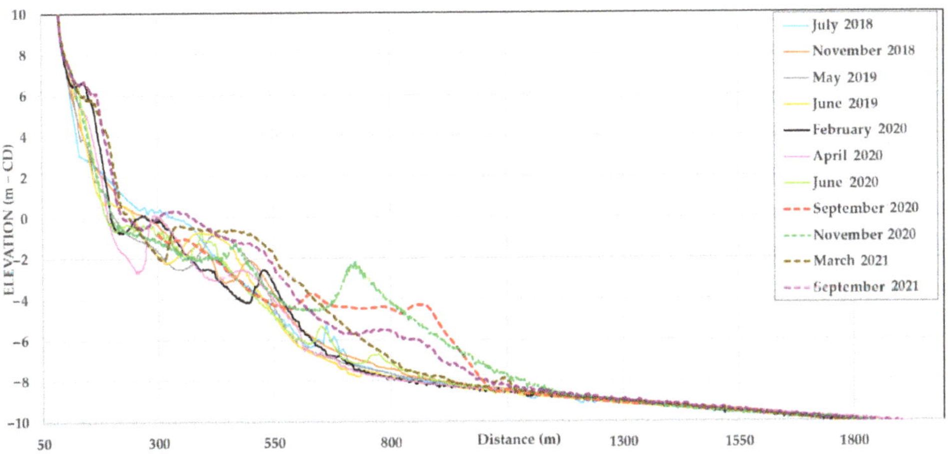

Figure 7. Beach Shoreface Profiles (BSP) at Costa Nova between July 2018 and September 2021. The red dotted line of September 2020 represents the final construction profile of the shoreface nourishment (elevations referred to CD—Chart Datum).

A complex longshore bar system dominates inner shoreface morphology and morphological changes. Inner bars (single or multiple), extend up to 500 m offshore, and develop between—4 m (CD) and 0 m (CD), the elevation offset between bar crest and trough reaching up to 2 m. After the SN, a plateau is evident in September 2020, from which a prominent bar is evident in November 2020 (800 m offshore). The inland migration of this bar is highlighted in Mar (500 m offshore) and September 2021 (350 m offshore).

The shoreface nourishment (illustrated by BSP September 2020 in Figure 7) created a 500 m wide, flat summited sand mound, with a broadly trapezoidal shape between −4 m CD and −8 m CD. This resulted in seaward displacement of 100 m and 300 m of the −8 m CD and −6 m CD contour lines, respectively (Figure 8a).

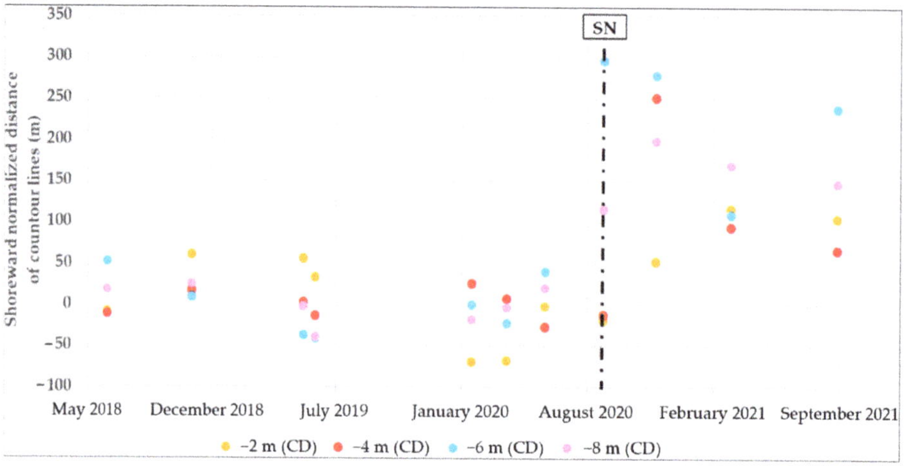

(**a**)

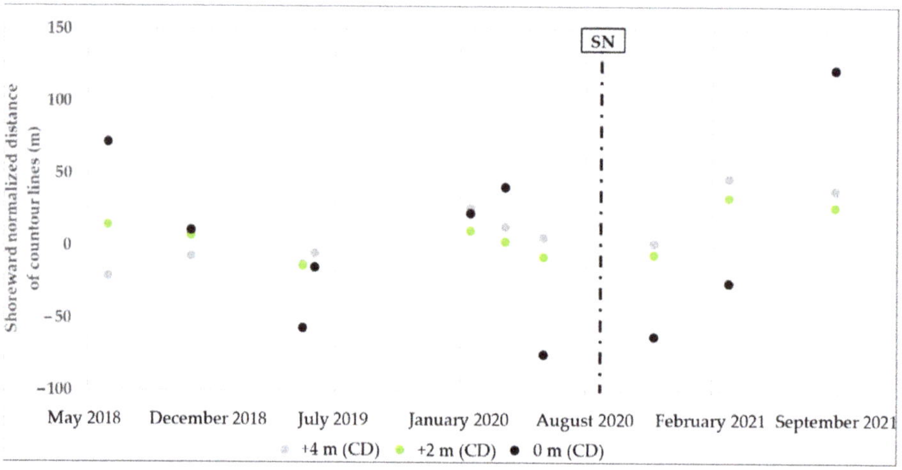

(**b**)

Figure 8. Temporal variation in the position of contour lines in the shoreface (**a**) and subaerial beach (**b**).

About two months later (November 2020) (Figure 7) the cross-section profiles show a landward translation and significant reshaping of the fill, with the development of two asymmetrical bars: a larger one atop the artificial sand mound and a smaller one merging with the pre-existing bar system. The seaward slope of the mound became milder, mimicking the pre-nourishment shape, and extended further seaward of the post-construction profile until the DoC. This slope adjustment is manifested by additional 80 m seaward displacement of the −8 m CD, and limited opposite (landward) translation of 20 m at −6 m CD.

Seven months after the SN, almost all sand in the artificial mound moved landward, feeding and building a wider and shallower nearshore bar system. This behaviour is mirrored by a landward displacement of depth contour lines−8 m CD, −6 m CD, and −4 m CD (Figure 8a). Simultaneously, it was observed significant subaerial beach accretion by ca. 40 m seaward advance of the beach face (cf. contour lines + 2 m CD and + 4 m CD) (Figure 8b).

Profile changes between March 2021 and September 2021 (Figure 7) are less pronounced landward of −1 m CD, where limited landward displacement of the nearshore bar and beach face retreat occurred. Seaward of −1 m CD, lowering of the profile was observed, whereas sand accumulated between −5 m CD and −8 m CD.

Profile changes in BSP may be taken as representative of sediment and morphological cross-shore dynamics in the area influenced by the SN. They suggest that SN at Costa Nova evolved rapidly, and essentially consisted of reshaping and landward translation of placed sand mound over the shoreface, the nourishment effects eventually affecting the subaerial beach in about seven months.

4.4. Topo-Bathymetric Surveys (TBS) and Multi-Beam Hydrographic Survey (MBHS)

Analyses of TBS and MBHS allowed for the evaluation of volume changes and sediment budget within the SNA and SD. Moreover, it provides additional insights on sediment dispersion driven by both longshore and cross-shore process.

Comparison of TBS June 2020 (pre-nourishment survey) and MBHS September 2020 (post-construction final survey) in the SNA indicate a volume increase of 2.087×10^6 m^3. This is 12% less than the volume of material dredged from the borrow area, as reported by the Port Authority (2.375×10^6 m^3). This difference may be explained by: (i) permanent loss (8%) to the offshore of the finer size fractions of sediment (<0.125 mm); and (ii) sediment transfer (4%) from SNA to the SD, caused by waves and currents over four months, during which numerous dumps added to produce the post nourishment measured volume. In agreement, a volume of 2.185×10^6 m^3 is considered a fair estimate of the amount of compatible sand placed at the SNA. Sediment budget calculations for the entire SD includes an additional amount of 295,000 m^3.

Morphological changes from pre- to post-nourishment (June 2020 to March 2021) reveal a heterogeneous spatial pattern, with alternating accumulation and erosion patches (Figure 9), showing larger longshore continuity. Bed level changes range from −6 m to +5 m, with the largest changes occurring in the bar system and subaerial beach. The SNA, together with updrift and downdrift adjacent regions, show higher accumulation and increased spatial continuity, in contrast with the southern half of the SD where patches of accumulation and erosion are smaller and more fragmented.

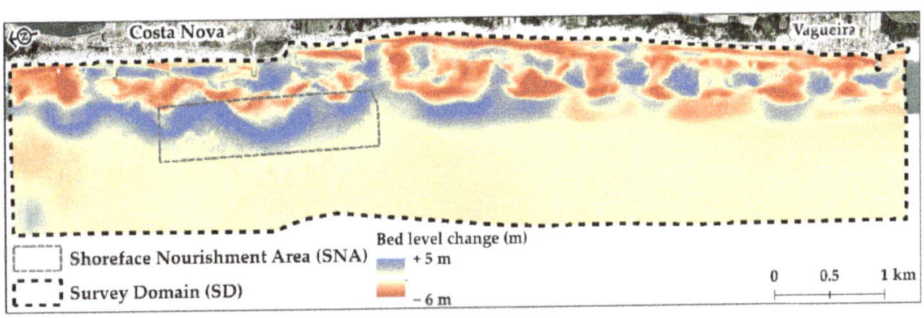

Figure 9. Morphological changes along the SD with respect to the pre-nourishment situation (June 2020) and after-nourishment (March 2021).

The Multiple Monitoring Cell (MMC) approach was applied to SD by splitting this area in twenty rectangular cells, organized in four cross-shore rows and five longshore columns, as illustrated in Figure 10. Cross shore rows broadly correspond to the following domains: (i) subaerial beach; (ii) bar system; (iii) depth range of SNA; (iv) depth range between seaward boundary of SNA and offshore limit of SD.

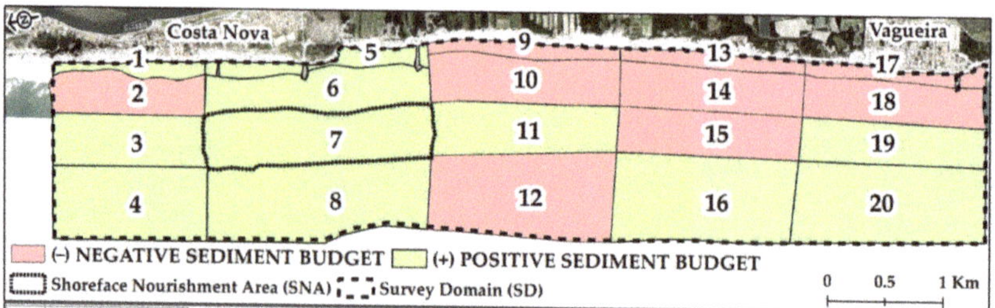

1	44,191 m³	5	37,846 m³	9	−250,000 m³	13	−30,545 m³	17	−144,446 m³
2	−73,918 m³	6	217,784 m³	10	−321,990 m³	14	−46,472 m³	18	−64,090 m³
3	384,035 m³	7	1,302,412 m³	11	623,647 m³	15	−49,457 m³	19	21,663 m³
4	33,187 m³	8	44,167 m³	12	−83,943 m³	16	46,407 m³	20	25,278 m³

Figure 10. Multiple Monitoring Cell (MMC) approach with sediment budget along the SD between June 2020 and March 2021.

Results shown in Figure 10 indicate that seven months after the nourishment, the SNA (cell 7) retained 1.3 $\times$ 10^6 m^3 (60% of the dumped volume).

Cells 5, 6, 7, and 8 present a positive sediment budget in the entire beach domain, comprising the dry-beach and nearshore up to -10 m CD. South of SNA, namely, in cells 9, 10, 13, 14, 17, and 18, concentrated landward of -4 m CD (in the inner surfzone), they all have a negative sediment budget.

Within the survey domain (SD), the June 2020 to March 2021 sediment budget is positive, and of 1.73 $\times$ 10^6 m^3. Considering that an additional volume of 295,000 m^3 (September 2020 to March 2021) was added to the initial measured volume of 2.185 $\times$ 10^6 m^3 (deposited between June 2020 and September 2020), a loss of 750,000 m^3 (30 %) was observed in this period.

Regarding subaerial beach response to the nourishment in the adjacent beaches (i.e., Costa Nova) in front of the SNA and further south until Vagueira Beach, Figure 11 shows the horizontal displacement of the + 1 m (CD) contour line, and its variation in terms of advance or retreat between June 2020 and March 2021.

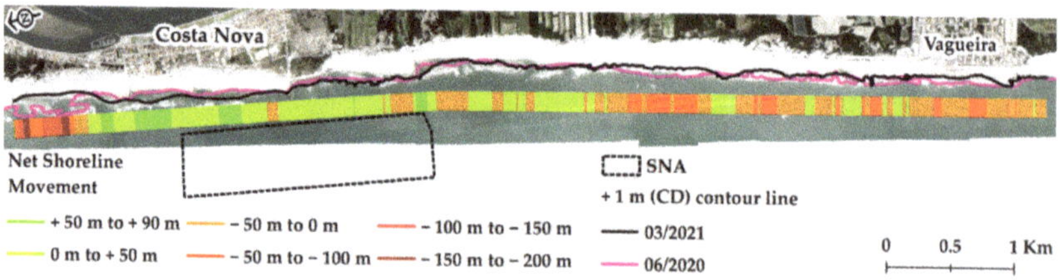

Figure 11. Horizontal displacement of the + 1 m (CD) contour line and its variation in terms of advance or retreat.

Results presented in Figure 11 confirm the above mentioned regarding subaerial beach variation, with a global positive net shoreline movement up to + 90 m within 0.5 km to the north and in front of the SNA. This trend changes further south, with a retreat of the +1 m contour line position, in average between -30 m to -60 m. The entire SD this contour line has an average displacement of + 9 m $\pm$ 44.5 m, which illustrates well the spatial variability of subaerial beach in response to the SN, because the standard deviation is much larger than the mean value.

5. Discussion

Herein, we describe the impacts of a high-magnitude shoreface nourishment in the morphological evolution and behaviour of a wave-dominated, high-energy beach–dune system. We add a sediment budget approach to previous work targeting the same area [20]. This has improved the understanding of the processes governing post-nourishment sediment dispersion within the survey domain. In addition, we provide a more detailed description of cross-shore changes along the nourished profile and evaluate the time scale required for the beach to acquire a new equilibrium profile following the placement of a significant sand volume.

5.1. Cross-Shore Processes

Coastal monitoring data show that cross-shore changes occur rapidly, with short-term landward migration of the shoreface nourishment. Right after the intervention, the cross-shore shape of the sand mound skewed in onshore direction, by decreasing its seaward slope and increasing crest elevation, together with the formation of two bars that eventually merged with the pre-existing ones. Seven months after the intervention, the seaward slope of the SN is milder, more similar to the pre-construction native profile. This is interpreted as resulting from the dominance of onshore transport accompanied by progressive depletion of the sediment source at the seaward toe of the nourishment. This in agreement with observed post-nourishment profile responses in the Netherlands [13] that required two to four years to replicate the native configuration, whereas the timescale inferred in this study is much smaller (up to seven months). Concomitantly, cross-shore processes induced rapid subaerial increase in beach volume and width in front of the placement area (Costa Nova beach), while the dune profile remained unchanged (Figures 6 and 7).

The observed rate of morphological responses is interpreted as mirroring the energetic wave conditions offshore Costa Nova, which are higher than the ones verified in the Netherlands [15] and California [19], where similar changes occur at larger timescales.

The magnitude of beachface seaward translation, represented by changes in beach width (Figure 6) measured between July 2020 and April 2021 is 76 m; the same estimate using October 2020 and January 2021 surveys is 60 m. These figures are in close agreement with the maximum seaward displacement (60 m) of the +3.05 m CD contour line observed between June 2020 and January 2021 in a nearby profile [20]. Moreover, our data suggest that beachface progradation slowed down or ceased after March 2021 (cf. Figure 7 for changes between March 2021 and September 2021). The magnitude of these changes is in close agreement with those yielded by solving Bruun's Rule (ca. 67 m), under the assumption that dumped sediment has been redistributed solely by cross-shore processes, aiming at restoring the equilibrium profile (replicating Bruun's equilibrium beach profile) up to the DoC. This similarity adds arguments to interpret earlier stages of post-nourishment morphodynamics as dominated by cross-shore processes. Following a convergence to the condition of equilibrium, additional increase in subaerial beach width is not expected.

5.2. Longshore Processes

The patterns of morphological changes illustrated in Figures 7 and 8 indicate that a significant volume of sediment ($\approx$ 600,000 m^3) accumulated immediately southward of the SNA (cell 11 in Figure 10), suggesting that longshore currents over the shoreface nourishment fed the seaward slope of the pre-existing bar system. Feeding tends to fade out further southward (see Figures 9 and 10), due to the combined effects of diffusion and low rate of advection characterizing longshore sediment transport.

The bar system and subaerial beach downdrift of the SNA show a spatially consistent pattern of erosion, with remarkably high values immediately south and landward of SNA (cells 9 and 10 in Figure 10). This behaviour is interpreted by blockage or decrease in longshore drift landward of the SNA, induced by the "reef effect" of the dumped sand mound, as suggested by [43]. Lower wave heights shoreward of the SNA due to wave dissipation reduces longshore sediment supply to the domain located immediately downdrift. Thus, a longshore drift gradient is created shortly after the intervention, leading to downdrift erosion. Smoothening of the morphological disturbance imposed by the nourishment will lead to the fading out of the "reef effect" and related erosive signal. These results suggest that subaerial domain of beaches facing and located downdrift the SNA are out of phase: while the former experience rapid accretion, the latter experience temporary enhanced erosion; this trend can only be reversed at larger time scales in tune with increasing importance of the longshore processes.

5.3. Sediment Budget and Nourishment Lifetime

Sediment budget calculations from June 2020 to March 2021 within the SD were based on the following assumptions:

(i) The seaward boundary was set seaward of the DoC and considered closed to sediment transport;
(ii) The landward boundary was set at the foredune toe in March 2021 and beach–dune sediment exchanges, as well as variations in coastline position, were considered negligible;
(iii) The northern boundary was located 1.5 km southward of the Aveiro inlet southern jetty. Sediment input through this boundary is unknown, but expected to be of smaller magnitude, in agreement with: (i) retention efficiency of the northern Aveiro jetty [23] and outer shoals of the Aveiro inlet; (ii) local reversal of longshore drift direction over the domain between the northern boundary of SD and the southern jetty due to wave diffraction induced by the north jetty and refraction over the outer inlet shoals;
(vi) The southern boundary is coincident with the southern limit of SD and was considered open to sediment transport.

Data collected in this study indicate that between June 2020 and March 2021 the sediment budget of the SD was positive and of about1.73 $\times$ 10^6 m^3. Considering that the shoreface nourishment was the main sediment source to the SD, corresponding to 2.48 $\times$ 10^6 m^3, a loss of 0.75 $\times$ 10^6 m^3 can be inferred. As the southern SD limit is the only open boundary this figure should correspond to a rough estimate of net longshore drift at that location. This conclusion is further supported by the independent assessments of potential net longshore drift performed in the scope of this work and by [22]. The former points to a value of 0.87 $\times$ 10^6 m^3 (cf. Section 4.1 and Figure 5) and the latter provides an estimate of 0.85 $\times$ 10^6 m^3, which was obtained by adding up June to March monthly averages of longshore drift magnitude over a twenty-year period. The agreement between the computed losses and the potential longshore drift provides confidence in the results of the sediment budget.

Linear extrapolation into the future of the mean annual net longshore drift indicates that, despite the large magnitude of this intervention, the permanence of placed sand within the SD is of 2–3 years.

High magnitude of littoral drift suggests that large scale and frequent (e.g., every other year) renourishment operations may be required if updrift sand placement is to be maintained as single adaptation strategy regarding beach erosion along the sediment cell, which extends for 50 km south of the Aveiro inlet. However, large uncertainties remain regarding the optimal design of shoreface nourishments, including not only magnitude and frequency, but also plan and cross-section shape of the sand mound. Altogether, these parameters govern morphodynamic feedback of the construction over wave forcing and potential impacts on the adjacent coast.

6. Conclusions

This work addresses the impacts of a high-magnitude shoreface nourishment in the morphological evolution and behaviour of a wave-dominated, high-energy beach–dune system. Results show that cross-shore processes dominate the early stages of morphological evolution with significant onshore sand transport. The beach aligned with the sand mound rapidly (in about seven months) acquires a condition of equilibrium, replicating Bruun's equilibrium beach profile that corresponds to an increase in subaerial beach width.

Longshore transport of sediment sourced in the placement explains the intense growth of the downdrift adjacent bar system, which diffuses and fades out southward. At the subaerial beach southward of the SNA, longshore drift was temporarily influenced by the "reef effect" of the sand mound, enhancing the previous erosive trend. This trend is expected to reverse in time, as longshore processes gain relevance in net sediment transfers over the SD.

Observed rates of change at Costa Nova are significantly higher than the ones reported for similar interventions undertaken in lower wave energy coasts of the Netherlands and California. This highlights the importance of wave energy in regulating the timescale of coastal readjustment to large magnitude shoreface nourishments. In consequence, great caution should be exercised when extrapolating behaviour-based models to predict morphological evolution under conditions other than those for which they were developed.

Current coastal protection strategies define beach and shoreface nourishment as a standard procedure to mitigate coastal erosion in critical areas. This is considered as an adaptation measure under present climate change scenario, including sea level rise. New insights provided by this work are expected to support decision-making regarding similar high-magnitude interventions foreseen in other areas along the high-wave energy Portuguese western coast and elsewhere. This gains relevance considering that shoreface nourishments are much more cost-effective in replenishment of large sand volumes than traditional beach nourishments (≈half-price per m^3).

Additionally, results of this study highlight the importance of nourishment design to optimize mitigation of beach erosion in high-magnitude drift-dominated coasts. Further investigation of morphodynamic feedback of shoreface nourishments over wave forcing and assessment of potential impacts on the adjacent coast is required.

Accurate and repeatable coastal monitoring data have proven to be essential to perform this evidence-based analysis, highlighting the need to maintain systematic coastal monitoring programmes through time, like the current Portuguese COaStal MOnitoring Program (COSMO).

Author Contributions: Conceptualization, C.A.P.; formal analysis, C.A.P., R.T. and C.A.; funding acquisition, C.A.P. and P.B.; investigation, C.A.P.; methodology, C.A.P., R.T. and C.A.; validation, C.A.P., R.T. and C.A.; writing—original draft, C.A.P.; writing—review & editing, C.A.P., R.T., C.A., P.B., P.A.S., D.M. and J.P.-B. All authors have read and agreed to the published version of the manuscript.

Funding: Some of the collected data were co-funded by POSEUR (Ref. POSEUR-02-1809-FC-000051—Remoção dos Inertes da Zali do Porto de Aveiro para Reforço do Cordão Litoral a Sul da Costa Novas and are gratefully acknowledged. The Portuguese Coastal Monitoring Program (COSMO), developed and implemented by the Portuguese Environment Agency (APA), is co-funded by the Operational Program for Sustainability and Efficiency in the Use of Resources (POSEUR) (ref. POSEUR-02-1809-FC-000004) and are gratefully acknowledged. We acknowledge financial support of FCT through project UIDB/50019/2020—IDL. This work is a contribution to project SANDTRACK (POCI-01-0145-FEDER-031779) funded by FEDER, through COMPETE2020—Programa Operacional Competitividade e Internacionalização (POCI), and by national funds (OE), through FCT/MCTES. We acknowledge financial support to CESAM by FCT/MCTES (UIDP/50017/2020+UIDB/50017/2020+LA/P/0094/2020), through national funds.

Institutional Review Board Statement: Not applicable.

Informed Consent Statement: Not applicable.

Data Availability Statement: Not applicable.

Acknowledgments: C.A.P. acknowledge the company consortium (GEOGLOBAL and ATLATICLAND) for the acquisition and processing of topographic and hydrographic data under the COSMO programme service contract. Aveiro Port Authority (APA, S.A.), namely, Fátima Alves and Carla Garrido, is gratefully acknowledged for providing post-construction multi-beam survey and sediment characterization. C.A.P. also acknowledges André Inácio from APA for supporting data analyses and calculations for Figures 6, 8 and 11. Images from Figure 3 were extracted from a video produced by JAN de NUL and INERSEL for Aveiro Port Authority/Portuguese Environment Agency during the shoreface nourishment intervention and are gratefully acknowledged. POSEUR is gratefully acknowledged for co-funding of the COSMO Program (ref. POSEUR-02-1809-FC-000004).

Conflicts of Interest: The authors declare no conflict of interest.

References

1. Vousdoukas, M.; Ranasinghe, R.; Mentaschi, L.; Plomaritis, T.; Athanasiou, P.; Luijendijk, A. Sandy coastlines under threat of erosion. *Nat. Clim. Chang.* **2020**, *10*, 260–263. [CrossRef]
2. Luijendijk, A.; Hagenaars, G.; Ranasinghe, R.; Baart, F.; Donchyts, G.; Aarninkhof, S. The State of the World's Beaches. *Sci. Rep.* **2018**, *8*, 6641. [CrossRef] [PubMed]
3. Ranasinghe, R. Assessing climate change impacts on open sandy coasts: A review. *Earth-Sci. Rev.* **2016**, *160*, 320–332. [CrossRef]
4. Hinkel, J.; Nicholls, R.; Tol, R.; Wang, Z.B.; Hamilton, J.M.; Boot, G.; Vafeidis, A.; McFadden, L.; Ganopolski, A.; Klein, R.J. A global analysis of erosion of sandy beaches and sea-level rise: An application of DIVA. *Glob. Planet. Chang.* **2013**, *111*, 150–158. [CrossRef]
5. Masselink, G.; Russell, P.E. Impacts of Climate Change on Coastal Erosion. MCCIP Annual Report Card 2010-11. *MCCIP Sci. Rev.* **2013**, 71–86. [CrossRef]
6. Andrade, C.; Pires, H.O.; Taborda, R.; Freitas, M.C. Projecting future changes in wave climate and coastal response in Portugal by the end of the 21st century. *J. Coast. Res.* **2007**, *SI 50*, 253–257.
7. Lobeto, H.; Menendez, M.; Losada, I.J. Future behavior of wind wave extremes due to climate change. *Sci. Rep.* **2021**, *11*, 7868. [CrossRef]
8. Rangel-Buitrago, N.; Neal, W.; Bonetti, J.; Anfuso, G.; Jonge, V.N. Vulnerability assessments as a tool for the coastal and marine hazards management: An overview. *Ocean. Coast. Manag.* **2020**, *189*, 105134. [CrossRef]
9. Dean, R. *Beach Nourishment: Theory and Practice*; Advanced Series on Ocean Engineering—Volume 18; World Scientific: Singapore, 2002; p. 399.
10. De Schipper, M.A.; Ludka, B.C.; Raubenheimer, B.; Luijendijk, A.P.; Schlacher, T.A. Beach nourishment has complex implications for the future of sandy shores. *Nat. Rev. Earth Environ.* **2020**, *2*, 70–84. [CrossRef]
11. Hamm, L.; Capobianco, M.; Dette, H.H.; Lechuga, A.; Spanhoff, R.; Stive, M.J.F. A summary of European experience with shore nourishment. *Coast. Eng.* **2002**, *47*, 237–264. [CrossRef]
12. Stive, M.J.F.; Schipper, M.A.; Luijendijk, A.P.; Aarninkhof, S.G.J.; van Gelder-Maas, C.; van Thiel de Vries, J.S.M.; Vries, S.; Henriquez, M.; Marx, S.; Ranasinghe, R. A new alternative to saving our beaches from Sea-Level rise: The Sand Engine. *J. Coast Res.* **2013**, *29*, 1001–1008. [CrossRef]
13. Huisman, B.J.A.; Walstra, D.J.R.; Radermacher, M.; de Schipper, M.A.; Ruessink, B.G. Observations and Modelling of Shoreface Nourishment Behaviour. *J. Mar. Sci. Eng.* **2019**, *7*, 59. [CrossRef]
14. Pinto, C.A.; Silveira, T.M.; Teixeira, S.B. Beach nourishment practice in mainland Portugal (1950–2017): Overview and retrospective. *Ocean. Coast. Manag.* **2020**, *192*, 105211. [CrossRef]
15. Gravens, M.; Ebersole, B.; Walton, T.; Wise, R. Beach fill design. In *Coastal Engineering Manual. Part V. Coastal Project Planning and Design*; Chapter IV. Engineer Manual 1110-2-1100 U.S.; Ward, D., Ed.; Army Corps of Engineers: Washington, DC, USA, 2006.
16. Pinto, C.; Pires, B.; Penacho, N.; Silveira, T.; Mil-Homens, M.; Terrinha, P.; Rosa, M.; Brito, P.; Neres, M.; Magalhães, V. Caracterização de Manchas de Empréstimo Para Alimentação Artificial de Praias—Projeto CHIMERA. In Proceedings of the 5a Conferência Sobre Morfodinâmica Estuarina e Costeira, Lisboa, Portugal, 24–26 June 2019; pp. 75–76. (In Portuguese)
17. Teixeira, S.B.; Pinto, C.A.; Rosa, M. Off-Shore Sources of Beach-Fill Material in Quarteira (Algarve-Portugal). In Proceedings of the 5a Conferência Sobre Morfodinâmica Estuarina e Costeira. Livro de Resumos, Lisboa, Portugal, 24–26 June 2019; pp. 85–86.
18. Mil-Homens, M.; Brito, P.; Magalhães, V.; Rosa, M.; Neres, M.; Silva, M.; Salgueiro, E.; Drago, T.; Rodrigues, A.; Guerra, M.T.; et al. Integrated geophysical and sedimentological datasets for assessment of Offshore Borrow Areas: The CHIMERA project (Western Portuguese Coast). *Geol. Soc. Lond. Spec. Publ.* **2020**, *505*. [CrossRef]
19. Mesa, C. Nearshore berm performance at Newport Beach, California, USA. *Coast. Eng. Proc.* **1996**, *1*, 4636–4649.
20. Mendes, D.; Pais-Barbosa, J.; Baptista, P.; Silva, P.A.; Bernardes, C.; Pinto, C. Beach Response to a Shoreface Nourishment (Aveiro, Portugal). *J. Mar. Sci. Eng.* **2021**, *9*, 1112. [CrossRef]
21. Santos, F.D.; Lopes, A.M.; Moniz, G.; Ramos, L.; Taborda, R. *Grupo de Trabalho do Litoral: Gestão da Zona Costeira: O desafio da mudança*; Santos, F.D., Penha-Lopes, G., Lopes, A.M., Eds.; FCUL/FCT/CE3C/BASE: Lisboa, Portugal, 2017; 396p, ISBN 978-989-99962-1-2. (In Portuguese)
22. Silva, P.A.; Monteiro, N.; Oliveira, T.; Abreu, T.; Coelho, C.; Lima, M.; Carvalho, R. *Estudo de Viabilidade da Transposição Aluvionar das Barras de Aveiro e da Figueira da Foz. Relatório Intercalar 2. Tarefa 3—Aveiro*; Universidade de Aveiro/Agência Portuguesa do Ambiente: Amadora, Portugal, 2021; 65p. (In Portuguese)
23. Coelho, C.; Afonso, A.; Bernardes, C.; Silva, P.A.; Baptista, P.; Roebeling, P.; Fernández, S.; Abreu, T.; Oliveira, T.; Ferreira, M.; et al. *Estudo de Viabilidade da Transposição Aluvionar das Barras de Aveiro e da Figueira da Foz. Sumário Executivo*; Universidade de Aveiro/Agência Portuguesa do Ambiente: Amadora, Portugal, 2021; 26p. (In Portuguese)

24. Silva, A.N.; Taborda, R.; Bertin, X.; Dodet, G. Seasonal to Decadal Variability of Longshore Sand Transport at the Northwest Coast of Portugal. *J. Waterw. Port Coast. Ocean Eng.* **2012**, *138*, 464–472. [CrossRef]
25. Vicente, C.; Clímaco, M.; Bertin, X. *Agitação Marítima e Transporte Sólido. Litoral na Costa de Aveiro. Relatório 164/2013—DHA/NEC*; LNEC: Lisboa, Portugal, 2013; 33p. (In Portuguese)
26. Silva, R.; Baptista, P.; Veloso-Gomes, F.; Coelho, C.; Taveira-Pinto, F. Sediment grain size variation on a coastal stretch facing the North Atlantic (NW Portugal). *J. Coast. Res.* **2009**, *SI56*, 762–766.
27. Bettencourt, P.; Ângelo, C. *Faixa Costeira Centro Oeste (Espinho-Nazaré): Enquadramento Geomorfológico e Evolução Recente*; Geonovas n Especial (A Geologia e o Ambiente): Lisboa, Portugal, 1992; pp. 7–30. (In Portuguese)
28. Vicente, C.; Clímaco, M. *Trecho de costa do douro ao Cabo Mondego. Caraterização geral do Processo Erosivo. Relatório 253/2012—DHA/NEC*; LNEC: Lisboa, Portugal, 2012; 56p. (In Portuguese)
29. Fernández-Fernández, S.; Baptista, P.; Bernardes, C.; Silva, P.; Fontán-Bouzas, A.; López-Olmedilla, L.; Ferreira, C. Variação da Linha de Costa em Praias Arenosas: Aveiro (Portugal). In *Livro de resumos do IX Congresso sobre Planeamento e Gestão das Zonas Costeiras dos Países de Expressão Portuguesa*; APRH: Lisboa, Portugal, 2019. (In Portuguese)
30. Pinto, C.A. *Enquadramento Metodológico Para a Demarcação Das Faixas de Salvaguarda à Erosão Costeira (Nível I e II) em Litoral Baixo e Arenoso—Aplicação ao Troço Costeiro Ovar—Marinha Grande*; Relatório Técnico DLPC n. 1/2015; Agência Portuguesa do Ambiente: Lisboa, Portugal, 2015; 11p. (In Portuguese)
31. Vicente, C.; Clímaco, M. *Trecho de Costa a Sul de Espinho. In Simulação Numérica do Processo Erosivo e de Alternativas de Intervenção*; Relatório 101/2014—DHA/NEC; LNEC: Lisboa, Portugal, 2014; 68p. (In Portuguese)
32. Bernardes, C.; Bernardes, C.; Fernández, S.; Santos, F.; Baptista, P.; Silva, P.A.; Abreu, T.; Coelho, C.; Lima, M.; Carvalho, R. Estudo de Viabilidade da Transposição Aluvionar das Barras de Aveiro e da Figueira da Foz. Relatório Intercalar 1. In *Tarefa 1—Análise da Evolução Histórica da Embocadura da Laguna de Aveiro*; Universidade de Aveiro/Agência Portuguesa do Ambiente: Amadora, Portugal, 2020; 59p. (In Portuguese)
33. Rato, D.; Relatório, I.I. Análise Textural de Sedimentos; Technical Report; SandTrack Project: 2019. Available online: http://sandtrack.web.ua.pt/pt/outputs/reports/02_Relatorio_II_Analise_textural_sedimentos.pdf (accessed on 28 June 2021). (In Portuguese)
34. Pinto, C.A.; Penacho, N.; Pires, B. Programa de Monitorização da Faixa Costeira de Portugal Continental (COSMO): Da Conceção à Implementação. In *X Congresso sobre Planeamento e Gestão das Zonas Costeiras dos Países de Expressão Portuguesa*; de Janeiro, R., Ed.; APRH: Lisboa, Portugal, 2021; 3p. (In Portuguese)
35. USACE. *Shore Protection Manual*; Department of the Army, U.S. Corps of Engineers: Washington, DC, USA, 1984; p. 20314.
36. Barbaro, G.; Foti, G.; Sicilia, L.; Malara, G. A formula for the calculation of the longshore sediment transport including spectral effects. *J. Coast. Res.* **2014**, *30*, 961–966. [CrossRef]
37. Hallermeier, R.J. A Profile Zonation for Seasonal Sand Beaches from Wave Climate. *Coast. Eng.* **1981**, *4*, 253–277. [CrossRef]
38. Roest, B.; de Vries, S.; de Schipper, M.; Aarninkhof, S. Observed Changes of a Mega Feeder Nourishment in a Coastal Cell: Five Years of Sand Engine Morphodynamics. *J. Mar. Sci. Eng.* **2021**, *9*, 37. [CrossRef]
39. Himmelstoss, E.A.; Henderson, R.E.; Kratzmann, M.G.; Farris, A.S. *Digital Shoreline Analysis System (DSAS) Version 5.0 User Guide (No. 2018-1179)*; US Geological Survey: Reston, Virginia, 2018.
40. Ferreira, O. Caraterização Dos Principais Fatores Condicionantes do Balanço Sedimentar e da Evolução da Linha de Costa Entre Aveiro e o Cabo Mondego. Master's Thesis, Universidade de Lisboa, Lisboa, Portugal, 1993; 404p. (In Portuguese)
41. Coelho, C. Riscos de Exposição de Frentes Urbanas Para Diferentes Intervenções de Defesa Costeira. Ph.D. Thesis, Universidade de Aveiro, Aveiro, Portugal, 2009. (In Portuguese)
42. Nicholls, R.J.; Birkemeier, W.A.; Lee, G.H. Evaluation of Depth of Closure Using Data from Duck, NC, USA. *Mar. Geol.* **1998**, *148*, 179–201, ISSN 0025-3227. [CrossRef]
43. Van Duin, M.; Wiersma, N.; Walstra, D.; van Rijn, L.; Stive, M. Nourishing the shoreface: Observations and hindcasting of the Egmond case, The Netherlands. *Coast. Eng.* **2004**, *51*, 813–837. [CrossRef]

Journal of
*Marine Science
and Engineering*

MDPI

Article

Beach Response to a Shoreface Nourishment (Aveiro, Portugal)

Diogo Mendes [1,*,†], Joaquim Pais-Barbosa [2], Paulo Baptista [1], Paulo A. Silva [2], Cristina Bernardes [1] and Celso Pinto [3]

[1] Departamento de Geociências e Centro de Estudos do Ambiente e do Mar (CESAM), Campus de Santiago, Universidade de Aveiro, 3810-193 Aveiro, Portugal; renato.baganha@ua.pt (P.B.); cbernardes@ua.pt (C.B.)
[2] Departamento de Física e Centro de Estudos do Ambiente e do Mar (CESAM), Campus de Santiago, Universidade de Aveiro, 3810-193 Aveiro, Portugal; joaquim.paisbarbosa@ua.pt (J.P.-B.); psilva@ua.pt (P.A.S.)
[3] Agência Portuguesa do Ambiente, I.P., Rua da Murgueira, 9, Zambujal, Alfragide, 2610-124 Amadora, Portugal; celso.pinto@apambiente.pt
* Correspondence: dsmendes@ua.pt
† Current address: HAEDES, Casais do Arrocho, 2025-452 Azóia de Cima, Portugal.

Abstract: In Aveiro (NW coast of Portugal), a coastal monitoring programme was carried out in sequence of a shoreface nourishment intervention (over than 2 M m^3) performed in 2020. In this programme, almost one year of biweekly subaerial topographies and quarterly bathymetric surveys have been collected along a 10 km coastal stretch between June 2020 and June 2021. In this study, topographic and bathymetric surveys were analysed to assess the expectation that if the shoreface nourishment is located in sufficiently shallow water depths, its landward movement will feed adjacent beaches and, consequently, increase the subaerial beach volume. Results show that the subaerial beach volume is well correlated with the 1.05 m (above MSL) isoline displacement through time. While the seaward limit of the shoreface nourishment moved landwards about 200 m, the shoreline proxy (isoline of 1.05 m) displayed a maximum seaward displacement of 60 m. The displacement of the shoreline proxy was highly variable in space, along the 10 km coastal stretch, and also in time, during storm events. During such events, both landward and seawards displacement of the shoreline proxy took place, depending on the spatial position. Moreover, while beaches close to the initial shoreface nourishment intervention displayed faster accretion patterns than those located farther away, the well-defined onshore movement of the shoreface nourishment did not result in a considerable beach volume increase. The achieved results were also compared against case studies of shoreface nourishments with similar volumes performed worldwide.

Keywords: beach nourishment; field observations; storm; beach accretion

Citation: Mendes, D.; Pais-Barbosa, J.; Baptista, P.; Silva, P.A.; Bernardes, C.; Pinto, C. Beach Response to a Shoreface Nourishment (Aveiro, Portugal). *J. Mar. Sci. Eng.* **2021**, *9*, 1112. https://doi.org/10.3390/jmse9101112

Academic Editor: Rodger Tomlinson

Received: 1 September 2021
Accepted: 6 October 2021
Published: 13 October 2021

1. Introduction

Beach nourishment constitutes a nature-based engineering solution commonly employed by coastal managers on sandy beaches [1]. It comprises the placement of large quantities of good quality sand on the beach to advance it seaward [2]. This advance is of key importance for tourism and recreation because it provides enough space on the dry beach for this type of activities. In general, urbanized beaches, such as those located or backed up by buildings and infrastructures, are the most prone to be improved with sand nourishments because a municipality tax revenue can be used for such interventions (e.g., [2]). The sand can be either placed on the subaerial beach or in the subtidal beach, as an underwater mound. While the former is usually referred in the literature as a beach nourishment, the latter can be referred as profile nourishment [2], berm nourishment [3], nearshore berm [4] or shoreface nourishment [5]. In this study, shoreface nourishment was used to designate the placement of sand in the subtidal zone of a beach profile.

Following Dean [2], a shoreface nourishment has two advantages compared to beach nourishment. First, the dredging-dumping operation is less expensive (e.g., [6]). Second, it is associated with less restrictive policies regarding sediment quality characteristics.

Unlike beach nourishments, shoreface nourishments can use sediments that are dredged from nearby navigation channels, subtidal bars or offshore deposits. Though, there are some examples that sediments dredged from navigation channels are used for beach nourishment [7]. As an example, a large beach nourishment intervention (1.5×10^6 m^3) was performed in Dunkirk, France, using sediments dredged from a nearby navigation channel but with a median grain size (D_{50}) coarser than those of the native beach [8].

Beach nourishment has been employed throughout the world. In the US, beach nourishment has been used since 1923 with an exponentially growth in sand volume placement by the end of the last century [9]. In Europe, beach nourishment interventions started after 1950. In the last decades, it has been a gradual change from the use of hard to soft coastal protection/defense techniques both for short-term and long-term coastal planning [10]. In Australia, beach nourishment interventions are generally smaller in scale but more frequent and mainly begin in spring to promote beach accretion [11]. In China, beach nourishment was first introduced in 1990 and the number of beach nourishment interventions has also show an exponential increase between 1990 and 2010 [12]. Despite the overall use of beach nourishment on sandy beaches, this solution can also be used in complex reef environments associated with irregular bathymetries [13,14].

In Portugal, the first beach nourishment intervention was performed in 1950 [7]. A recent review and compilation of beach nourishment practice in Portugal has shown that the main objectives are shoreline stability and erosion mitigation [7]. Nourishment interventions have been mainly performed with sediments dredged from maintenance channels' dredging. Moreover, there has been an increasing tendency to use soft engineering techniques in opposition to hard engineering solutions [7]. The largest shoreface nourishment intervention ever made in Portugal, before that reported in this study, took place in Aveiro in 1996 where a sand volume of 1.7×10^6 m^3 was deposited nearshore. Sediments were dredged from the navigation channel of the Aveiro lagoon and were placed southwards at Costa Nova beach. Shoreline stability was the objective of this intervention. Previous works conducted in Aveiro associated with shoreface nourishments observed cross-shore volume variations of up to 1500 m^3/m (e.g., [15]), which suggests that this coastal stretch is morphologically very dynamic.

Following Brutsché et al. [4], pioneer shoreface nourishment interventions, undertaken between 1930 and 1940 in the US, raised some doubts about their overall effectiveness. Besides the observed sediment accretion near the shoreface nourishment, sediment erosion took place near the shoreline. These initial results postponed the use of shoreface nourishments until late 1960. Later on, the effectiveness of shoreface nourishments started to become documented. As an example, observations indicated that a shoreface nourishment located in Durban (South Africa) provided shelter to beaches [16]. Those beaches that were located on the lee side of the shoreface nourishment experienced up to 25% less erosion than those that were not. Nowadays, shoreface nourishments constitute a viable solution from a technical perspective [2]. However, stakeholders and the general public can still be reluctant to this type of solutions because they cannot be easily seen from the dry beach. Even though, the large number of shoreface nourishment interventions after 1940, that were documented in [4], provides some confidence that they are becoming better accepted.

Beach nourishment design, construction and subsequent monitoring is well documented in the literature (e.g., [2]). On the contrary, shoreface nourishment interventions are less well understood. In particular, shoreface nourishments can behave in two ways [2]. The shoreface nourishment can be placed in sufficiently deep water depths so that it remains there in time, usually referred as a stable berm. The stable berm main objective is to reduce storm damage relative to the level of damage that would have resulted without the nourishment. The shoreface nourishment can also be placed in sufficiently shallow water depths so that it moves landwards, usually referred as a feeder berm. The main goal of a feeder berm is to feed adjacent beaches with sand. The hypothesis is if a shoreface nourishment is placed in sufficiently shallow depths, its onshore movement will continuously promote beach accretion, through an increase in the subaerial beach width or volume. In this study, field observations associated with a coastal monitoring program were analysed to assess the spatial and time evolution of a large (2×10^6 m^3) shoreface nourishment intervention that was deployed between −10 m and −6 m (Mean Sea Level, hereafter MSL) water depths in Costa Nova (Aveiro, Portugal). In particular, this dataset was used to test the hypothesis just mentioned.

This paper is structured as follows. A review of previous works on shoreface nourishment interventions with a considerable volume (>0.8×10^6 m^3) is performed in Section 2. The characterization of the study site and of the shoreface nourishment intervention, together with a description of the field data collection, processing and analysis is presented in Section 3. The results of the shoreface nourishment subtidal evolution and of the subaerial beach shoreline proxy evolution are presented in Section 4. In Section 5, a discussion is performed in light of the hypothesis that shoreface nourishment interventions that moved landwards can increase the subaerial beach width or volume. The discussion in Section 5 also compares the results analysed in this work with other shoreface nourishments performed elsewhere. Conclusions are summarized in Section 6.

2. Previous Works on Shoreface Nourishments

Five shoreface nourishment interventions were reviewed in this Section. They were chosen from a recent review on shoreface nourishments [4] and from a review of beach nourishment experience in Europe [10]. The five chosen shoreface nourishments were selected based on two criterion. First, shoreface nourishments had an overall volume > 0.8×10^6 m^3, to be comparable with that reported in the present study (2×10^6 m^3). Second, the chosen shoreface nourishments had ready-available and well-documented general characteristics, such as shoreface nourishment length (L) and placement water depths (h) (Table 1). It was noted that a shoreface nourishment of 8.2×10^6 m^3 was performed in Anglet, France [10] but, unfortunately, its general characteristics were not available. The shoreface nourishment of 2.0×10^6 m^3 performed in Rio de Janeiro, Brazil, [17] was also excluded because the project not only included a shoreface nourishment but also a beach nourishment. Consequently, the overall beach response was not solely due to the shoreface nourishment. Moreover, the interesting case study in Denmark [18], where a comparison between a beach and a shoreface nourishment was conducted, was also excluded because the shoreface nourishment volume was 0.25×10^6 m^3.

Table 1. General characteristics of shoreface nourishments in previous works.

Location	V ($\times 10^6$ m^3)	V/L (m^3/m)	h (m, MSL)
California, US (1992) [19]	1.0	1070	−9 to −4
Terschelling, NL (1993) [6]	2.1	480	−7 to −4
Terheijde, NL (1997) [6]	0.9	530	−8 to −5
Egmond, NL (1999) [6]	0.9	390	−8 to −5
Wassenaar, NL (2002) [6]	2.5	410	−8 to −5

In 1992, a shoreface nourishment of about 1.0×10^6 m^3 was built at Newport Beach, California, US [19]. The shoreface nourishment was placed in water depths between −4 to −9 m (MSL). Offshore wave conditions along the California coastline are characterized by an averaged offshore significant wave height (H_{m0}) that ranges from 1.75 m to 3.5 m in summer and in winter, respectively. Large swell waves generated in the Pacific Ocean are common at this site with an averaged peak wave period (T_p) of 12.3 s [20]. During storms, H_{m0} can reach or exceed 5 m offshore but it is effectively reduced due to Channel Islands sheltering effect close to Newport Beach [20]. The wave-induced sediment transport is towards the southeast direction [19] and the tidal range is about 2 m during spring tides. Overall, this nourishment moved onshore likely due to wave action. While the outer limit of the shoreface nourishment moved about 180 m onshore, the MSL contour moved offshore (i.e., beach width increase) about 30 m in 2.5 yr (Figure 2 in [19]). Moreover, the analysis of profile surveys displayed no indication of an alongshore shoreface nourishment movement [19].

All the other shoreface nourishments in Table 1 were performed in The Netherlands. The wave climate along the Dutch coast is associated with an average significant wave height of 1.0 m during summer, which increases to 1.7 m during winter [21]. Winter storms are typically associated with an H_{m0} of 4 to 5 m and a T_p of 10 s. The net sand transport rate ranges between 0.25×10^6 m^3 up to 0.6×10^6 m^3 [22]. The Terschelling shoreface nourishment displayed a complex behaviour with both offshore and onshore movement of the intermediate bar after nourishment (Figure 7d in [6]). The other shoreface nourishments performed in the Dutch coast (Terheijde, Egmond and Wassenaar) displayed an overall onshore movement of about 100 m in 4.5 yr for Wassenaar, about 100 m in 2 yr for Egmond and about 150 m in 3 yr for Terheijde. The MSL contour varied less than about 20 m for all four shoreface nourishments.

The values associated with onshore and offshore shoreface nourishments' displacement are summarized in Table 2. The shoreface nourishment onshore displacement increases for a more energetic wave climate. The MSL contour seaward displacement is much smaller (<20%) than the outer part of the shoreface nourishment landward movement, independently of the offshore wave conditions.

Table 2. Shoreface nourishment onshore displacement (S_{on}), MSL contour offshore displacement (s_{off}), and respective migration rates.

Location	S_{on} (m)	s_{off} (m)	Rate S_{on} (m/yr)	Rate s_{off} (m/yr)
California, US (1992) [19]	180	30	72	12
Terschelling, NL (1993) [6]	-	-	-	-
Terheijde, NL (1997) [6]	150	20	50	7
Egmond, NL (1999) [6]	100	20	50	10
Wassenaar, NL (2002) [6]	100	20	22	4

3. Case Study and Field Observations

3.1. Brief Description of the Coastal Settings

The study area is located south of the Aveiro harbour entrance (Barra), and extends up to 10 km towards south until Vagueira beach (Figure 1). The coastline orientation is approximately 15° N. Coastal defense works along the study area (Figure 1) comprise two breakwaters near the Aveiro harbour entrance (white), six groynes (blue) and two revetments (orange). The beach is delimited by groynes at Costa Nova (between Barra and G5) and it is backed-up by dune systems southwards of G5. At Vagueira, the beach is both confined by one revetment and one groyne. The reader is referred to [23] for a more in-depth characterization of the study area.

Based on an analysis of field measurements collected by deep-water wave buoys offshore mainland Portugal [24], the offshore H_{m0} has a monthly-averaged value of 1.7 ± 0.7 m during maritime summer and increases to 3.1 ± 1.3 m during maritime winter. The monthly-averaged values of mean wave period oscillate between 6.0 ± 1.0 s and 8.0 ± 1.7 s along the year, with large swell waves easily reaching 20 s of T_p during winter. The most frequent mean wave direction is from NW, with some sea states displaying a mean wave direction from W. On average, five to fifteen coastal storms (H_{m0} threshold of 4.5 m) hit the study site per year [25]. Tidal range varies from approximately 1.2 m to 3.6 m during neap and spring tides, respectively. The most frequent northwestern wave conditions can promote a net potential sediment transport rate of about 1×10^6 m^3/yr directed towards south [26].

According to numerical modelling results [27], the coastal stretch south of Aveiro after 25 years of reduced sediment supply, critical situations of imminent sand-spit disruption are expected, as well as an ultimate linkage between the sea and the lagoon. The scenarios of sea level rise (SLR) are less important than the scenarios of wave-climate change after

25 years. A slight increase in the relative frequency of higher waves would have greater effects than a pessimistic scenario of the SLR rate.

Morphodynamically, the beach profile can be characterized as intermediate [28]. A well-defined subtidal bar is often present on the shoreface which induces partial depth-induced breaking of incoming waves. On the foreshore, the beach slope is mild during low tide and steep during high tide. Intertidal beach sediments are typically composed by fine to medium sand with a median grain size between 0.4 mm to 0.6 mm [29].

3.2. *Shoreface Nourishment Intervention*

The shoreface nourishment intervention comprised the dumping of sediments, dredged from the Zona de Atividades Logísticas e Industriais (ZALI) deposition area located inside the Aveiro harbour, between groynes 3 and 5 (G3 and G5 in Figure 1). These groynes are located at Costa Nova beach. Consequently, the beach response to the shoreface nourishment at Costa Nova is of particular interest because it is located closer to the intervention. The intervention was materialized by an underwater mound with a length of 1900 m and a width of 500 m. The shoreface nourishment was placed in water depths between −6 m to −11 m (MSL). At the end of the intervention, the mean water depth at the flat top of the underwater mound was −6 m (MSL). Sediments that were dumped at that location were compatible with native sediments but slightly coarser because the former were dredged in part from the Aveiro harbour entrance channel [30]. According to the Aveiro harbour authority, dredging works have started in May and last until August 2020. The analysis of the available multibeam surveys performed on June 2020 and on August 2020 allowed to detect a volume increase of 1.9×10^6 m^3 at that location. This calculation was performed based on the August 2020 survey coverage area with reference to the depth of closure which is −10 m (MSL) (see Section 4.1).

3.3. *Methodology*

3.3.1. Data Collection

Bathymetric surveys were performed with the research vessel Nereide from Centro de Estudos do Ambiente e do Mar (CESAM). This vessel can be equipped with a single beam or a multi-beam echo-sounder. The multi-beam survey was performed in an area pre-defined by the Portuguese Environment Agency (hereafter APA). The single beam survey was performed along the study area through cross-shore transects with a 500 m spacing (Figure 1). The vertical accuracy of the single-beam surveys is approximately 0.10 m. Part of the surf zone was not surveyed due to wave conditions at the study site. The bathymetric surveys were performed on July 2020, on September 2020 and on January 2021 (see Figure 4, below).

Topographic surveys were performed using a quad-bike equiped with the INSHORE monitoring system [31,32]. This allows a horizontal and vertical accuracy better than 0.05 m (see Baptista et al. [31]). Topographic surveys were performed approximately biweekly from September 2020 to June 2021 along the 10 km coastal stretch (from Barra to Vagueira in Figure 1). These surveys were performed during low-water on spring tides and each took between 3 h to 4 h to be completed. Each topographic survey was performed through cross-shore beach transects with a spacing between 30 m to 80 m and through longshore transects that accounted for the water line, berm crest and dune base, with a maximum spacing of about 30 m. These topographic surveys were used to extract the 1.05 m isoline that was analysed in Section 4 (Figures 5–7, see below).

At approximately 55 km northwest of the study site, offshore wave conditions were obtained from the Copernicus Marine Service (CMEMS, https://marine.copernicus.eu/, accessed on 28 June 2021). The wave parameters were H_{m0}, T_p and the mean wave direction ($MDir$). The numerically generated wave data was carefully compared against buoy and satellite observations (see CMEMS report and assessment [33]).

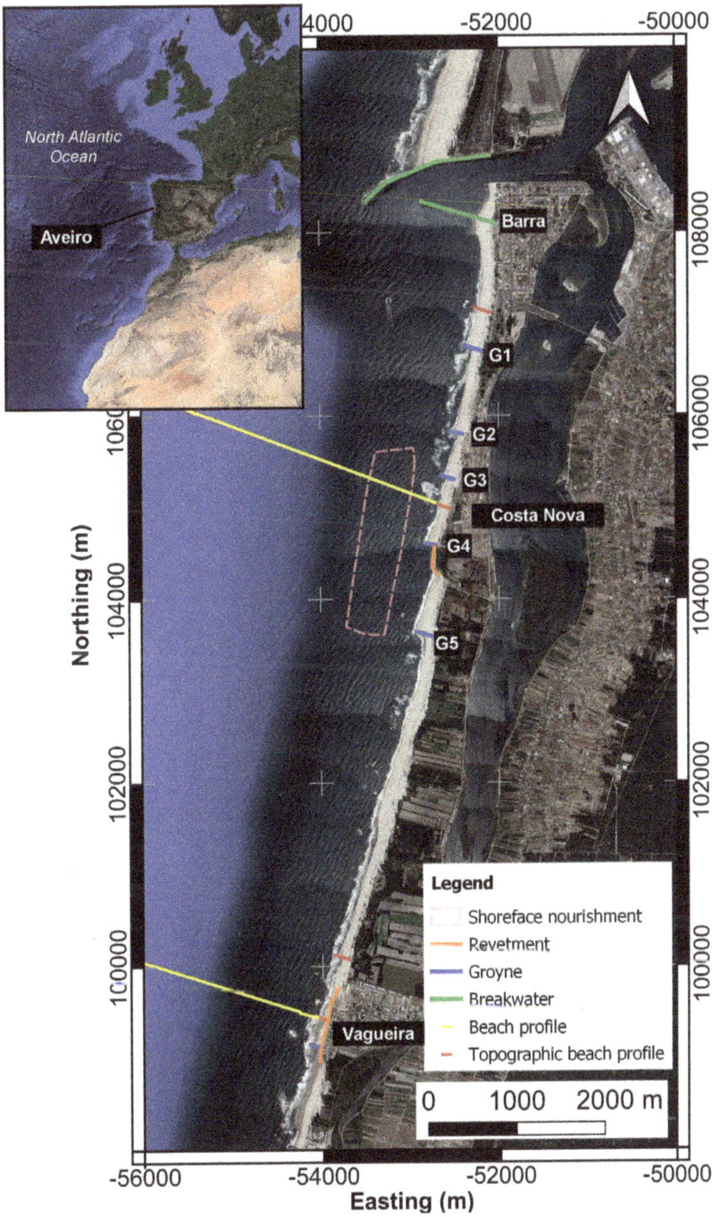

Figure 1. Study area. Coastal defense works include two breakwaters at the Aveiro entrance channel, five groynes at the northern part of the study area and another at Vagueira beach and two revetments one close to G4 and the other at Vagueira beach. The area associated with the shoreface nourishment intervention was between G3 and G5. Beach profiles and topographic beach profiles were obtained within the COSMO programme. Coordinates are referred to the ETRS89 system.

3.3.2. Data Processing and Analysis

For the bathymetric surveys, the acquired data was processed with Caris software (version 9.5). The obtained XYZ coordinates were interpolated in the QGIS software to produce a digital elevation model (DEM) with a horizontal resolution of 5 m using a Triangular Irregular Network (TIN) interpolation method. For the topographic surveys, the acquired data was processed with a set of dedicated software, namely the Trimble Business Center (version 5.32, www.trimble.com, accessed on 10 October 2021) to GPS data process and the Matlab (version R2020a) to estimate the inclination angles (i.e., the attitude) of the GPS antennas in the quad-bike for each sample period. The obtained XYZ coordinates were interpolated in the ArcGIS software to produce a DEM with a horizontal resolution of 1 m using a Kriging interpolation method.

The analysis conducted in this study used a set of cross-shore transects with a spacing of 50 m that were extracted from the generated DEM between Barra and Vagueira. These transects were perpendicular to the main coastline orientation (15° N). The extraction of cross-shore transects with a 50 m spacing allowed to analysed in more detail the spatial variations of the beach response in time along the coastal stretch.

3.3.3. Additional Data Sources

To complement the analysis conducted in this study, additional data sources were used. These sources were all the available topo-bathymetric surveys, beach profile surveys and topographic beach surveys obtained between July 2018 and November 2020 under the Portuguese COaStal MOnitoring Programme (COSMO), developed and implement by APA (https://cosmo.apambiente.pt/, accessed on 28 June 2021) [34]. Moreover, a bathymetric survey provided by the Aveiro harbour authority was also used in this study. The latter was performed right after the shoreface nourishment intervention on August 2020 and it covered the area in Figure 1 (dashed white line).

4. Results

4.1. Depth of Closure and Beach Volume Based on a Shoreline Elevation Proxy

The beach profile surveys, obtained before the shoreface nourishment through the COSMO programme, were used to estimate the elevation associated with the depth of closure in the study area. The beach profile surveys were conducted with a wave runner in the submerged part, thereby surveying the entire surf zone, and by foot on the intertidal and aerial part. In Figure 1, the guide lines of the submerged part of the beach profile survey are displayed in yellow. The available profiles were located near Costa Nova beach (between G3 and G4) and at Vagueira beach.

Figure 2 shows the beach profile surveys for Vagueira (left) and Costa Nova (right). The bottom panels display the standard deviation of the beach profiles. At Vagueira, the standard deviation rapidly increases from its offshore value of less than 0.20 m to values larger than about 0.50 m (Figure 2e). In more details, the standard deviation has an inflexion point for distance equal to 650 m. This inflexion point is associated with an elevation of −10 m (MSL). At Costa Nova, it is clear that the November 2020 profile has an influence of the shoreface nourishment (Figure 2b,d) and it was not used in the standard deviation calculation. A similar analysis performed on Costa Nova beach profile also suggests that the elevation −10 m (MSL) is associated with very small morphological changes (less than 0.20 m) (Figure 2d). Therefore, the analysis of the COSMO profiles allowed to estimate a depth of closure of −10 m (MSL) for this study site. Note that the shoreface nourishment intervention led to morphological modifications up to −11 m (MSL) at Costa Nova (November 2020 in Figure 2a,b) but this was due to the intervention and not due to the natural beach behaviour (see Figure 4 of [35]).

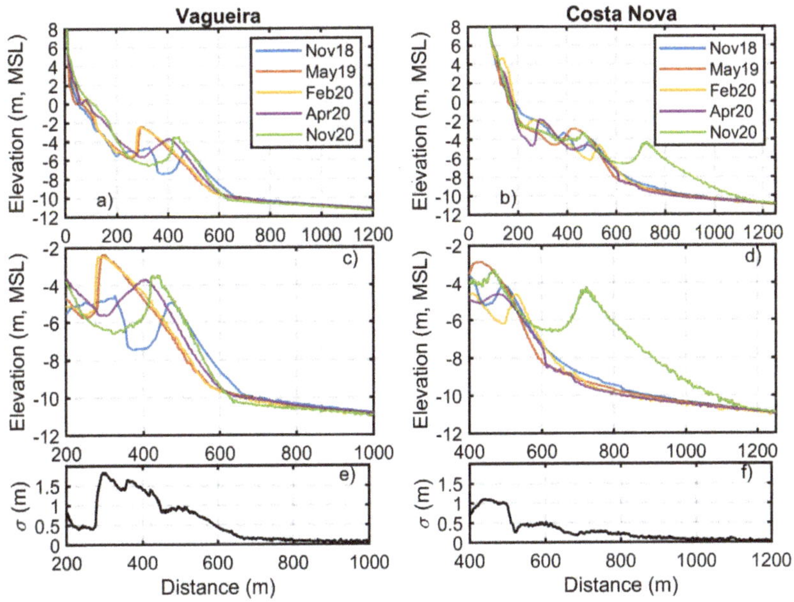

Figure 2. Beach profiles at Vagueira (panel (**a**)) and at Costa Nova (panel (**b**)) beaches (yellow lines in Figure 1) between 2018 and 2020. Zoom-in of the submerged beach profiles (panels (**c**,**d**)). Standard deviation of beach profile elevations for each cross-shore position (panels (**e**,**f**)).

Also available through the COSMO programme were four subaerial topographic profiles that can be found along the coastal stretch depicted in Figure 3 (see also red lines in Figure 1). The profiles were collected between August 2018 and October 2020 every 3 months. Therefore, the seasonal subaerial beach variability throughout is well captured. These field measurements were used to understand how well the displacement of an elevation in time can represent the beach volume temporal evolution. Beach profiles display an elevation (about 5 m, MSL) above which the morphological variations are smaller than 0.3 m (top panels in Figure 3). An exception is the beach profile Vagueira 2 because it is located in front of a revetment. The beach volume (per unit width) for Barra, Costa Nova and Vagueira 1 was determined as the area above MSL until the point in which the profiles converge (5 m, MSL), see Figure 3. For Vagueira 2, the beach volume (per unit width) was calculated as the volume above MSL that is delimited by the revetment toe (distance = 25 m, Figure 3). Next, it was assessed how well the displacement of several isolines (ranging from 0 m to 5 m) through time can describe the temporal beach volume variations. The middle panels of Figure 3 show the correlation coefficient between those two quantities. In general, the onshore-offshore displacements of isolines between 1 and 3 m are well correlated with the beach volume variations. For the Vagueira 1 beach, this is not the case and there is a well-defined maximum of the correlation coefficient around 1 m. Motivated by this maximum and since other beach profiles have also correlation coefficients higher than about 0.9 for 1 m, the following shoreline proxy was used: the 1.05 m (MSL) contour, for two reasons. First, this contour is capable of describing well the temporal variations of beach volume for the four beach profiles located across the coastal stretch (bottom panels of Figure 3). Second, the 1.05 m (MSL) contour is the averaged value of the high-tide at Aveiro tidal gauge, which corresponds to the shoreline [36]. Therefore, the displacement of the 1.05 m (MSL) contour will be used to characterize the subaerial beach response to the shoreface nourishment in time and in space.

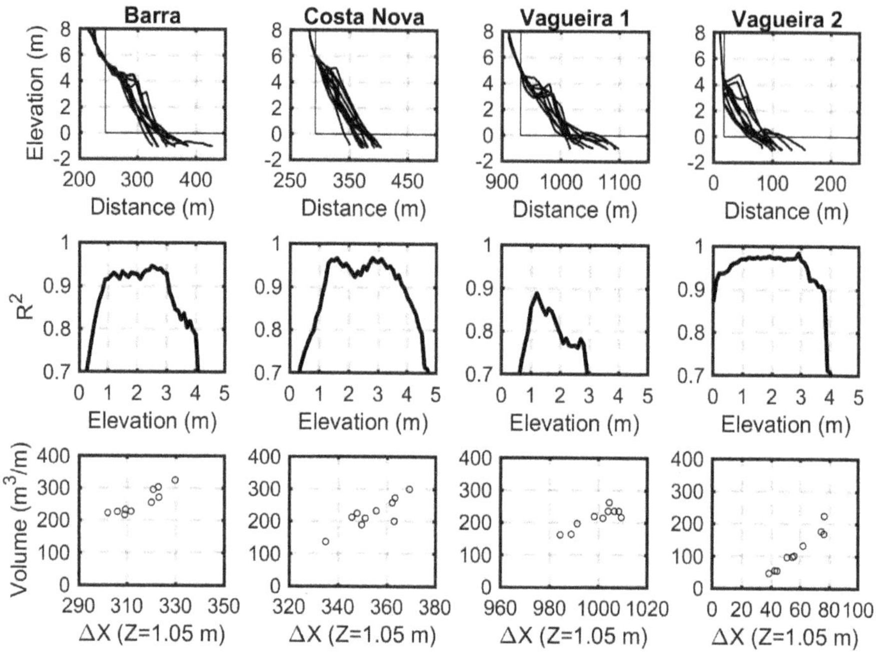

Figure 3. Topographic beach profiles at Barra, Costa Nova, Vagueira 1 and Vagueira 2 beaches (red lines in Figure 1, from North to South) (**top** panels). Correlation coefficient between beach volume (per unit width) and the displacement of elevation contours (**middle** panels). Scatter diagrams between beach volume and the displacement of the 1.05 m (MSL) contour (**bottom** panels).

4.2. Subtidal Beach Response to the Shoreface Nourishment

Figure 4 shows the temporal evolution of the −10 m and of the −8 m (MSL) contours along the coastal stretch between June 2020 (previous to the nourishment) and January 2021. The initial sediment deposition zone, between G3 and G5, is obtained from the August 2020 multibeam survey provided by the Aveiro harbour administration.

Since the closure depth was disturbed by the shoreface nourishment intervention, its onshore or offshore displacement is associated with the shoreface nourishment movement through time. As an example, if the isobathymetric contour of −10 m (MSL) does not change in time, it means that the shoreface nourishment is relatively stable. On the opposite, if the −10 m (MSL) contour moves landward, it means that the shoreface nourishment moved towards the coast. To complement the analysis, the −8 m (MSL) contour was also used. The latter is expected to have a larger displacement in space than the −10 m (MSL). Other less deep contours were also envisaged but they were more influenced by the bar movement that occurred southward and northward of the initial shoreface nourishment area. Consequently, the shallower isobathymetric contours will not be considered hereafter.

Regarding the −10 m (MSL) contour, its offshore displacement ranged between 250 m at the southward end to 120 m at the northward end of the initial deposition zone (between June and September 2020). Between September 2020 and January 2021, while the onshore displacement at the northward end was smaller than 50 m, the onshore displacement reached 120 m at the southern end. Differences between the −10 m (MSL) contour between G1 and G2 are likely associated with the ebb-tidal delta shoal dynamics of the Aveiro inlet and to the different survey technologies (i.e., multibeam vs single-beam), and will not be discussed hereafter. The horizontal differences between the −10 m (MSL) contours are smaller than 50 m at 1600 m southwards (Y = 102,000 m in Figure 4a). Therefore, the influence of the shoreface nourishment on the displacement of the −10 m (MSL) contour south of the initial deposition zone had an extent of about 1600 m.

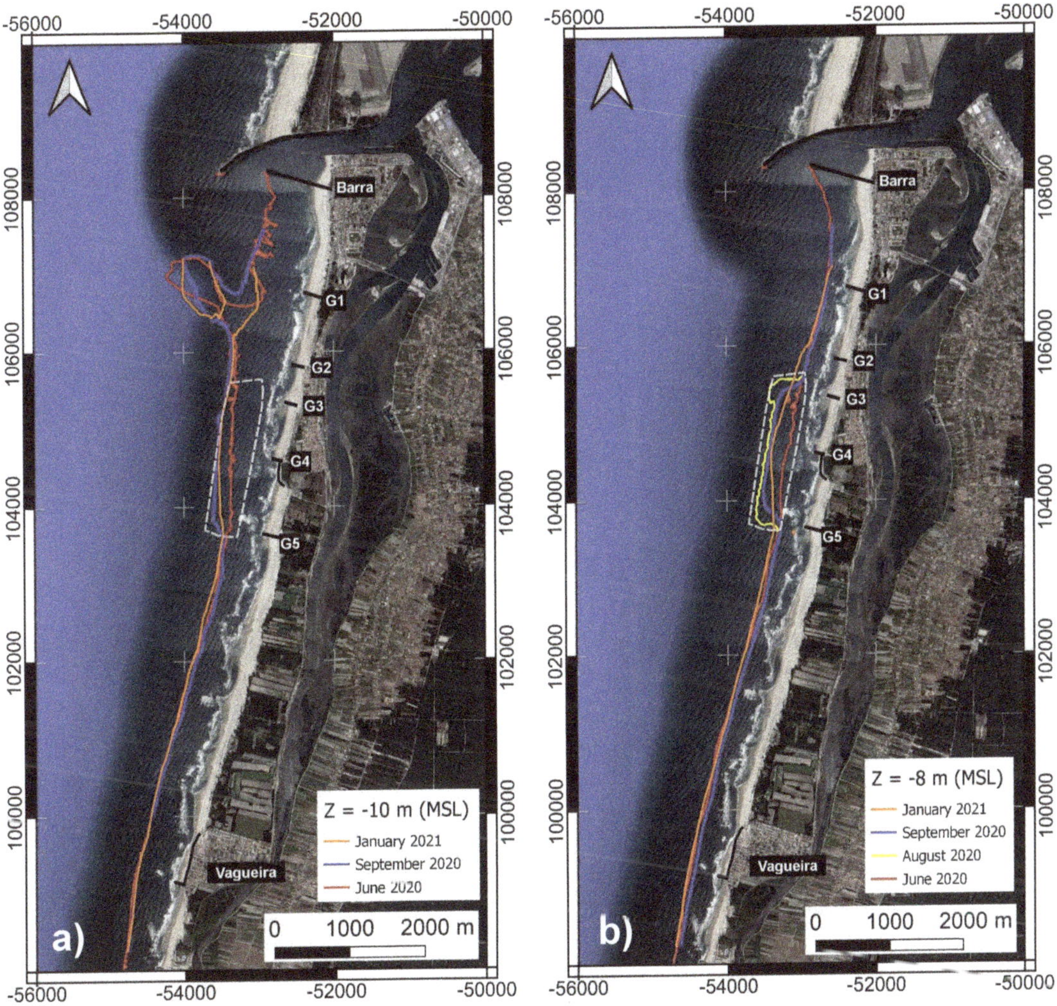

Figure 4. Time evolution of the bathymetric contour associated with −10 m (MSL) (panel (**a**)) and with −8 m (MSL) (panel (**b**)). Coastal defense works over the study area (black) and the area associated with the shoreface nourishment intervention was between G3 and G5 (dashed white). Coordinates are referred to the ETRS89 system.

The patterns displayed by the −8 m (MSL) contour are more pronounced than those associated with the −10 m (MSL) contour. The offshore displacement between June and August 2020 ranged between 280 m to 350 m at the initial deposition zone. After one month (September 2020), the −8 m (MSL) contour shifted landwards between 50 m at the central area (in front of G4) and 120 m at the end points (in front of G3 and G5). The largest landward migration at the end points is in part explained by the very sharp contours that are rapidly eroded. On January 2021, the overall shape of the shoreface nourishment resembles a gaussian function (Figure 4b, January 2021), denoting the spreading of the nourishment in the longshore direction (Figure 4b). The onshore displacement of the −8 m

(MSL) contour was 30 m between September 2020 and January 2021 at the central part (in front of G4). This onshore displacement increased farther away, reaching about 100 m in front of G3 and of G5. The influence of the shoreface nourishment was felt until about 600 m towards north (200 m north of G3). At south, the onshore or offshore displacements of the −8 m (MSL) contour can be associated with the subtidal bar movement, as seen in Figure 2a, thereby preventing to isolate the effect of the nourishment. The latter aspect is visible in front of Vagueira beach where the September 2020 −8 m (MSL) contour is not only located 60 m landward than the January 2021 contour but also about 40 m landward than the June 2020 contour.

4.3. Subaerial Beach Response to the Shoreface Nourishment

The analysis of the beach response to the shoreface nourishment was based on the 1.05 m (MSL) contour (see Figure 3 and Section 4.1). Figure 5 shows the spatial and temporal displacement of the 1.05 m (MSL) contour along the coastal stretch relative to its position on the beginning of June 2020, together with the the time series of wave parameters. While warm colors indicate an offshore displacement (i.e., beach volume increase), cold colors are associated with an onshore displacement (i.e., beach volume decrease). The horizontal black lines represent the five groynes (G1 to G5 in Figure 1) located at Costa Nova beach. The absence of colored circles means no data.

In general, three main features can be observed. First, the dominance of a landward displacement (blue circles) for distance between 3000 m and 4000 m. This displacement, that can reach locally −60 m, is located at the downdrift side of the shoreface nourishment. In more details, the shoreface nourishment was performed between G5 and G3. The retreat of the 1.05 m (MSL) contour relative to the June 2020 can be associated with modification of local wave parameters which drive divergency of the longshore sediment transport at the end parts of the shoreface nourishment. This erosion pattern was pointed out by [5] where the downdrift part can experience shoreline retreat in analogy to a detached breakwater. Second, the large seawards displacement (red circles) for distance equal to 2800 m between September and November 2020, and also on February and on April 2021. It is speculated that this seawards displacement is associated with the natural beach behaviour which increases in volume until late summer. Unfortunately, a comparison with older surveys is not possible because they are not available with the high-temporal and spatial resolution as those presented in this study. Third, the mild offshore displacement of the 1.05 m (MSL) contour located between G4 and G2. This offshore displacement through time is likely induced by the shoreface nourishment. The beaches delimited between G1 and G4 do experience the expected response due to the longshore drift until November 2020, which is from North to South at this study area during summer (see Figure 7 in [26]). Between September and October 2020, those beaches are associated with a seawards displacement northward of each groyne and with an onshore displacement southward (i.e., downdrift side) of each groyne. From November 2020 onward, this pattern changes and the onshore displacement is northward of each groyne, while the offshore displacement occurs at the downdrift side. This inversion in the expected beach erosion and accumulation associated with groynes is clearly observed for beaches between G2 and G5. Moreover, the mean wave direction become more confined to the WNW and W sectors (see January to February 2021 in Figure 5d). The maximum landward displacement of the shoreline proxy was −60 m at distance between 3000 m and 4000 m between January and April 2021. The maximum seawards displacement was +60 m at distance equal to 2800 m between September and October 2020.

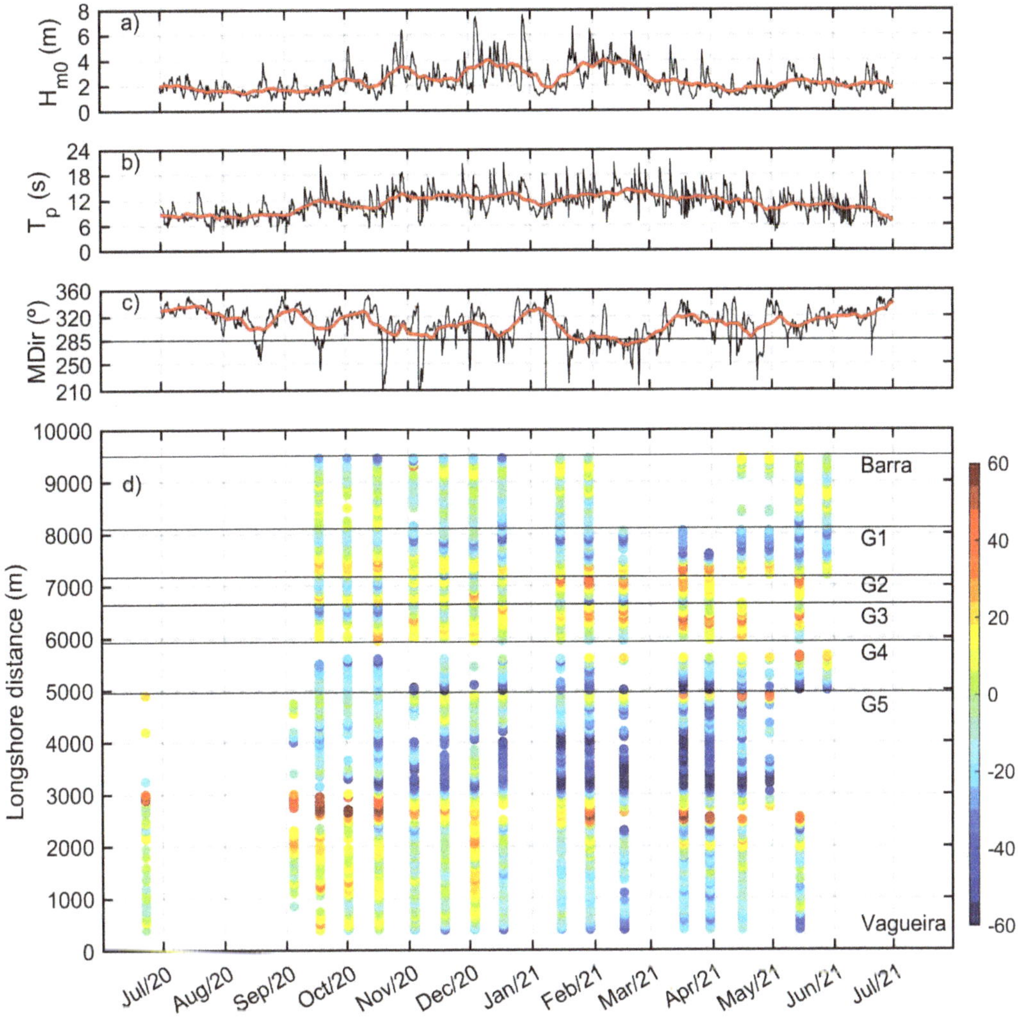

Figure 5. Time series of offshore significant wave height (**a**), peak period (**b**) and mean wave direction (**c**) obtained from CMEMS (https://marine.copernicus.eu/, accessed on 28 June 2021), and 15-day moving average (red). Time-series of the relative 1.05 m (MSL) shoreline proxy advance seawards (warm colors) or retreat landwards (cold colors) (**d**) along the coastal stretch from Vagueira (0 m) to Barra (9500 m). The relative advance or retreat is in comparison with the shoreline proxy position surveyed on June 2020 (reference situation before shoreface nourishment). Horizontal black lines refer to groynes and to breakwater.

4.4. Subaerial Beach Response to Storms and Subsequent Recovery

The beach response to storms was analysed based on the relative displacement of the shoreline proxy (1.05 m, MSL) in relation to the previous topographic survey, as shown in Figure 6. As an example, the shoreline proxy displacement on 17 December 2020 is obtained as the difference between 17 and 3 December 2020 surveys.

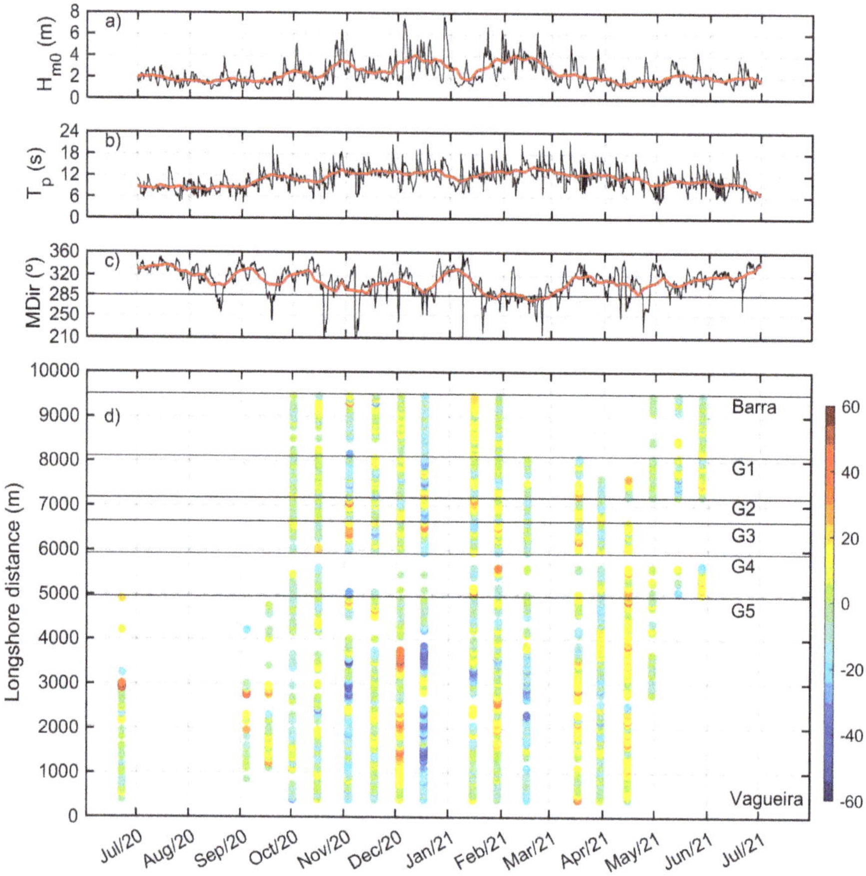

Figure 6. Time series of offshore significant wave height (**a**), peak period (**b**) and mean wave direction (**c**) obtained from CMEMS (https://marine.copernicus.eu/, accessed on 28 June 2021), and 15-day moving average (red). Time-series of the relative 1.05 m (MSL) shoreline proxy advance seawards (warm colors) or retreat landwards (cold colors) (**d**) along the coastal stretch from Vagueira (0 m) to Barra (9500 m). The relative advance or retreat is in comparison with the shoreline proxy position surveyed on the previous survey. Horizontal black lines refer to groynes and to breakwater.

Looking at Figure 6d, the largest differences between consecutive surveys took place on December 2020. More specifically, the shoreline proxy moved seawards on 3 December and moved landwards on 17 December. Regarding the landward displacement, this occurred after storm Dora, which hit mainland Portugal on 4 December 2020. However, the beach response was very variable across the coastal stretch. The largest changes are more visible on natural beaches (south of G5) than on beaches close to coastal structures (between Barra and G5), which are under the influence of the shoreface nourishment. The offshore H_{m0} time-series clearly displays a very large increase on that day (up to 8 m) with an incident mean wave direction of about 320° N. The shoreline proxy retreat reach up to −50 m in some areas. The seaward displacement downdrift of G3 and G2 after storm Dora is likely associated with the presence of the shoreface nourishment. At that locations, the shoreline proxy advanced approximately 10 m. Regarding the seaward displacement that occurred on 3 December (south of G5), this beach accretion was likely associated with a reduction of the offshore H_{m0}, together with large T_p. The 15-day moving average window suggests that H_{m0} decreased from 4 m to about 2 m. This reduction accompanied by large wave periods have most likely promoted the beach accretion (i.e., seaward displacement of the shoreline proxy). Again, the shoreline proxy advance is not uniform along the coastal stretch.

Beach accretion after storm Dora is displayed in Figure 7d. In this Figure, each shoreline proxy relative position is compared with the shoreline proxy position on 3 December (before storm Dora). Two major patterns emerged in Figure 7. First, the shoreline proxy 1.05 m (MSL) rapidly moved landwards at the downdrift side of G2 and of G3 from January 2021 onward. Second, the shoreline proxy continued to moved landwards on beaches backed up by dunes (those located southwards of G5). At those locations, there are no clear patterns of total recovery because the circles are still associated with negative values. The shoreline proxy close to Vagueira beach started to display some values close to 0 m on March 2021. This indicates that the 1.05 m (MSL) contour has recovered its 3 December relative position at that location, which can be in part ascribed to the Vagueira beach natural behaviour.

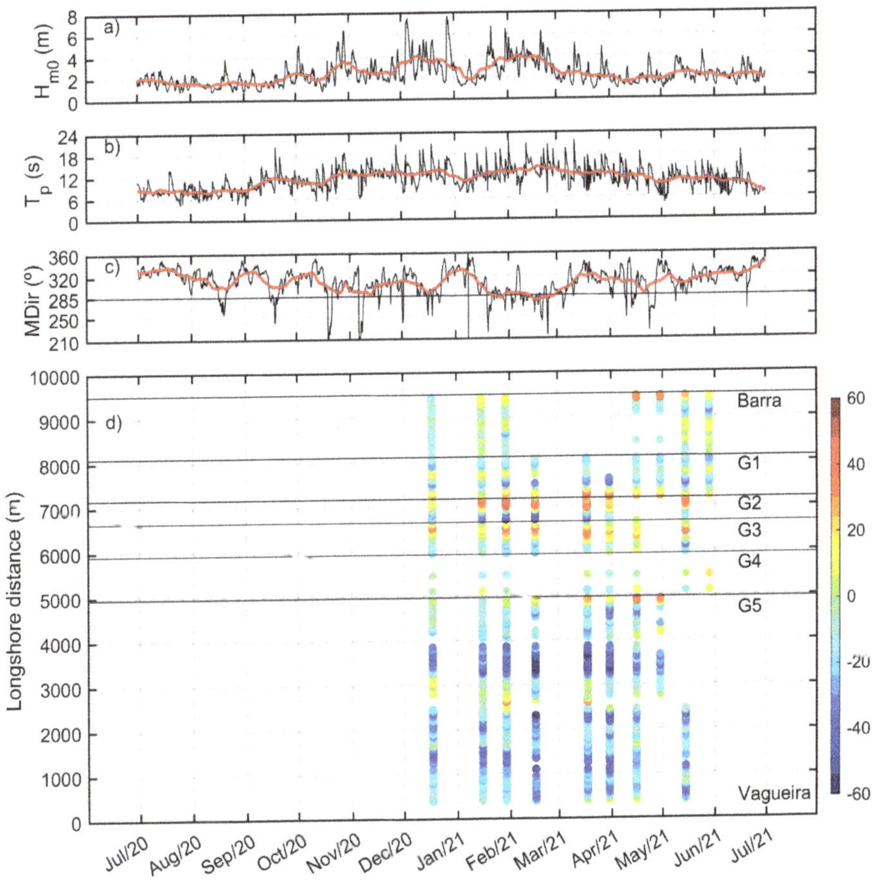

Figure 7. Time series of offshore significant wave height (**a**), peak period (**b**) and mean wave direction (**c**) obtained from CMEMS (https://marine.copernicus.eu/, accessed on 28 June 2021), and 15-day moving average (red). Time-series of the relative 1.05 m (MSL) shoreline proxy advance seawards (warm colors) or retreat landwards (cold colors) (**d**) along the coastal stretch from Vagueira (0 m) to Barra (9500 m). The relative advance or retreat is in comparison with the shoreline proxy position surveyed on 3 December 2020 (reference situation before storm Dora). Horizontal black lines refer to groynes and to breakwater.

5. Discussion

5.1. General Characteristics of the Aveiro Shoreface Nourishment in Comparison with Previous Works

The review of previous works presented in Section 2 (see Table 1) suggests that shoreface nourishments performed at locations with both larger significant wave heights and wave periods have a V/L value (ratio between nourishment volume and shoreface nourishment length) about a factor of 2 greater than those performed at locations characterized by milder and local generated waves. The shoreface nourishment movement landwards is performed with a migration rate about 50% larger than those in milder environments. In both mild or energetic wave conditions, the MSL contour displacement ranges between 13% to 20% of the shoreface nourishment onshore migration. Moreover, for the shoreface nourishment in California, Figure 2 in [19] suggests that the increase in beach volume above MSL is much smaller than the reduction of shoreface nourishment volume. For the Dutch shoreface nourishments, ref. [37] mentioned that the nourishment only acted partially as a feeder berm.

The Aveiro shoreface nourishment had an overall sediment volume (V) of about 2×10^6 m^3 and a length (L) = 1900 m, which gives a V/L = 1050 m^3/m. Its −8 m (MSL) contour onshore displacement, ranged from 80 m to 230 m between August 2020 and January 2021. Although the observations available in this study do not coverage one year, the initial onshore displacement rate (rate S_{on}) is between 200 m/yr to 500 m/yr. The local wave conditions at Aveiro are similar to those offshore because there are neither a sheltering effect promoted by islands nor dissipation by bottom friction over the continental shelf (about 80 km at Aveiro). In general, the Aveiro shoreface nourishment shares similar characteristics with that performed in California, US [19]. Wave conditions at California are associated with a larger H_{m0} than those at Aveiro (3.5 m in winter compared to 2.8 m) but the local wave conditions at California are likely milder due to the Channel Islands sheltering effect. Therefore, the large onshore displacement of Aveiro shoreface nourishment (200–500 m/yr), when compared to California (72 m/yr), can be attributed to the more energetic local wave conditions.

Regarding the Dutch nourishments [6], they were performed with a V/L ratio about a factor of 2–3 smaller than both Aveiro and California nourishments (Table 1). Their onshore migration rates varied for each nourishment, ranging from about 20 m/yr, for Wassenaar, to 50 m/yr, for both Egmond and Terheijde. These rates are smaller when compared with 72 m/yr for California and with 200–500 m/yr for Aveiro. In general, the wave climate along The Netherlands is milder and characterised by local generated waves. Even if the wave climate is different from that in Aveiro and California, often more energetic and characterized by swell waves, the differences between wave climates alone cannot explain the differences in the onshore migration rates. This is because for the same wave climate and with similar V/L ratios, the onshore migration rate of Wasenaar nourishment (22 m/yr) was about half of the rate of Egmond nourishment (50 m/yr). We suggest that other factors, such as tidal-induced velocities, beach or surfzone slopes, sediment grain sizes, coastal geomorphology and man-made structures (e.g., harbours) can contribute to explain these differences on the overall shoreface nourishment evolution.

The maximum seaward displacement of the 1.05 m (MSL) contour was about 60 m when compared to its reference position on June 2020. Although this displacement varied along the study area (see Figure 5), this gives an approximate offshore migration rate of 60 m/yr. Despite the fact that this offshore migration rate is larger than that of other shoreface nourishments, it is still about 30% of the onshore movement of the Aveiro shoreface nourishment (200 m/yr). Therefore, although the onshore and offshore displacements associated with the Aveiro shoreface nourishment are more pronounced than the shoreface nourishments performed elsewhere, the shoreline proxy advance seawards only accounted for 30% of the smallest shoreface nourishment outer limit advance landward. This suggests that the Aveiro shoreface nourishment behaved only partially as a feeder berm. In other words, the Aveiro shoreface nourishment did not contribute exclusively to a subaerial beach volume increase. The remaining part of the sediment volume associated with the shoreface nourishment have likely dispersed alongshore driven by the littoral drift. The latter suggestion is supported by the gaussian shape of the shoreface nourishment −8 (MSL) contour in January 2021 (Figure 4b).

5.2. Response of the Shoreline Proxy to the Shoreface Nourishment

The results presented above clearly highlighted distinct behaviours of the shoreline proxy relative to its cross-shore position before the shoreface nourishment intervention (June 2020, in Figure 5) and after the major storm Dora (December 2020, in Figure 6). The results suggest that the most clear shoreline proxy displacement onshore occurs at the downdrift side of the shoreface nourishment intervention (cold colours in Figure 5d). This behaviour has been pointed out by Van Duin et al. [5]. In our case study, the extent of this effect was about 2000 m southwards of the initial deposition zone. Since the length of the Aveiro shoreface nourishment is 1900 m, it is suggested that the shoreline retreat associated with this shoreface nourishment is about the same length as the shoreface nourishment itself.

The longshore spreading and evolution of the shoreface nourishment is about their initial length (1900 m). The results show a southward displacement of the −10 m and of the −8 m isolines up to 1600 m (Figure 4). Towards north, the shoreface nourishment evolution is influenced by the ebb delta shoal of Aveiro inlet but it extends up to 600 m northwards. Based on the available bathymetric surveys, it can be expected that the shoreface longshore spreading is about its initial length during the first year of morphological evolution.

Beaches located closer to the shoreface nourishment were benefited. This effect was present in three comparisons: to its relative position on June 2020; after storm Dora; and also during subsequent beach recovery. As an example, during storm Dora beaches located near the shoreface nourishment (between G3 and G5) even experienced a shoreline proxy advance seawards. This effect is likely offered by the sheltering effect of the shoreface nourishment which is capable of reducing wave amplitude during storms. Moreover, beaches closer to the shoreface nourishment also experienced recovery much quicker than those farther away. Looking at Figure 7d, the shoreline proxy associated with beaches located between G4 and G2 displayed positive values after storm Dora (>20 m, yellow and red markers) from January until June 2021. On the opposite, the shoreline proxy associated with beaches located southwards of G5 displayed negative values (<−20 m, cyan and blue markers). There are some locations where this pattern was not so clear, such as beaches located updrift of G3 (where shoreline proxy retreat reached −40 m between February and March 2021) and also beaches located downdrift of G5 (where the shoreline proxy advanced up to 40 m on mid April 2021). Though, the general pattern is that the relative position of the shoreline proxy accreted more between G4 and G2 than at beaches located southwards of G5, where a shoreline proxy retreat was observed. Therefore, beaches located closer to the shoreface nourishment (between G4 and G2) achieved their position before storm Dora (warm colours in Figure 7d) quicker than beaches located southward of G5 (cold colours in Figure 7d).

On a seasonal scale (from September 2020 to April 2021), beaches located closer to the shoreface nourishment experienced a milder but steady advance seawards (see warm colours between G4 and G2 in Figure 5d). An interesting result is that beaches delimited by groynes have experienced a different morphodynamic pattern than that expected from the longshore drift. The expected pattern occurred until October 2020 on beaches located between G2 and G4, with erosion downdrift (cold colours in Figure 5d) and accretion updrift (warm colours in Figure 5d). From November 2020 on, but more clear in January and February 2021, this pattern changed and beach accretion occurred on the downdrift side of a groyne (warm colours in Figure 5d), while beach erosion took place on the updrift side of a groyne (cold colours in Figure 5d). This inversion is possibly linked to the nearshore circulation patterns induced by the shoreface nourishment [5], and also with the directional wave conditions. Future numerical modelling efforts may be used to understand the reason for this observed inversion.

The results presented in this study suggest that the shoreface nourishment was beneficial to the subaerial beach in different ways. Although the results presented above also allow to better understand the beach response to a shoreface nourishment at Aveiro, more coastal monitoring is definitely desirable to draw more firm conclusions. The use of simple techniques to assess shoreface nourishment expected evolution can be of interest for preliminary designs [38]. Additionally, the use of video cameras can be included to complement topographic and bathymetric measurements [39]. Moreover, the application of numerical models (either more simple, such as one-line models, or more complex, such as coastal area models) can be envisaged in future (e.g., [40]). These type of applications would not only allow to verify numerical models but also to test different shoreface nourishment options and configurations by varying its width, length, volume and distance to shoreline, and also to perform cost-benefit analysis of coastal protection strategies [41].

6. Conclusions

In this study, the beach response to a shoreface nourishment was analysed based on topographic and bathymetric surveys performed over one year along 10 km in Aveiro (Portugal). The analysis was based on the evolution of the −10 m and −8 m (MSL) contours and of the 1.05 m (MSL) contour. The former is associated with the displacement of the shoreface nourishment outer limit and the latter is associated with the temporal variation of the beach volume above MSL, for this study area.

While the onshore displacement of the −10 m (MSL) and of the −8 m (MSL) contour ranged between 80 m and 230 m, the maximum offshore displacement of the 1.05 m (MSL) contour was 60 m. Therefore, our results suggest that the shoreface nourishment intervention only acted partially as a feeder berm. This is in agreement with shoreface nourishment of similar volume magnitudes performed elsewhere.

The beach response (the onshore or offshore displacement of the 1.05 m contour) was highly variable in time and along the 10 km study area. Beaches located closer to the shoreface nourishment (delimited by groynes) exhibit a more stable behaviour, are not so vulnerable to wave conditions than those located south of G5 and display a large seaward displacement (i.e., beach volume increase). During storm Dora (December 2020), beaches located closer to the shoreface nourishment even display an offshore advance (i.e., beach volume increase). Moreover, the beach response is also quick on that beaches. A drawback associated with the shoreface nourishment was the beach retreat at the downdrift side of the nourishment. This effect occurred until a distance that is about the shoreface nourishment length.

Author Contributions: Conceptualization, D.M.; data collection, D.M.; methodology, D.M.; software, D.M.; formal analysis, D.M., J.P.-B., P.B., P.A.S., C.B. and C.P.; resources, P.B.; writing—original draft preparation, D.M.; writing—review and editing, J.P.-B., P.B., P.A.S., C.B. and C.P.; project administration, P.B. and C.P.; funding acquisition, P.B. and C.P. All authors have read and agreed to the published version of the manuscript.

Funding: Some of the collected data were co-funded by POSEUR (Ref. POSEUR-02-1809-FC-000051—Remoção dos Inertes da Zali do Porto de Aveiro para Reforço do Cordão Litoral a Sul da Costa Nova) and are gratefully acknowledged. Data were also obtained under the COSMO Programme-Coastal Monitoring Programme of Continental Portugal, of the Portuguese Environment Agency, co-funded by the Operational Program for Sustainability and Efficiency in the Use of Resources (POSEUR), https://cosmo.apambiente.pt, accessed on 28 June 2021.

Acknowledgments: The authors acknowledge Rita Cavalinhos (CESAM field technician) and Paulo Rosa (CESAM skipper) for their fantastic work during the bathymetric surveys. Rita Cavalinhos is also acknowledged for her work during the multibeam surveys pre-processing phase. Fábio Santos is also acknowledged for his help during the bathymetric surveys. The Aveiro harbour authority is warmly acknowledged for providing the multibeam survey. Comments and suggestions provided by two anonymous reviewers are acknowledged. Thanks are due to FCT/MCTES for the financial support to CESAM (UIDP/50017/2020+UIDB/50017/2020), through national funds. SANDTRACK project (PTDC/CTA-GEO/31779/2017) funded by FEDER, through COMPETE2020 - Programa Operacional Competitividade e Internacionalização (POCI), and by national funds (OE), through FCT/MCTES.

Conflicts of Interest: The authors declare no conflict of interest.

References

1. de Schipper, M.A.; Ludka, B.C.; Raubenheimer, B.; Luijendijk, A.P.; Schlacher, T.A. Beach nourishment has complex implications for the future of sandy shores. *Nat. Rev. Earth Environ.* **2020**, *2*, 70–84. [CrossRef]
2. Dean, R. *Beach Nourishment: Theory and Practice*; Advanced Series on Ocean Engineering—Volume 18; World Scientific: Singapore, 2002; p. 399.
3. Brutsché, K.E.; Wang, P.; Beck, T.M.; Rosati, J.D.; Legault, K.R. Morphological evolution of a submerged artificial nearshore berm along a low-wave microtidal coast, Fort Myers Beach, west-central Florida, USA. *Coast. Eng.* **2014**, *91*, 29–44. [CrossRef]
4. Brutsché, K.E.; McFall, B.C.; Bryant, D.B.; Wang, P. *Literature Review of Nearshore Berms—ERDC/CHL SR-19-2*; Technical Report; USACE: Washington, DC, USA, 2019.
5. van Duin, M.J.P.; Wiersma, N.R.; Walstra, D.J.R.; van Rijn, L.C.; Stive, M.J.F. Nourishing the shoreface: observations and hindcasting of the Egmond case, The Netherlands. *Coast. Eng.* **2004**, *51*, 813–837. [CrossRef]
6. Huisman, B.J.A.; Walstra, D.J.R.; Radermacher, M.; de Schipper, M.A.; Ruessink, B.G. Observations and Modelling of Shoreface Nourishment Behaviour. *J. Mar. Sci. Eng.* **2019**, *7*, 59. [CrossRef]
7. Pinto, C.A.; Silveira, T.M.; Teixeira, S.B. Beach nourishment practice in mainland Portugal (1950–2017): Overview and retrospective. *Ocean Coast. Manag.* **2020**, *192*, 105211. [CrossRef]
8. Spodar, A.; Héquette, A.; Ruz, M.H.; Cartier, A.; Grégoire, P.; Sipka, V.; Forain, N. Evolution of a beach nourishment project using dredged sand from navigation channel, Dunkirk, northern France. *J. Coast. Conserv.* **2018**, *22*, 457–474. [CrossRef]
9. Elko, N.; Briggs, T.R.; Benedet, L.; Robertson, Q.; Thomson, G.; Webb, B.M.; Garvey, K. A century of U.S. beach nourishment. *Ocean Coast. Manag.* **2021**, *199*, 105406. [CrossRef]
10. Hanson, H.; Brampton, A.; Capobianco, M.; Dette, H.; Hamm, L.; Laustrup, C.; Lechuga, A.; Spanhoff, R. Beach nourishment projects, practices, and objectives—A European overview. *Coast. Eng.* **2002**, *47*, 81–111.
11. Cooke, B.C.; Jones, A.R.; Goodwin, I.D.; Bishop, M.J. Nourishment practices on Australian sandy beaches: A review. *J. Environ. Manag.* **2012**, *113*, 319–327. [CrossRef] [PubMed]
12. Cai, F.; Dean, R.G.; Liu, J. Beach nourishment in China: Status and Prospects. *Coast. Eng. Proc.* **2010**, *1*, 1–12. [CrossRef]
13. Habel, S.; Fletcher, C.H.; Barbee, M.; Anderson, T.R. The influence of seasonal patterns on a beach nourishment project in a complex reef environment. *Coast. Eng.* **2016**, *116*, 67–76. [CrossRef]
14. Muñoz-Perez, J.J.; Gallop, S.L.; Moreno, L.J. A Comparison of Beach Nourishment Methodology and Performance at Two Fringing Reef Beaches in Waikiki (Hawaii, USA) and Cadiz (SW Spain). *J. Mar. Sci. Eng.* **2020**, *8*, 266. [CrossRef]
15. Marinho, B.; Coelho, C.; Larson, M.; Hanson, H. Monitoring the evolution of nearshore nourishments along Barra-Vagueira coastal stretch, Portugal. *Ocean Coast. Manag.* **2018**, *157*, 23–39. [CrossRef]
16. Zwaborn, J.; Fromme, G.; FitzPatrick, J. Underwater mound for the protection of Durban's beaches. *Coast. Eng. Proc.* **1970**, *1*, 62. [CrossRef]
17. Vera-Cruz, D. Artificial Nourishment of Copacabana Beach. *Coast. Eng. Proc.* **1972**, *1*, 1451–1463. [CrossRef]
18. Laustrup, C.; Madsen, H.T.; Sorensen, P.; Broker, I. Comparison of beach and shoreface nourishment Torsminde Tang, Denmark. *Coast. Eng. Proc.* **1996**, *1*, 2927–2940. [CrossRef]
19. Mesa, C. Nearshore berm performance at Newport Beach, California, USA. *Coast. Eng. Proc.* **1996**, *1*, 4636–4649. [CrossRef]
20. Wilson, J.H.; Beyene, A. California Wave Energy Resource Evaluation. *J. Coast. Res.* **2007**, *23*, 679–690. [CrossRef]
21. Wijnberg, K.M. Environmental controls on decadal morphologic behaviour of the Holland coast. *Mar. Geol.* **2002**, *189*, 227–247. [CrossRef]
22. van Rijn, L.C. Sediment transport and budget of the central coastal zone of Holland. *Coast. Eng.* **1997**, *32*, 61–90. [CrossRef]
23. Coelho, C. Riscos de Exposição de Frentes Urbanas para Diferentes Intervenções de Defesa Costeira. Ph.D. Thesis, University of Aveiro, Aveiro, Portugal, 2005. (In Portuguese)
24. Mendes, D.; Oliveira, T.C. Deep-water spectral wave steepness offshore mainland Portugal. *Ocean Eng.* **2021**, *236*, 109548. [CrossRef]
25. Oliveira, T.C.; Neves, M.G.; Fidalgo, R.; Esteves, R. Variability of wave parameters and Hmax/Hs relationship under storm conditions offshore the Portuguese continental coast. *Ocean Eng.* **2018**, *153*, 10–22. [CrossRef]
26. Silva, A.N.; Taborda, R.; Bertin, X.; Dodet, G. Seasonal to Decadal Variability of Longshore Sand Transport at the Northwest Coast of Portugal. *J. Waterw. Port Coast. Ocean Eng.* **2012**, *138*, 464–472. [CrossRef]
27. Coelho, C.; Silva, R.; Veloso-Gomes, F.; Taveira-Pinto, F. Potential effects of climate change on northwest Portuguese coastal zones. *ICES J. Mar. Sci.* **2009**, *66*, 1497–1507. [CrossRef]
28. Ferreira, Ó. Morfodinâmica de Praias Expostas: Aplicação ao Sector Costeiro Aveiro-Cabo Mondego. Ph.D. Thesis, University of Algarve, Aveiro, Portugal, 1998. (In Portuguese)

29. Silva, R.; Baptista, P.; Veloso-Gomes, F.; Coelho, C.; Taveira-Pinto, F. Sediment grain size variation on a coastal stretch facing the North Atlantic (NW Portugal). *J. Coast. Res.* **2009**, *SI56*, 762–766.
30. Rato, D. *Relatório II—Análise Textural de Sedimentos*; Technical Report; SandTrack Project: 2019. Available online: http://sandtrack.web.ua.pt/pt/outputs/reports/02_Relatorio_II_Analise_textural_sedimentos.pdf (accessed on 28 June 2021). (In Portuguese)
31. Baptista, P.; Bernardes, C.; Cunha, T. The validation analysis of the INSHORE system—A precise and efficient coastal survey system. *Environ. Monit. Assess.* **2011**, *179*, 589–604. [CrossRef]
32. Baptista, P.; Cunha, T.; Bernardes, C.; Gama, C.; Ferreira, Ó.; Dias, A. A Precise and Efficient Methodology to Analyse the Shoreline Displacement Rate. *J. Coast. Res.* **2011**, *27*, 223–232. [CrossRef]
33. Toledano, C.; Dalphinet, A.; Lorente, P.; de Alfonso, M.; Ghantous, M.; Aouf, L.; García Sotillo, M. *Qualitity Information Document Atlantic-Iberian Biscay Irish IBI Production Centre IBI ANALYSIS FORECAST WAV 005 005*; Technical Report; Copernicus Marine Environment Monitoring Service: 2020. Available online: https://catalogue.marine.copernicus.eu/documents/QUID/CMEMS-IBI-QUID-005-005.pdf (accessed on 28 June 2021).
34. Pinto, C.; Penacho, N.; Pires, B. Programa de Monitorização da Faixa Costeira de Portugal Continental (COSMO): Da concepção à implementação. In Proceedings of the X CPGZCPEP, Rio de Janeiro, Brazil, 6–10 December 2021; submitted. (In Portuguese)
35. Palalane, J.; Fredriksson, C.; Marinho, B.; Larson, M.; Hanson, H.; Coelho, C. Simulating cross-shore material exchange at decadal scale. Model application. *Coast. Eng.* **2016**, *116*, 26–41. [CrossRef]
36. Carapuço, M.M.; Taborda, R.; Silveira, T.M.; Psuty, N.P.; Andrade, C.; Freitas, M.C. Coastal geoindicators: Towards the establishment of a common framework for sandy coastal environments. *Earth-Sci. Rev.* **2016**, *154*, 183–190. [CrossRef]
37. Hoekstra, P.; Houwman, K.T.; Kroon, A.; Ruessink, B.G.; Roelvink, J.A.; Spanhoff, R. Morphological development of the Terschelling shoreface nourishment in response to hydrodynamic and sediment transport processes. *Coast. Eng. Proc.* **1996**, *1*, 2897–2910. [CrossRef]
38. Bain, R.; McFall, B.; Krafft, D.; Hudson, A. Evaluating Transport Formulations for Application to Nearshore Berms. *J. Waterw. Port Coast. Ocean Eng.* **2021**, *147*, 04021031. [CrossRef]
39. Jóia Santos, C.; Andriolo, U.; Ferreira, J.C. Shoreline Response to a Sandy Nourishment in a Wave-Dominated Coast Using Video Monitoring. *Water* **2020**, *12*, 1632. [CrossRef]
40. Baptista, P.; Coelho, C.; Pereira, C.; Bernardes, C.; Veloso-Gomes, F. Beach morphology and shoreline evolution: Monitoring and modelling medium-term responses (Portuguese NW coast study site). *Coast. Eng.* **2014**, *84*, 23–37. [CrossRef]
41. Lima, M.; Coelho, C.; Veloso-Gomes, F.; Roebeling, P. An integrated physical and cost-benefit approach to assess groins as a coastal erosion mitigation strategy. *Coast. Eng.* **2020**, *156*, 103614. [CrossRef]

Journal of
Marine Science
and Engineering

MDPI

Article

Morphological Analysis of a Nearshore Nourishment along the Atlantic Coast of New Jersey, USA

Sean P. McGill [1,*], Brian D. Harris [1], Brian C. McFall [1], Douglas R. Krafft [1], Rachel L. Bain [1], Nicholas R. Olsen [1], Ian W. Conery [2] and Monica A. Chasten [3]

[1] Coastal and Hydraulics Laboratory, U.S. Army Engineer Research and Development Center, Vicksburg, MS 39180, USA
[2] Coastal and Hydraulics Laboratory-Field Research Facility, U.S. Army Engineer Research and Development Center, Duck, NC 27949, USA
[3] U.S. Army Corps of Engineers, Philadelphia, PA 19107, USA
* Correspondence: sean.p.mcgill@usace.army.mil; Tel.: +1-601-634-2622

Abstract: Nearshore nourishment is a common coastal flood risk management technique that can be constructed beneficially by using dredged sediment from navigation channels. A nearshore nourishment project was completed during the summer of 2021 in Harvey Cedars, NJ, USA, with 67,500 m^3 of dredged sediment from Barnegat Inlet placed along approximately 450 m of beach in a depth of 3–4 m. In situ instruments were installed to monitor hydrodynamic conditions before and after dredged material placement, and nine topographic and bathymetric surveys were conducted to monitor nearshore morphological response to the nourishment. Shoreline location was extracted from satellite imagery using CoastSat software to compare historical trends to the shoreline response after construction. Seven months after construction, 40% of the nearshore nourishment was transported from the initial footprint and the centroid of the nourishment migrated towards shore and alongshore (north). The sheltering capacity of the nearshore berm appears to have captured an additional 58% of the placed volume from the longshore transport system and the beach width onshore of the placement increased by 10.9 m. Measured data, satellite imagery analysis, and rapid predictions all indicate that the nearshore nourishment at Harvey Cedars had a positive impact on the adjacent beach.

Keywords: nearshore nourishment; sediment transport; shoreline extraction; shoreface nourishment

Citation: McGill, S.P.; Harris, B.D.; McFall, B.C.; Krafft, D.R.; Bain, R.L.; Olsen, N.R.; Conery, I.W.; Chasten, M.A. Morphological Analysis of a Nearshore Nourishment along the Atlantic Coast of New Jersey, USA. *J. Mar. Sci. Eng.* **2022**, *10*, 1622. https://doi.org/10.3390/jmse10111622

Academic Editor: Carlos Daniel Borges Coelho

Received: 8 October 2022
Accepted: 29 October 2022
Published: 2 November 2022

1. Introduction

Nearshore nourishments are a common flood risk management technique used extensively around the world [1–6]. In the United States (US), nearshore nourishment projects are a common technique to beneficially use dredged sediment from navigation channels. From 2015 to 2020, the US Army Corps of Engineers (USACE) placed 28 Mm^3 in nearshore littoral systems [7], which allows natural forces to redistribute the material and nourish the broader beach profile [8,9]. The primary objective of these nearshore beneficial use projects is to retain the sediment in the littoral system. Nearshore nourishment can beneficially use dredged sediment that is not suitable for a direct beach placement because the fine material is naturally winnowed from the placement by wave action [10,11]. Additional flood risk management benefits include nourishing the beach profile, breaking waves farther offshore [12], and extending the life of co-located subaerial beach nourishments [13].

Subaerial beach nourishment has been the leading form of coastal protection in the US for the last four decades [14] and commonly uses sediment from offshore sources. Direct subaerial nourishment is more expensive than nearshore nourishment [8], and sheltering the subaerial nourishment from erosive storm conditions using a co-located nearshore nourishment may have significant financial benefit. Consequently, some nearshore nourishments are placed in concentrated footprints as nearshore berms, which may have the capacity to break waves farther offshore and reduce their erosive potential [12,15]. This potential to dissipate wave energy farther offshore has inspired research into co-located nearshore nourishments and subaerial beach nourishments [16].

Prior research suggests that the morphologic change around shallow nearshore nourishments results from two processes: (1) capturing alongshore sediment transport in the lee of the berm as waves break farther offshore, and (2) onshore transport driven by shoaling waves [17]. Both processes manifest as erosion in the placement footprint and accretion on the lee side of the placement. Placement monitoring via topographic and bathymetric surveys without direct sediment transport measurements can make it difficult to distinguish between these two nearshore processes in dynamic coastal environments. In one example where transport direction could be resolved (the nearshore nourishment tracer study in the controlled environment of the large-scale sediment transport facility [18]), the measurements suggested that alongshore transport can dominate the removal of material from the placement footprint in some settings [19]. In the field, nearshore nourishments are generally observed to migrate onshore, with a smaller number of placements remaining stable [20–24].

Although the nearshore placement of dredged sediment is common practice, the critical questions of how often the sediment will be mobilized, where the sediment will go, and how the shoreline will respond are often site-specific and remain poorly understood [22,24]. Nearshore nourishments have been studied with numerical modeling [25–27], physical modeling [28,29], and field measurements [10,14,22,30–32]. Although each method has its strengths and weaknesses, field monitoring is ideal in locations where projects are likely to be repeated because it allows for adaptive management of the project and improved outcomes of subsequent placements [33,34]. Monitoring nearshore nourishment projects commonly involves topographic and bathymetric surveys [35], hydrodynamic measurements [9,36], and shoreline response analysis from aerial imagery [37].

In the summer of 2021, sediment was dredged by the USACE Philadelphia District from the Barnegat Inlet federal navigation channel and used beneficially to construct a nearshore berm on the historically erosive beach at Harvey Cedars, NJ, USA [38,39]. To improve the understanding of the morphological evolution of nearshore nourishments and the adjacent beach response, this nearshore nourishment project was extensively monitored using a combination of in situ sensors and a series of topographic and bathymetric surveys. Satellite observations in conjunction with a trained machine learning algorithm were also utilized to track shore position at a higher temporal resolution and larger spatial extent than offered by traditional methodologies [40]. Knowledge gained from this monitoring will support future projects and advance the practice of nearshore nourishment.

2. Materials and Methods

2.1. Study Site

Harvey Cedars is a borough in Ocean County, NJ, USA, located along the barrier island of Long Beach Island approximately 110 km south of New York City, NY, USA (Figure 1a,b). Harvey Cedars has approximately 3 km of beach along the Atlantic Ocean. Wave Information Studies (WIS) hindcasts between 1976 and 1995 found Harvey Cedars had an average 0.9 m wave height and 6.4 s period from the southeast with most wave periods ranging from 5.0 to 9.0 s [41]. The site has a semi-diurnal tide range of 1.3 m [42]. A maximum wave height of 8.33 m and a 2% wave height of 3.22 m were recorded at the National Oceanic and Atmosphere Administration (NOAA) Station 44091 between 2015 and 2022 offshore of the project site. Harvey Cedars was selected to receive a nearshore nourishment because of an erosional hotspot identified along a stretch of shoreline within the Federal Coastal Storm Risk Management (CSRM) Project, which stretches for 26 km along Long Beach Island, NJ, USA. The Harvey Cedars beachfill portion of CSRM was originally completed in July 2010 [38] using 2.4 Mm^3 of sediment, expanding the beach and burying a groin field (Figure 1c). Since that time, an emergency beachfill in 2013 used 1.2 Mm^3 of sediment to repair damages to the beach and dune system after Hurricane Sandy, and 0.9 Mm^3 of sediment was used for a periodic nourishment in 2018.

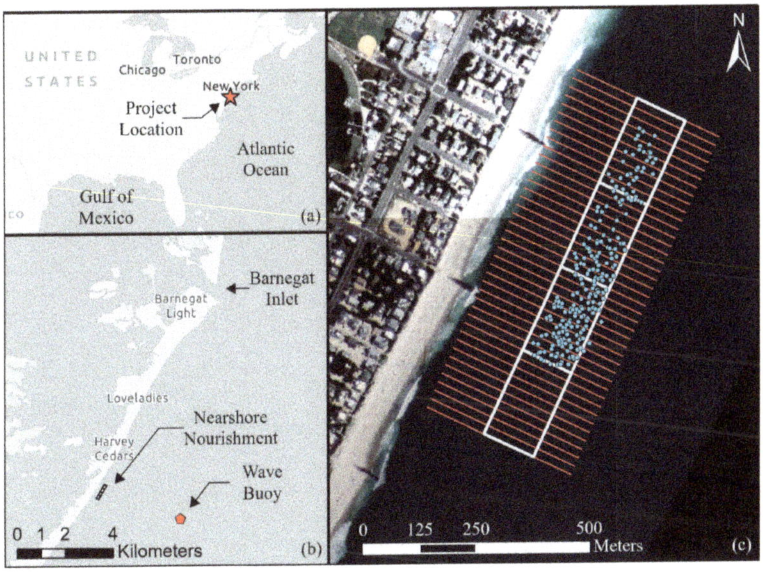

Figure 1. (**a**) Project location, (**b**) Barnegat Inlet and Harvey Cedars, and (**c**) the design extent of the nourishment divided into four 150 m wide sections. Vessel heading lines spaced at 15 m shown in red and blue circles denote drop locations. (**c**) contains Landsat/Copernicus imagery from Google Earth.

In July and August 2021, the USACE Philadelphia District placed 67,500 m^3 of dredged sediment in the nearshore of Harvey Cedars, NJ, USA, approximately 100 m from shore ([38,39]; see Figure 1c). The sediment was dredged from Barnegat Inlet, a federal navigation channel located approximately 10 km north of the placement site. The dredged sediment consisted of sandy material with a median grain size of 0.42 mm. Barnegat Inlet was one of the 10 beneficial use pilot projects that the Water Resource and Development Act (WRDA) Section 1122 program established across the nation and was carried out by USACE Philadelphia District in partnership with the NJ Department of Environmental Protection. The nearshore nourishment was constructed using the USACE's split hull hopper dredge *Murden* to remove sediment from shoals in the channel and transport dredged material south to Harvey Cedars. The *Murden* has a hull capacity of 380 m^3 and could complete a trip, which includes dredge time and travel to and from the site, in approximately 3 h. The design of the nearshore berm consisted of four 150 m wide shore-parallel boxes with their landward side at the −2.75 m NAVD88 contour line (Figure 1c). Vessel heading lines were spaced every 15 m, and the dredge was instructed to "nose in" along the placement line and then split the hopper to place the sediment as shallow as possible given the tide. The average placement depth was 3–4 m. The dredge began construction at the north end of the southernmost box and placed multiple loads per line before progressing north. Approximately halfway through construction, the dredge reduced the number of drops per line to develop a longer berm [39], resulting in the northern half of the berm being less prominent when compared to the southern end.

2.2. In Situ Instruments

A variety of hydrodynamic data were collected at 3 locations in the area around the nearshore nourishment with 3 distinct water depths to measure the forcing of the morphological response to nearshore nourishment. From 23 June 2021 to 10 May 2022, an RBR Virtuoso pressure sensor was deployed 630 m north of the nearshore nourishment in 1.5 m of water. Another pressure sensor was deployed landward of the nearshore nourishment at the same time in a similar water depth, but the data could not be recovered. Pressure measurements from the surviving RBR sensor were converted to bulk wave statistics to quantify nearshore wave characteristics in the immediate vicinity of the nearshore nourishment. The instrument was installed in close proximity to buried groins that periodically became exposed during the project monitoring and may have influenced the nearshore wave characteristics. The measurements from the RBR sensor are described in greater detail in the Results section. From 7 December 2021 to 15 April 2022, an ADCP measured water level and vertical velocity profiles in 8 m of water on the seaward edge of the nearshore nourishment at 4 Hz for 34 min each hour. These data were collected for future numerical model validation at the site. A Sofar Spotter buoy collected hourly wave spectra approximately 3 km offshore of the site in 15.2 m of water (Figure 1b). The buoy collected wave data between 1 July 2021 and 17 December 2021, after which it became unmoored. After 17 December 2021, wave data from NOAA Station 44091 were substituted to understand wave conditions offshore of the nourishment and calculate sediment transport rates. Information about sensor deployments is summarized in Table 1.

Table 1. Sensor deployment information.

Sensor	Location	Approximate Depth	Timing
RBR Virtuoso (Pressure)	630 m North of the nourishment	1.5 m	23 June 2021 to 10 May 2022
Nortek Signature 1000 (ADCP)	Seaward of the nourishment	8.0 m	7 December 2021 to 15 April 2022
Sofar Spotter buoy (Spectral wave info)	3 km offshore of nourishment	15.2 m	1 July 2021 to 17 December 2021

2.3. Site Evaluation for Nearshore Nourishment

Several commonly applied techniques have been previously developed to rapidly assess potential nearshore nourishment sites [24]. These include approximating the depth of closure [43–46], frequency sediment placed in the nearshore will be mobilized [47], the cross-shore transport direction [48], and comparing the potential site to historical projects [20]. All of these techniques can be applied in the interactive Sediment Mobility Tool web application [49] and the methods are detailed here.

The depth of closure is the depth beyond which repeated beach profiles compiled over time converge [50] or the depth along a beach profile where there is not significant sediment transport from nearshore processes [51]. Hallermeier [43,44] further defined an inner depth of closure and outer depth of closure. The outer depth of closure (h_i) is the depth at which waves cause minimal sediment transport and is calculated as

$$h_i = (\overline{H_s} - \sigma_s)\overline{T_s}\left(\frac{g}{5000\, d_{50}}\right)^{0.5} \tag{1}$$

where $\overline{H_s}$ is the mean annual significant wave height, σ_s is the significant wave height standard deviation, $\overline{T_s}$ is the average period associated with $\overline{H_s}$, g is the gravitational constant, and d_{50} is the median sediment grain size. The inner depth of closure (h_l) is the seaward boundary for the littoral zone where the bed experiences increased significant stresses and sediment transport from wave near breaking and nearshore circulation. The inner depth of closure is the appropriate limit for the shoreline response and beach nourishment projects [52]. The equation to calculate the inner depth of closure (h_l) is given by Hallermeier [43] as

$$h_l = 2.28H_e - 68.5\left(\frac{H_e^2}{gT_e^2}\right) \tag{2}$$

where H_e is the effective wave height, or the wave conditions exceeded only 12 h per year (largest 0.137% waves in a year), and T_e is the wave period associated with H_e. Birkemeier [45] evaluated Equation (2) using high fidelity bathymetric profiles from the USACE

Field Research Facility in Duck, NC, USA from June 1981 to December 1982, and found the more appropriate coefficients for the equation for this site to be

$$h_l = 1.75H_e - 57.9\left(\frac{H_e^2}{gT_e^2}\right) \tag{3}$$

Hands and Allison [20] compared the Hallermeier inner (Equation (2)) and outer (Equation (1)) depth of closure values at 11 historical nearshore berm sites to analyze the relationship between the equations and whether the placements were stable or active. Stable berms maintained most of the placed volume for years, whereas the active berms dispersed within months. All the berms built shallower than the inner depth of closure were active, and the one nearshore berm built deeper than the outer depth of closure was stable. This analysis was expanded by McFall et al. [24] to include a total of 20 historical projects, and similar trends emerged.

Another technique to estimate how active a nearshore nourishment will be at a site is to calculate how often placed sediment is expected to be mobilized. This is calculated by the Sediment Mobility Tool in two different ways [47]. One method uses bed shear stress and linear wave theory using the technique described by Soulsby [53]. The critical bed shear stress to initiate sediment mobility is calculated as

$$\tau_{cr} = \theta_{cr} g(\rho_s - \rho) d_{50} \tag{4}$$

where θ_{cr} is the Shields parameter, ρ_s is the sediment density, and ρ is the water density. The maximum bed shear stress is calculated using the bed shear stress induced by the waves and current as

$$\tau_m = \tau_c\left[1 + 1.2\left(\frac{\tau_w}{\tau_c + \tau_w}\right)^{3.2}\right] \tag{5}$$

and

$$\tau_{max} = \left[(\tau_m + \tau_w cos\phi)^2 + (\tau_w sin\phi)^2\right)]^{1/2} \tag{6}$$

where τ_m is the mean bed shear stress, τ_c is the current induced bed shear stress, τ_w is the wave induced bed shear stress, and ϕ is the angle between the wave and current direction. To quantify the average normalized difference of the maximum shear stress and the critical threshold, the mean mobility score (M) is calculated as

$$M = \overline{\left(\frac{\tau_{max} - \tau_{cr}}{\tau_{cr}}\right)} \tag{7}$$

The second method to calculate the sediment mobility uses the near-bottom velocity and nonlinear stream function wave theory using a procedure described by Ahrens and Hands [21]. The critical near-bottom velocity (u_{cr}) for sediment with grain size diameter less than 2 mm is calculated as

$$u_{cr} = \sqrt{8g\gamma d_{50}} \tag{8}$$

where $\gamma = (\rho_s - \rho)/\rho$. The maximum wave induced near-bottom velocity for the wave crest and wave trough are calculated as

$$u_{maxcrest} = \left(\frac{H}{T}\right)\left(\frac{h}{L_o}\right)^{-0.579} \exp\left[0.289 - 0.491\left(\frac{H}{h}\right) - 2.97\left(\frac{h}{L_o}\right)\right] \tag{9}$$

and

$$u_{maxtrough} = -\left(\frac{H}{T}\right) \exp\left[1.966 - 6.70\left(\frac{h}{L_o}\right) - 1.73\left(\frac{H}{h}\right) + 5.58\left(\frac{H}{L_o}\right)\right] \tag{10}$$

where H is the wave height at the placement site, T is the wave period, h is the water depth, and L_0 is the offshore wavelength given by $L_0 = (g\ T^2)/2\ \pi$. The maximum near-bottom velocity was taken as $u_{\max} = \max(|u_{maxcrest}|, |u_{maxtrough}|)$. The mean mobility score for this method is calculated as

$$M_u = \overline{\left(\frac{u_{\max} - u_{cr}}{u_{cr}}\right)} \tag{11}$$

For the site evaluation at this project, the Sediment Mobility Tool was applied using wave characteristics from the offshore WIS Station 63137 transformed to the nearshore depth of 4 m using Snell's Law and conservation of energy flux. The sediment mobility analysis was applied to hourly hindcasts for 10 years (1 January 1990–31 December 1999).

2.4. Surveys

Nine topographic and bathymetric surveys were performed throughout the duration of this project. These surveys consisted of 26 lines of single-beam and RTK-GPS transects at 75 m spacing, combined with periodic multi-beam surveys in the immediate area surrounding the nearshore placement. Table 2 shows a summary of completed surveys and methods for the project. The location of the MHHW contour at 0.61 m NAVD88 was extracted at each of the transects for each survey. The survey data were then combined to create Digital Elevation Models (DEMs) for site analysis. A change map based on pre- and post-placement elevations was used to create a mask for the initial placement's extent. Additional change maps between successive survey dates were created, and the berm mask was used to extract the placement area for sediment centroid calculations and volume change analysis.

Table 2. Survey information.

Survey Type	Date	Period	Notes
Single-Beam	13 May 2021	Pre-Nourishment	-
Multi-Beam	22 July 2021	During Construction	-
Multi-Beam	28 July 2021	During Construction	-
Multi-Beam	9 August 2021	During Construction	-
Single- and Multi-Beam	25 August 2021	Post Placement	6 days post construction
Single- and Multi-Beam	13 October 2021	Post Placement	55 days post construction
Single- and Multi-Beam	9 December 2021	Post Placement	112 days post construction
Single-Beam	22 March 2022	Post Placement	215 days post construction
Single-Beam	17 June 2022	Post Placement	302 days post construction

2.5. Deflation Code

A method for generating order-of-magnitude estimates of sediment loss from a nearshore placement site was previously developed by Bain et al. [54]. Time series measurements of wave height, period, and direction are used to estimate longshore and cross-shore transport rates. The longshore transport (Q_y) is calculated using the equation by Shaeri et al. [55] as

$$\frac{Q_y T_p}{H_b^3} = \frac{3 \times 10^{-4}}{(1-a)} \frac{\rho_w}{\rho_s - \rho_w} \left(\frac{H_b}{L_0}\right)^{-0.9} \left(\frac{H_b}{d_{50}}\right)^{0.2} sin^{0.5}(2\theta_b) \tag{12}$$

where T_p is the peak wave period, H_b is the breaking wave height, and a is the porosity that is set to 0.4. The coefficient in Equation (12) was calibrated using 47 longshore transport datasets from sites around the world [55]. The cross-shore transport (Q_x) is calculated per unit width using a modified method from Dronkers [56] as

$$Q_x = \alpha[-\lambda m\langle|u_w|^3\rangle + \langle|u_w|^2 u_w\rangle(1-k)]cos\theta_{crest} \tag{13}$$

where α and λ are empirical coefficients, m is the bed slope, u_w is the near-bottom horizontal velocity, k is the critical velocity scaling term and is given as $k = \min[u_{cr}/u_w^{max}, 1]$, and θ_{crest} is the wave angle over the crest of the nearshore berm. Angular brackets $\langle\,\rangle$ indicate averaging over a wave period. The negative term in Equation (13) represents the gravity-driven offshore transport, whereas the positive term represents the wave-driven onshore transport. Hudson et al. [57] used a series of placements at the Columbia River mouth to optimize the empirical parameters in Equation (13) as $\alpha = 3 \times 10^{-5}$ and $\lambda = 1.7$, which are retained as constants in the present study. Superimposing the longshore and cross-shore sediment transport rates predicted by Equations (12) and (13) and integrating over a specified time period yields an estimate of the total volume of sediment removed from the original placement footprint. The validation in Bain et al. [54] achieved acceptable order-of-magnitude deflation rate predictions at 11 historical nearshore placements along the Atlantic, Pacific, and Gulf Coasts of the United States with the same empirical parameters as Hudson et al. [57] and Shaeri et al. [55].

To test this methodology for the Harvey Cedars placement, the sediment transport algorithm was forced using Spotter Buoy wave measurements from 8 July 2021 until 9 December 2021. Wave data from NOAA Station 44091 (32 km northeast of the placement site at 25.6 m depth) were used to force the model from 9 December 2021 until 22 March 2022 due to the unmooring of the Spotter Buoy in mid-December. Conditions were not modeled following the March survey due to missing data from NOAA Station 44091 beginning in April until June. For both sources of wave data, the offshore significant wave height was transformed to height across the nearshore profile based on conservation of energy flux (e.g., Komar [50]), and the offshore wave direction was transformed to direction at breaking using Snell's Law. Following the approach of Bain et al. [54], the placement's geometric parameters were treated as time-invariant and were based on the shape and position of the sediment mound during the first post-placement survey on 25 August 2021, as summarized in Table 3.

Table 3. Site parameters used to generate order-of-magnitude predictions of the sediment volume loss from the Harvey Cedars placement site.

Parameter	Value (Treated as Time-Invariant)
Shore angle	210°
Landward boundary of initial placement footprint	168 m
Cross-shore distance to initial placement crest	198 m
Seaward boundary of initial placement footprint	290 m
Water depth at landward placement boundary	3.0 m
Water depth at initial placement crest	2.5 m
Water depth at seaward placement boundary	7.9 m
Shore-parallel length of placement	427 m
Representative beach slope	0.03
d_{50} of placed sediment	0.42 mm
Water density	1025 kg/m^3
Sediment density	2650 kg/m^3
Sediment porosity	0.4

2.6. Satellite Analysis

The open-source CoastSat tool [40] was used to track shoreline position in the vicinity of the placement from January 2001 to May 2022 using satellite imagery collected from the Landsat 5 and Landsat 8 missions. Landsat 5 collected imagery from 1984 to 2013, whereas Landsat 8 began in 2013 and is still ongoing. Both missions have revisit periods of

16 days, and 30 m pixel resolution across Red, Green, Blue, Near Infrared, and Short-wave Infrared 1 bands. This software toolkit uses a trained multi-layer perceptron algorithm to classify pixels in multispectral imagery as land or water and calculates the normalized difference water index (NDWI) from observations downloaded from Google Earth Engine's public repository [40]. Otsu's thresholding algorithm is applied to the pairs of categorized and NDWI rasters [58]. These thresholds attempt to depict the land-water interface at the time of observation. Each pansharpened image, overlaid with corresponding shoreline detection, has been checked to ensure that only high-quality detections are used in analysis. The resulting collection of quality-checked shorelines can be used to track the shoreline position through time. Satellite-derived shorelines were filtered using a 45-day moving average with a 15-day step to capture the rapid shoreline response to the nourishment.

3. Results

3.1. Wave Analysis

Wave conditions during the monitoring period are shown in Figure 2. At the Spotter buoy (3 km offshore of the placement site in 15.2 m of water), the mean and median significant wave heights were $H_s = 0.91$ m and $H_s = 0.82$ m, respectively, with a maximum of $H_s = 3.12$ m (Figure 2a). The mean peak period was $T_p = 8.3$ s, and the median was $T_p = 7.9$ s. The most frequent offshore wave direction was between 90° and 120° relative to north (between 0° and 30° relative to shore-normal); however, waves between 120° and 180° relative to north (between −60° and 0° relative to shore-normal) were also common (Figure 2b). It should be noted that the Spotter buoy became unmoored after 17 December 2021, so the wave conditions in Figure 2a,b do not include data from winter storm waves.

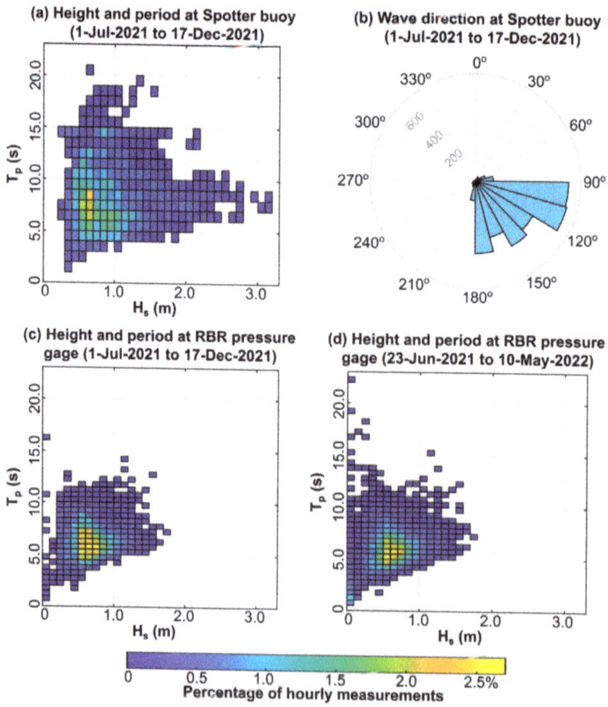

Figure 2. Distribution of offshore and nearshore wave conditions. (**a**) Wave height and period at the Spotter buoy, 3 km offshore of the placement site. (**b**) Wave direction at the Spotter buoy. The concentric circles indicate the number of hourly events, and all angles are relative to north. (**c**) Wave height and period at the RBR pressure gauge for the same time period shown in subplot (**a**). (**d**) Wave height and period at the RBR pressure gauge for the entire monitoring duration.

Nearshore wave measurements from the RBR pressure gauge (630 m north of the nourishment, in approximately 1.5 m of water) indicate that smaller-amplitude, higher-frequency waves are most common near the placement site. For the same time period recorded by the Spotter buoy, the mean, median, and maximum nearshore significant wave heights were H_s = 0.69 m, H_s = 0.65 m, and H_s = 1.73 m, respectively (Figure 2c). The mean peak period was T_p = 6.4 s with a median of T_p = 6.2 s. Over the full monitoring duration (23 June 2021 to 10 May 2022; Figure 2d), these values change to a mean H_s = 0.60 m, median H_s = 0.59 m, mean T_p = 6.1 s, and median T_p = 6.0 s.

3.2. Site Evaluation for Nearshore Nourishment

The Hallermeier inner and outer depths of closure were calculated to be 8.8 m and 14.0 m, respectively. The Birkemeier inner depth of closure was calculated as 6.7 m. The project placement depth is plotted in relation to the Hallermeier depth of closure values in Figure 3 for comparison with 20 historical nearshore nourishment projects. The placement depth was shallower than both the inner and outer depths of closure, indicating that this project is similar to highly active historical nearshore nourishment projects.

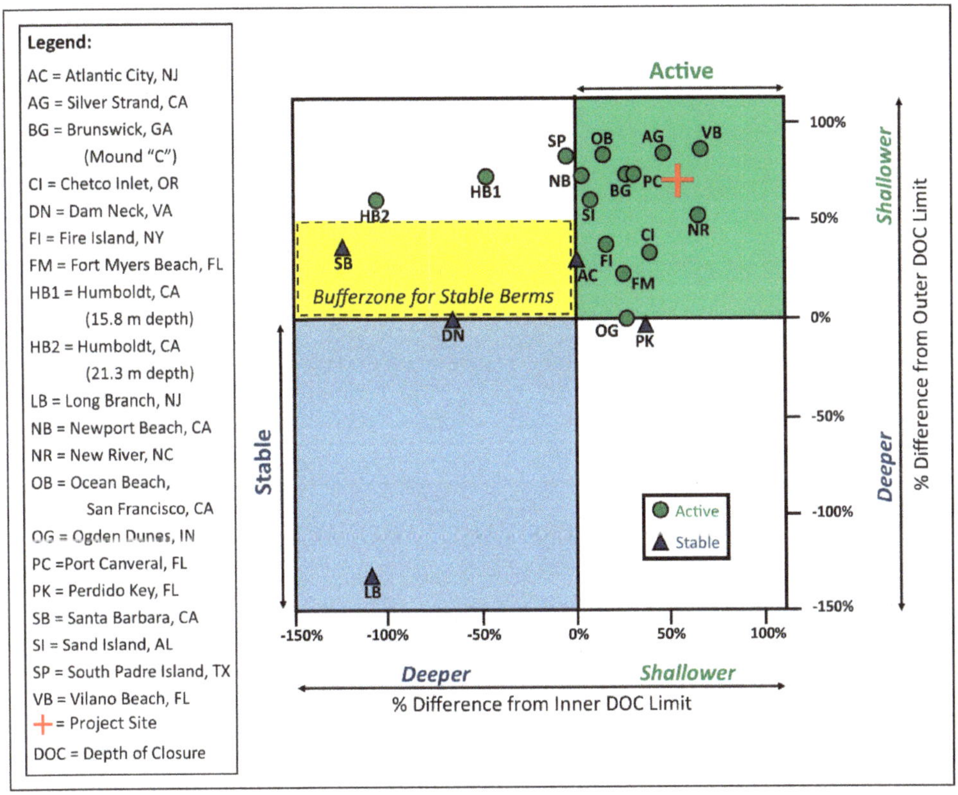

Figure 3. Relationship of the nearshore nourishment site to the Hallermeier depth of closure equations and compared to historical projects (adapted from McFall et al. [24] and Hands and Allison [20]).

The frequency of sediment mobility was also calculated. The median grain size of 0.42 mm is estimated to be mobilized by 91% of the waves using linear wave theory and 95% of the waves using stream function theory, as shown in the histograms in Figure 4. The mobility scores greater than 2 indicate a very active site. Mobility scores less than 1 tend to be stable projects [23]. These rapid site evaluation techniques indicate the project site will be very active, nourishing the beach profile. The placed sediment is predicted to be mobilized frequently because the placement was constructed at a relative depth and wave climate similar to historically active projects.

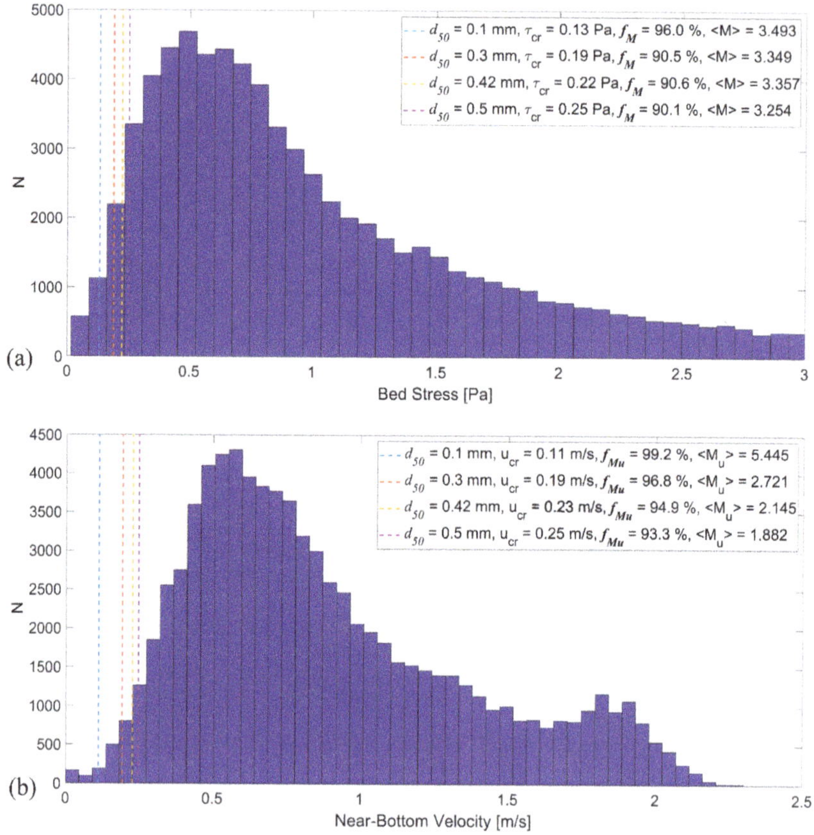

Figure 4. Histograms of the (**a**) bed shear stress and (**b**) near-bottom velocity calculated for the nearshore placement. The vertical dashed lines are the thresholds to initiate sediment mobility for the respective grain sizes.

3.3. Surveys

A total of 67,500 m^3 of sediment was placed in the nearshore during the construction of the nearshore berm, and 26,700 m^3 of sediment was transported away from the placement between the completion of construction in August 2021 and March 2022. Figure 5 shows the change in nearshore morphology during and after the construction of the berm. The centroid of the berm migrated toward the north throughout the surveys, with the exception of the June 2022 survey due to two Nor'easters that occurred in April and May. As seen in Figure 5b, sediment accretion occurred at the shoreline directly landward of the berm, as well as in the nearshore surrounding the initial placement. Minor avalanching of sediment on the seaward slope of the berm also occurred. Analysis of the three cross-shore transects drawn in Figure 5a shows the change in beach profiles throughout the surveys (Figure 6). Beach profile change was the most variable along the site transect post construction, with nearshore elevations increasing from the placement (Figure 6b). Between post-construction and the 22 March 2022 survey, the beach width above the MHHW line increased by 10.9 m before receding in the June 2022 survey.

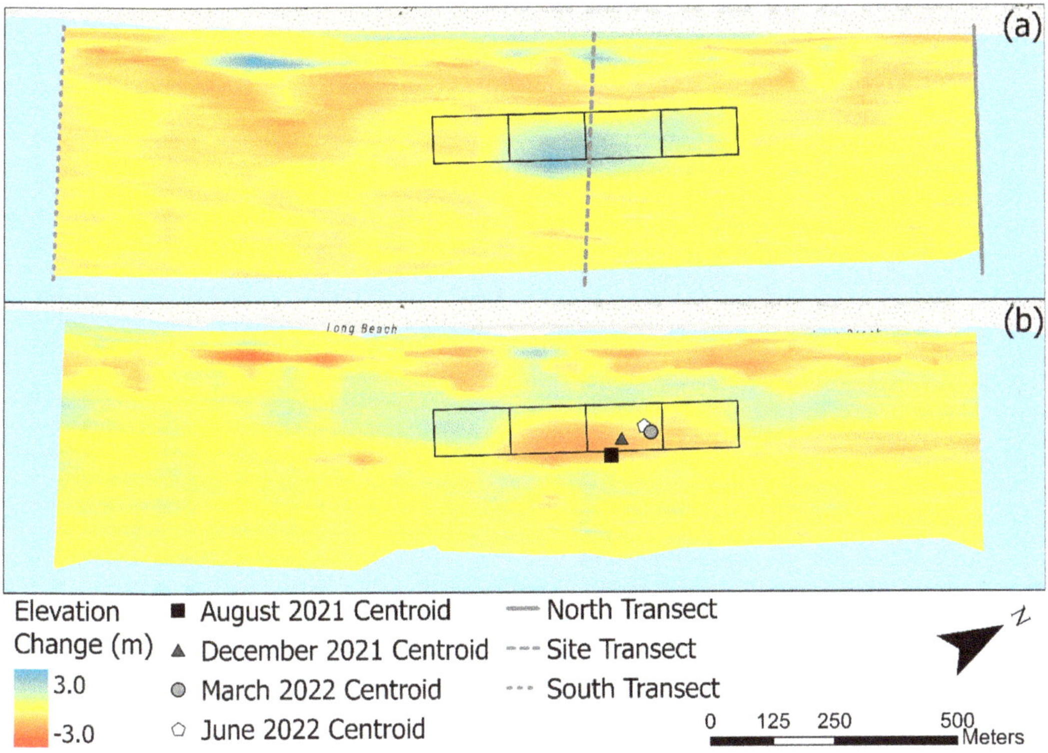

Figure 5. Change maps between pre- and post construction (**a**) and post construction and June 2022 survey (**b**). Within the placement area, the centroid of sediment was found and labeled for the surveys in August 2021, December 2021, March 2022, and June 2022.

3.4. Deflation Results

A time series of the predicted volumetric sediment transport rate at the placement site appears in Figure 7. Sediment loss within the placement footprint due to longshore transport remains near zero except during large wave events, when the breaking position is farther offshore (Figure 7a). Cross-shore removal of sediment from the placement site is predicted to occur more frequently (Figure 7b) but with a smaller maximum magnitude than calculated for the large longshore transport events that occurred in January and February 2022. Due to the shape of a wave under stream function wave theory (Equations (9) and (10)), the net transport over a single wave period tends to be onshore in shallow water. Instances of offshore transport predictions were considered minimal as they were 5 orders of magnitude smaller than onshore transport and were rounded to zero for our analysis. Time-integrating the volume loss rates in Figure 7 predicts that approximately 34,000 m^3 of sediment was removed from the placement footprint via longshore transport between 25 August 2021 and 22 March 2022, whereas approximately 31,400 m^3 was removed via cross-shore transport over the same time period for a total predicted volume loss of 65,400 m^3 (Table 4), i.e., almost the entire placed volume of sediment. The actual volume loss from the original footprint over the same timespan was 26,700 m^3, or a predictive error of 145% of the bulk volume removal rate.

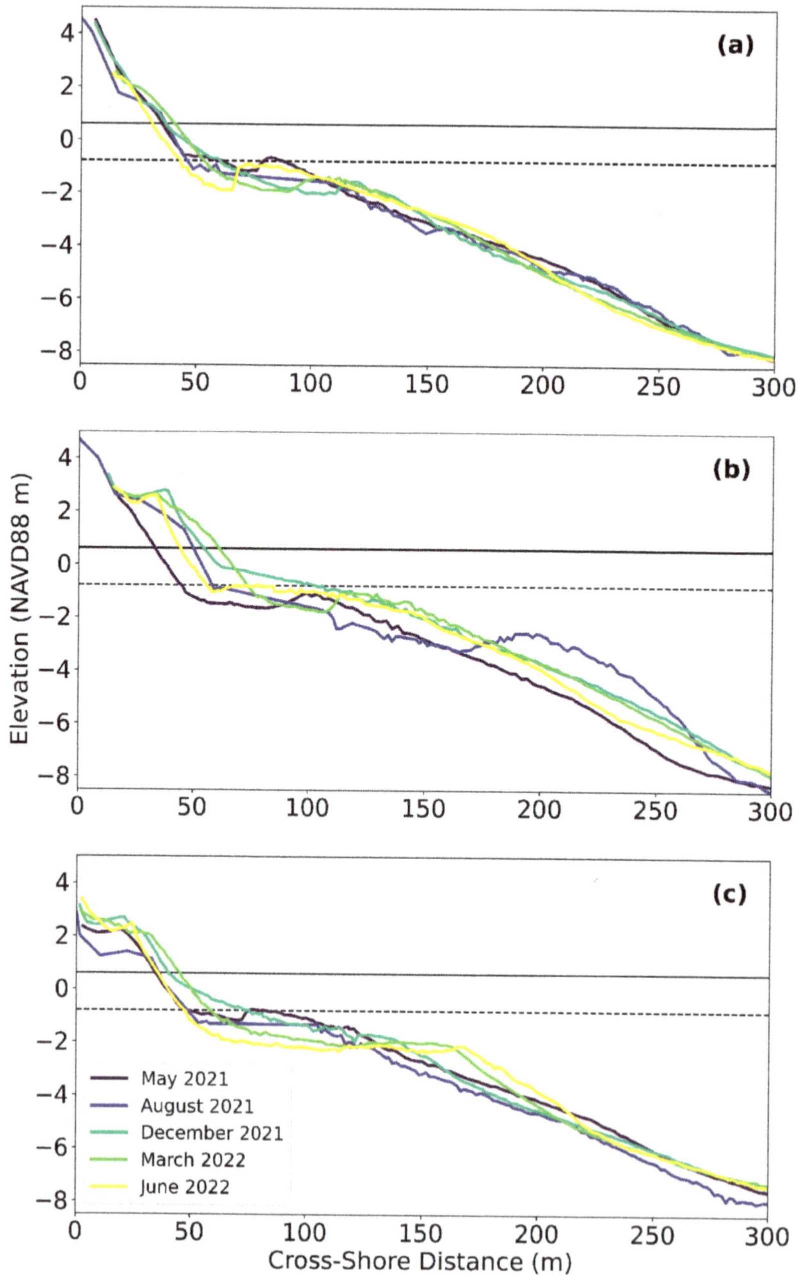

Figure 6. Beach profiles along the most northern transect (**a**), a transect within the placement area (**b**), and a transect south of the placement (**c**). MHHW (black solid line) is 0.61 m, and MLLW (black dashed line) is −0.79 m. The nearshore berm is seen in the August 2021 survey in (**b**) between 165 and 285 m.

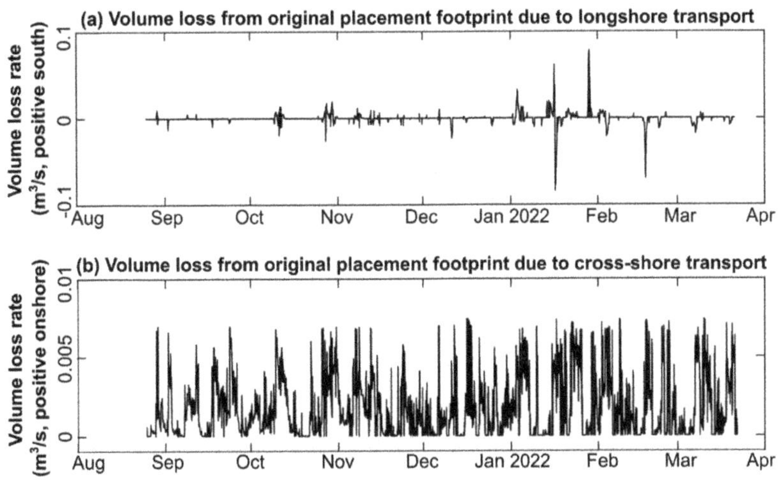

Figure 7. Time series of predicted volumetric sediment transport away from the original placement footprint based on the algorithm of Bain et al. [54]. In subplot (**a**), transport is directed alongshore towards the south when positive and alongshore towards the north when negative. In subplot (**b**), transport is directed onshore when positive and offshore when negative. Note that the cross-shore transport rate is for the entire shore-parallel length of the placement footprint (i.e., not unit width).

Table 4. Predicted volume of sediment lost from the original placement footprint between consecutive bathymetric surveys and the full monitoring duration. Values are rounded to the nearest hundred.

t_{start}	t_{end}	Predicted Volume Loss from Original Placement Footprint between t_{start} and t_{end} (m^3) *				
		Southward	Northward	Onshore	Offshore	Sum of All Directions
25 August 2021	13 October 2021	1800	1200	7100	0	10,100
13 October 2021	9 December 2021	2900	1800	7500	0	12,200
9 December 2021	22 March 2022	12,000	14,300	16,800	0	43,100
25 August 2021	22 March 2022	16,700	17,300	31,400	0	65,400

* From 25 August 2021 to 9 December 2021, the sediment transport model is forced using wave data from the Spotter Buoy. After 9 December 2021, the model is forced using wave data from NOAA Station 44091. Regardless of the data source, wave direction and height were transformed to nearshore values using Snell's Law and conservation of energy flux.

Bain et al. [54] validated their results for total (longshore + cross-shore) sediment removal from 11 historical placement sites but had insufficient survey data to evaluate whether the *direction* of predicted transport was accurate. To further evaluate the algorithm's ability to accurately predict relative proportions of longshore versus cross-shore sediment removal from a placement site, a grid surrounding the placement area was created and used to divide the bathymetric survey data (Figure 8). The change in sediment volume within each quadrant was then calculated for each pair of consecutive surveys. Between placement and March 2022, 16,000 m^3 of sediment accreted directly onshore, 9900 m^3 of sediment accreted directly offshore, 3200 m^3 accreted north of the placement and 36,900 m^3 accreted to the south of the placement (Figure 8). The surrounding area gained 145% more sediment than the placement area lost, indicating sediment from the longshore transport was retained in the project area. The previously noted predicted deflation of the nearshore nourishment (65,400 m^3) is coincidentally similar to the total volume gained in the areas surrounding the nourishment area (65,700 m^3), but the volume gain accounts for processes not accounted for in the predicted deflation.

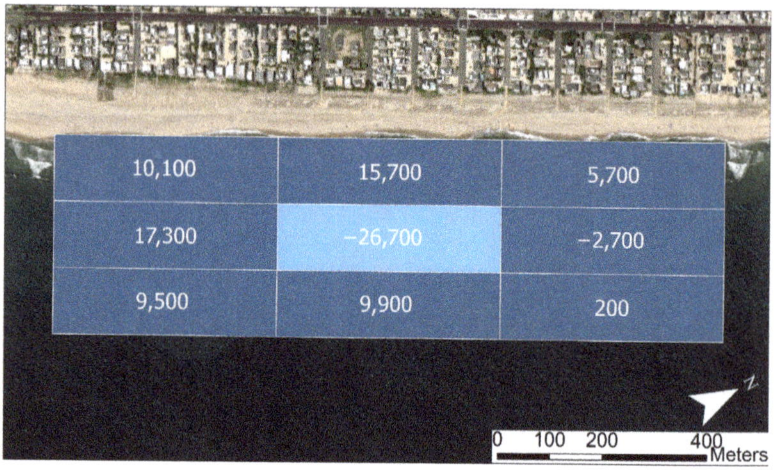

Figure 8. Sediment volume change grid between 25 August 2021 and 22 March 2022 surveys in m^3. The initial placement area is in the center of the grid, indicated by lighter shading.

3.5. Shoreline Change

Shoreline change in the study area was investigated with a combination of satellite imagery and surveys. Satellite images were analyzed from 2000 onwards to investigate long-term shoreline position, showing two major trends separated by the first of four nourishment projects. From 2000 until 2010, beach width remained stable with some seasonal variability. Starting with the first major recent nourishment in 2010, beach width increased up to 120 m at the site transect (Figure 9b). After project construction, subaerial equilibration was accelerated by Hurricanes Irene (2011) and Sandy (2012), and the satellite data shows the beach returned to pre-project width after approximately 36 months following a shoreline retreat rate of −39.3 m/year (Figure 9b). Hurricane Sandy triggered an emergency nourishment which returned the beach to roughly the 2010 post-nourishment width. Following the most recent beachfill in 2018, the satellite-derived shoreline appears to be remaining stable south of the site, but retreated up to 50 m at the site and 62 m north before the 2021 nearshore nourishment (Figure 9b). Following the 2021 nearshore nourishment at Harvey Cedars, surveys between October 2021 and March 2022 show the nearby shoreline generally advanced, with erosion in the southern half of the placement footprint (Figure 9a). Elevation data indicates that the shoreline retreated over much of the surveyed area between March and June 2022, generally eroding past the recent accretion (Figure 9a). The universal erosion shown in the June shoreline was due to a Nor'easter impacting the study site from 7 May 2022 to 12 May 2022.

Shore position was derived from satellite imagery at 49 transects at and near the placement area from the year preceding the placement to present to investigate alongshore spatial patterns more thoroughly (Figure 10). These changes are unfiltered and relative to 3 October 2020 shore position. The alongshore area in the immediate vicinity of the placement falls between the two black lines overlaying the image. All change values were subtracted by the average values of all transects to isolate local variation from trends seen across the region. CoastSat analysis generally captures the shoreline advance that the survey data indicated across much of the placement area following project construction, with more accretion near the center of the placement area. Comparison of the shoreline from surveys to CoastSat-derived shoreline allows for error quantification at specific transects. The shoreline farther north was generally erosive in CoastSat analyses over the same time span, typically underestimating the shoreline change with an average error of 5.10 m. Satellite analysis south of the placement area indicates a mix of shoreline accretion and erosion in this timespan; that is, the shoreline is less accretionary relative to inside the placement area. This was the most accurate of the three transects, with a 3.69 m underestimation error skewed by the October 2021 and June 2022 surveys. Alongshore patterns of satellite-derived shoreline change appear to indicate more accretion landward of the nearshore nourishment than in the surrounding area, although this section also has the greatest error, overestimating shoreline change by an average of 9.96 m when compared to elevation surveys.

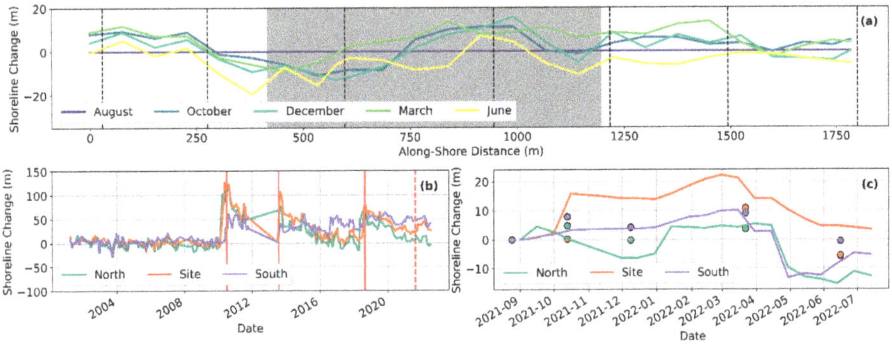

Figure 9. Change in mean shoreline extracted from the surveys (**a**), extracted shoreline from satellite imagery over the past 20 years (**b**), and comparing satellite extracted shoreline and survey extracted shoreline after construction (**c**). For the elevation surveys, shoreline change is relative to the post-construction August survey, groin locations are the dashed black lines, and the placement area is highlighted in grey. Previous subaerial beach nourishment projects are shown with solid red lines, with a dashed red line representing the 2021 nearshore nourishment (**b**), and DEM extracted shorelines are depicted as circles (**c**).

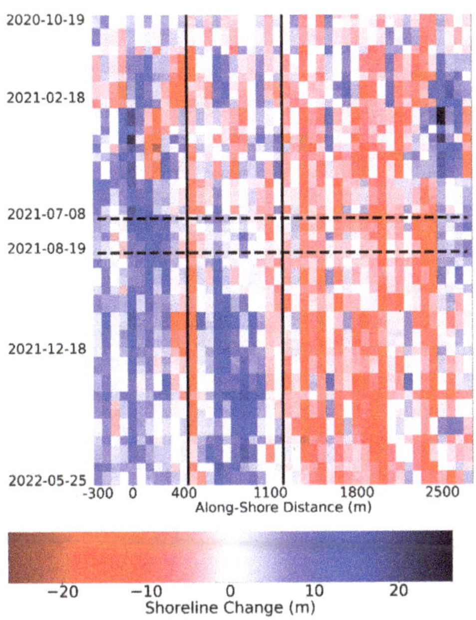

Figure 10. Change in raw shoreline extracted from the satellite imagery relative to the 3 October 2020 shoreline and normalized using the mean of all transects to remove tidal influences. The vertical solid lines show the along shore position of the nearshore nourishment, and the horizontal dashed lines frame the time of construction.

4. Discussion

The combination of collected data and analyses performed for the nearshore nourishment at Harvey Cedars can be compared to further describe the placement evolution and the effectiveness of predictive capabilities. Elevation survey data indicates that the nearshore berm deflated by 40% between 25 August 2021 and 22 March 2022. During this timespan, the nearshore berm appears to merge with the local sandbar system. Shoreline position from elevation surveys shows accretion along the northern half of the placement until the June 2022 survey. Shoreline retreat in the June 2022 survey could be related to the impacts of large Nor'easters in April and May. In addition to shoreline advance, a large volume of sediment had accreted south of and onshore of the initial placement footprint in the March 2022 survey. Centroid tracking at the nearshore berm indicates migration to the north in all but the June 2022 survey, implying that the accretion south of the placement was on the updrift side. Updrift and onshore accretion and downdrift erosion would generally match the Van Duin et al. [17] predicted sediment transport pattern of longshore trapping at nearshore berms. Although the nourishment footprint deflated 26,700 m^3, the survey area surrounding the nourishment gained 65,700 m^3, or an additional 145% more than was deflated, which is similar to the observed morphological response at Terschilling [59] at a much smaller scale (67,500 m^3 of placed sediment vs. 2 Mm^3).

Shoreline advance is also visible in CoastSat analysis of satellite imagery, demonstrating the satellite extracted shoreline approach is capable of monitoring the evolution of coastal management practices such as nearshore nourishment. The nearshore berm at Harvey Cedars is the first to be investigated with CoastSat, enabling more frequent measurements at a higher alongshore resolution than is feasible with traditional survey methods. Satellite-derived shorelines show the most shoreline advance in the lee of the nearshore berm, which is corroborated by the large volumes of accretion onshore of the initial placement in the elevation surveys. The general agreement between elevation surveys and CoastSat results highlight the value of this new, low-cost measurement technique. The errors associated with the CoastSat extracted shoreline are due to temporal and spatial differences in wave setup and runup across the study site and dataset. When comparing to the survey extracted shorelines, this error may be influenced by survey and interpolation errors, although for this analysis we considered these errors to be minimal. With these considerations, the shoreline errors ranged between 3.69 to 9.96 m, consistent with previous applications [60]. Additionally, CoastSat results over the past 20 years show that the area is generally erosive between beachfill projects, providing relevant context for the shoreline gain observed following the nearshore nourishment. The long satellite record could be used during future feasibility studies to quantify background, natural shoreline variability in the system.

Observations suggest that the nearshore nourishment at Harvey Cedars had a positive impact on the adjacent beach, addressing a number of the factors that motivated beneficial use of dredged sediment at this location. This project provides the opportunity to assess the rapid modeling techniques that are used for nearshore nourishment siting. Volume change and centroid calculations from elevation surveys corroborate the Sediment Mobility Tool prediction that the placement would be very active. Nearshore berm deflation predictions following Bain et al. [54] are intended to provide order of magnitude estimates with a target of less than 200% error. This target is met for bulk removal rate predictions, with an error of 145%. Additionally, the fraction of predicted transport following Bain et al. [54] that is directed onshore (48%) is very similar to the measured ratio of onshore accretion to nearshore berm deflation (59%).

A number of factors could contribute to the relatively large deflation over-predictions. The methodology of Bain et al. [54] does not modify the nearshore berm shape as it deflates or reduce the volume of sediment available for transport. Keeping the same crest elevation causes an increased deflation bias, and considering the placement as an infinite sediment source does not prevent the deflation prediction to exceed the measured placement volume. The prediction of approximately equal alongshore sediment transport to the north and south is not supported by measured elevation data. This unrealistic prediction could be related to the simple shore-parallel contour assumption omitting bathymetry impacts that were important to wave transformation in this case. While these simplifications appear to impact prediction accuracy, bulk deflation rates have less than 200% error, and the ratio between predicted onshore and alongshore transport may be reasonable, which emphasizes the value of this rapid technique.

5. Conclusions

In the summer of 2021, the USACE Philadelphia District beneficially used material dredged from Barnegat Inlet in a 67,500 m^3 nearshore nourishment at an erosional hotspot in Harvey Cedars, NJ, USA. This nearshore nourishment was extensively monitored to describe the morphology change of the placed feature and adjacent beach. Nine topographic and bathymetric surveys were conducted, covering pre-construction and the first 302 days following construction. Wave information was measured at 1.5 m, 8 m, and 15.2 m depths. These measurements indicate that the placement was active, and sediment was transported from the placement footprint. Elevation surveys show shoreline advance over much of the measured area, centroid movement onshore and generally to the north, and volume gain onshore and south of the initial placement footprint.

Between 25 August 2021 and 22 March 2022, elevation surveys indicate that the placement eroded by 40% of the original volume. The net sediment gain in the surveyed area is 39,000 m^3, which is the deflation volume (26,700 m^3) from the nourishment footprint subtracted from the volume gain in the surrounding area (65,700 m^3). This indicates the sheltering capacity of the nourishment for 7 months captured 58% of the placed volume from the longshore transport.

This is the first nearshore nourishment project to be investigated with satellite imagery using CoastSat. When compared to observed shoreline change, satellite-derived shoreline errors range from 3.69 to 9.96 m, which is consistent with past work, and results match elevation survey observations of more accretion onshore of the placement and to the south. The high temporal frequencies and spatial coverage at which the CoastSat analysis can be conducted, and the low cost make this a useful addition to capture shoreline change between elevation surveys. It could allow for new capabilities at locations that have not historically been able to monitor nearshore nourishments extensively.

Observed elevation changes also generally match predictions from the rapid assessment tools. Frequency of sediment mobilization and depth of closure estimates from the Sediment Mobility Tool both suggest that the placement should be active. The deflation predictions following Bain et al. [54] also imply that the placement should be active, but the predicted bulk deflation rates exceeded measured values by 145%. Agreement between predicted and observed behavior demonstrate the utility of these rapid tools while planning and designing nearshore nourishments.

Measured data, CoastSat analysis, and rapid predictions all indicate that the nearshore nourishment at Harvey Cedars had a positive impact on the adjacent beach. Data collected can be used to improve numerical modeling capabilities of nearshore nourishments. Future work can quantify long-term benefits of the nearshore nourishment, including the financial savings to the subaerial CSRM project by beneficially placing dredged sediment in the nearshore.

Author Contributions: Conceptualization, M.A.C., B.C.M. and D.R.K.; methodology, B.C.M., B.D.H., D.R.K., S.P.M. and R.L.B.; software, R.L.B., S.P.M., N.R.O. and B.C.M.; formal analysis, B.C.M., D.R.K., R.L.B. and S.P.M.; investigation, B.C.M., B.D.H., D.R.K. and S.P.M.; writing—original draft preparation, S.P.M., D.R.K., B.C.M. and B.D.H.; writing—review and editing, S.P.M., B.D.H., B.C.M., D.R.K., R.L.B., N.R.O., I.W.C. and M.A.C.; visualization, R.L.B., S.P.M., N.R.O., B.C.M. and B.D.H.; project administration, S.P.M., B.D.H., B.C.M. and D.R.K.; funding acquisition, M.A.C., B.C.M., D.R.K. and B.D.H. All authors have read and agreed to the published version of the manuscript.

Funding: This research was funded by the U.S. Army Corps of Engineers through the Regional Sediment Management (RSM) Program, Coastal Inlets Research Program (CIRP), and the Water Resource and Development Act (WRDA) Section 1122 program.

Institutional Review Board Statement: Not applicable.

Data Availability Statement: The data presented in this study are available on request from the corresponding author.

Acknowledgments: We would like to thank the Borough of Harvey Cedars and the NJ Department of Environmental Protection's Office of Coastal Engineering for their support and assistance with this project. We would like to thank the USACE Philadelphia District for assistance with surveying, historical data, and instrument deployment/recovery.

Conflicts of Interest: The authors declare no conflict of interest.

References

1. Zwamborn, J.A.; Fromme, G.A.W.; Fitzpatrick, J.B. Underwater mound for the protection of Durban's beaches. In Proceedings of the 12th International Conference on Coastal Engineering, Washington, DC, USA, 13–18 September 1970.
2. Vera-Cruz, D. Artificial nourishment of Copacabana beach. In Proceedings of the 13th International Conference on Coastal Engineering, Vancouver, BC, Canada, 10–14 July 1972.
3. Hanson, H.; Brampton, A.; Capobianco, M.; Dette, H.H.; Hamm, L.; Laustrup, C.; Lechuga, A.; Spanhoff, R. Beach nourishment projects, practices, and objectives—A European overview. *Coast. Eng.* **2002**, *47*, 81–111. [CrossRef]
4. Lodder, Q.J.; Sørensen, P. Comparing the Morphological Behaviour of Dutch–Danish Shoreface Nourishments. In Proceedings of the Coastal Management: Changing Coast, Changing Climate, Changing Minds, Amsterdam, The Netherlands, 8–9 September 2015.
5. Pinto, C.A.; Taborda, R.; Andrade, C.; Baptista, P.; Silva, P.A.; Mendes, D.; Pais-Barbosa, J. Morphological Development and Behaviour of a Shoreface Nourishment in the Portuguese Western Coast. *J. Mar. Sci. Eng.* **2022**, *10*, 146. [CrossRef]
6. Mendes, D.; Pais-Barbosa, J.; Baptista, P.; Silva, P.A.; Bernardes, C.; Pinto, C. Beach Response to a Shoreface Nourishment (Aveiro, Portugal). *J. Mar. Sci. Eng.* **2021**, *9*, 1112. [CrossRef]
7. Walker, B.M.; Krafft, D.R.; McFall, B.C.; McCaw, H.; Spurgeon, S.L. *Current State of Practice of Nearshore Nourishment by the United States Army Corps of Engineers*; ERDC/CHL SR-22-4; U.S. Army Corps of Engineers Engineer Research and Development Center: Vicksburg, MS, USA, 2022; 49p. [CrossRef]
8. Brand, E.; Ramaekers, G.; Lodder, Q. Dutch experience with sand nourishments for dynamic coastline conservation—An operational overview. *Ocean Coast. Manag.* **2022**, *217*, 106008. [CrossRef]
9. McFall, B.C.; Krafft, D.R.; McCaw, H.; Walker, B.M. *Metrics of Success for Nearshore Nourishment Projects Constructed with Dredged Sediment*; ERDC/TN RSM-21-3; U.S. Army Corps of Engineers Engineer Research and Development Center: Vicksburg, MS, USA, 2021; 13p. [CrossRef]
10. Wang, P.; Brutsché, K.E.; Beck, T.M.; Rosati, J.D.; Lillycrop, L.S. *Initial Morphologic Evolution of Perdido Key Berm Nourishment, Florida*; ERDC/CHL CHETN-IV-89; U.S. Army Corps of Engineers Engineer Research and Development Center: Vicksburg, MS, USA, 2013; 10p, Available online: http://hdl.handle.net/11681/2001 (accessed on 7 October 2022).
11. Brutsché, K.E.; Wang, P.; Beck, T.M.; Rosati, J.D.; Legault, K.R. Morphological evolution of a submerged artificial nearshore berm along a low-wave microtidal coast, Fort Myers Beach, west central Florida, USA. *Coast. Eng.* **2014**, *91*, 29–44. [CrossRef]
12. Brutsché, K.E.; McFall, B.C.; Li, H.; McNinch, J.E.; Ousley, J.D.; Engle, J.A.; Maglio, C.K. Strategic nearshore placement of dredged sediment at Vilano Beach, Florida. *Shore Beach* **2017**, *85*, 77–84.
13. Otay, E.N. Monitoring results of a nearshore disposal berm. In Proceedings of the Coastal Dynamics '95, Gdańsk, Poland, 4–8 September 1995.
14. Armstrong, S.B.; Lazarus, E.D.; Limber, P.W.; Goldstein, E.B.; Thorpe, C.; Ballinger, R.C. Indications of a Positive Feedback between Coastal Development and Beach Nourishment: Coastal Development Beach Nourishment. *Earth's Future* **2016**, *4*, 626–635. [CrossRef]
15. McFall, B.C.; Brutsché, K.E.; Ousley, J.D.; Maglio, C.K.; Engle, J.A. Innovative nearshore berm placement techniques at Vilano Beach, FL, and application of the sediment mobility tool. *World Dredg.* **2017**, *51*, 34–35.
16. Krafft, D.R.; Young, D.L.; Brutsché, K.E.; McFall, B.C.; Bruder, B.L. *Nearshore Placement Workshop 2019: Sediment Nourishment of the Nearshore Environment*; ERDC/CHL SR-20-02; U.S. Army Corps of Engineers Engineer Research and Development Center: Vicksburg, MS, USA, 2020; 28p. [CrossRef]
17. van Duin, M.J.P.; Wiersma, N.R.; Walstra, D.J.R.; Van Rijn, L.C.; Stive, M.J.F. Nourishing the shoreface: Observations and hindcasting of the Egmond case, The Netherlands. *Coast. Eng.* **2004**, *51*, 813–837. [CrossRef]
18. Hamilton, D.G.; Ebersole, B.A.; Smith, E.R.; Wang, P. *Development of a Large-Scale Laboratory Facility for Sediment Transport Study*; ERDC/CHL TR-01-22; U.S. Army Corps of Engineers Engineer Research and Development Center: Vicksburg, MS, USA, 2001; 187p. Available online: http://hdl.handle.net/11681/7560 (accessed on 7 October 2022).

19. Smith, E.R.; Mohr, M.C.; Chader, S.A. Laboratory experiments on beach change due to nearshore mound placement. *Coast. Eng.* **2017**, *121*, 119–128. [CrossRef]
20. Hands, E.B.; Allison, M.C. Mound migration in deeper water and methods of categorizing active and stable depths. In Proceedings of the Coastal Sediments '91, Seattle, WA, USA, 25–27 June 1991.
21. Ahrens, J.P.; Hands, E.B. Parameterizing Beach Erosion/Accretion Conditions. In Proceedings of the Coastal Engineering 1998, Cophenhagen, Denmark, 22–26 June 1998. [CrossRef]
22. Huisman, B.J.A.; Walstra, D.-J.R.; Radermacher, M.; de Schipper, M.A.; Ruessink, B.G. Observations and Modelling of Shoreface Nourishment Behaviour. *J. Mar. Sci. Eng.* **2019**, *7*, 59. [CrossRef]
23. Priestas, A.M.; McFall, B.C.; Brutsché, K.E. *Performance of Nearshore Berms from Dredged Sediments: Validation of the Sediment Mobility Tool*; ERDC/CHL TR-19-19; U.S. Army Corps of Engineers Engineer Research and Development Center: Vicksburg, MS, USA, 2019; 63p. [CrossRef]
24. McFall, B.C.; Brutsché, K.E.; Priestas, A.M.; Krafft, D.R. Evaluation Techniques for the Beneficial Use of Dredged Sediment Placed in the Nearshore. *J. Waterw. Port Coast. Ocean Eng.* **2021**, *147*, 04021016. [CrossRef]
25. Douglass, S.L. Estimating landward migration of nearshore constructed sand mounds. *J. Waterw. Port Coast. Ocean Eng.* **1995**, *121*, 247–250. [CrossRef]
26. Larson, M.; Ebersole, B.A. *An Analytical Model to Predict the Response of Mounds Placed in the Offshore*; ERDC/CHL CETN-II-42; U.S. Army Corps of Engineers Engineer Research and Development Center: Vicksburg, MS, USA, 1999; 13p. Available online: http://hdl.handle.net/11681/2124 (accessed on 7 October 2022).
27. Johnson, C.L.; McFall, B.C.; Krafft, D.R.; Brown, M.E. Sediment Transport and Morphological Response to Nearshore Nourishment Projects on Wave-Dominated Coasts. *J. Mar. Sci. Eng.* **2021**, *9*, 1182. [CrossRef]
28. Smith, E.R.; D'Alessandro, F.; Tomasicchio, G.R.; Gailani, J.Z. Nearshore placement of a sand dredged mound. *Coast. Eng.* **2017**, *126*, 1–10. [CrossRef]
29. Bryant, D.B.; McFall, B.C. Transport of nearshore dredge material berms. In Proceedings of the 6th International Conference on Application of Physical Modelling in Coastal and Port Engineering and Science, Ottawa, ON, Canada, 10–13 May 2016.
30. de Schipper, M.A.; de Vries, S.; Ranasinghe, R.; Reniers, A.J.H.M.; Stive, M.J.F. Morphological developments after a beach and shoreface nourishment at Vlugtenburg beach. In Proceedings of the NCK-days 2012: Crossing borders in coastal research, Enschede, The Netherlands, 13–16 March 2012.
31. Gijsman, R.; Visscher, J.; Schlurmann, T. A method to systematically classify design characteristics of sand nourishments. In Proceedings of the 36th International Conference on Coastal Engineering, Baltimore, MD, USA, 30 July–3 August 2018. [CrossRef]
32. Young, D.L.; Brutsché, K.E.; Li, H.; McFall, B.C.; Maloney, E.C.; McClain, K.E.; Bucaro, D.F.; LeRoy, J.Z.; Duncker, J.J.; Johnson, K.K.; et al. *Analysis of Nearshore Placement of Sediments at Ogden Dunes, Indiana*; ERDC/CHL TR-20-4; U.S. Army Corps of Engineers Engineer Research and Development Center: Vicksburg, MS, USA, 2020; 98p. [CrossRef]
33. de Looff, H.; Welp, T.; Snider, N.; Wilmink, R. Chapter 7: Adaptive Management. In *International Guidelines on Natural and Nature-Based Features for Flood Risk Management*; Bridges, T.S., King, J.K., Simm, J.D., Beck, Collins, G., Lodder, Q., Mohan, R.K., Eds.; U.S. Army Corps of Engineers Engineer Research and Development Center: Vicksburg, MS, USA, 2021.
34. Lodder, Q.; Jeuken, C.; Reinen-Hamill, R.; Burns, O.; Ramsdell, R., III; de Vries, J.; McFall, B.; IJff, S.; Maglio, C.; Wilmink, R. Chapter 9: Beaches and Dunes. In *International Guidelines on Natural and Nature-Based Features for Flood Risk Management*; Bridges, T.S., King, J.K., Simm, J.D., Beck, Collins, G., Lodder, Q., Mohan, R.K., Eds.; U.S. Army Corps of Engineers Engineer Research and Development Center: Vicksburg, MS, USA, 2021.
35. Tyler, Z.J.; McFall, B.C.; Brutsché, K.E.; Maloney, E.C.; Bucaro, D.F. *Physical Monitoring Methods for the Nearshore Placement of Dredged Sediment*; ERDC/TN RSM-18-6; U.S. Army Corps of Engineers Engineer Research and Development Center: Vicksburg, MS, USA, 2018; 13p. [CrossRef]
36. van Rees, C.B.; Naslund, L.; Hernandez-Abrams, D.D.; McKay, S.K.; Woodson, C.B.; Rosemond, A.; McFall, B.; Altman, S.; Wenger, S.J. A strategic monitoring approach for learning to improve natural infrastructure. *Sci. Total Environ.* **2022**, *832*, 155078. [CrossRef]
37. Arnold, D.E.; McFall, B.C.; Brutsché, K.E.; Maloney, E.C.; Bucaro, D.F. *Nearshore Placement Techniques in Southern Lake Michigan*; ERDC/CHL TR-18-3; U.S. Army Corps of Engineers Engineer Research and development Center: Vicksburg, MS, USA, 2018; 45p. [CrossRef]
38. USACE Philadelphia District. *WRDA 2016 Section 1122 Beneficial Use of Dredged Material Pilot Program: Beneficial Use Placement Opportunities in New Jersey Using Navigation Channel Sediments–Barnegat Inlet*. 2021. Available online: https://www.nap.usace.army.mil/Portals/39/docs/Civil/Coastal/Barnegat-Inlet-Section-1122-April-2021.pdf?ver=mDQWxLcR_kkLqf_2ndOvIg%3D%3D (accessed on 7 October 2022).
39. Harris, B.D.; McGill, S.P.; Krafft, D.R.; McFall, B.C.; Chasten, M.A.; Bain, R.L. Beneficial Use of Dredged Material for Beach Erosion Mitigation at Harvey Cedars, New Jersey. *World Dredg.* **2022**, *55*, 10–11.
40. Vos, K.; Splinter, K.D.; Harley, M.D.; Simmons, J.A.; Turner, I.L. CoastSat: A Google Earth Engine-enabled Python toolkit to extract shorelines from publicly available satellite imagery. *Environ. Model. Softw.* **2019**, *122*, 104528. [CrossRef]
41. Cialone, M.A.; Thompson, E.F. *Wave Climate and Littoral Sediment Transport Potential, Long Beach Island, New Jersey*; ERDC/CHL TR-00-21; U.S. Army Corps of Engineers Engineer Research and Development Center: Vicksburg, MS, USA, 2000; 75p. Available online: http://hdl.handle.net/11681/7451 (accessed on 7 October 2022).

42. Dally, W.R.; Osiecki, D.A. Evaluating the Impact of Beach Nourishment on Surfing: Surf City, Long Beach Island, New Jersey, U.S.A. *J. Coast. Res.* **2018**, *34*, 793–805. [CrossRef]
43. Hallermeier, R.J. Uses for a calculated limit depth to beach erosion. In Proceedings of the 16th International Conference on Coastal Engineering, Hamburg, Germany, 27 August–3 September 1978; pp. 1493–1512.
44. Hallermeier, R.J. A profile zonation for seasonal sand beaches from wave climate. *Coast. Eng.* **1981**, *4*, 253–277. [CrossRef]
45. Birkemeier, W.A. Field data on seaward limit of profile change. *J. Waterw. Port Coast. Ocean Eng.* **1985**, *111*, 598–602. [CrossRef]
46. Brutsché, K.E.; Rosati III, J.; Pollock, C.E.; McFall, B.C. *Calculating Depth of Closure Using WIS Hindcast Data*; ERDC/CHL CHETN-VI-45; U.S. Army Corps of Engineers Engineer Research and Development Center: Vicksburg, MS, USA, 2016; 9p. Available online: http://hdl.handle.net/11681/20259 (accessed on 7 October 2022).
47. McFall, B.C.; Smith, S.J.; Pollock, C.E.; Rosati, J.; Brutsché, K.E. *Evaluating Sediment Mobility for Sitting Nearshore Berms*; ERDC/CHL CHETN-IV-108; U.S. Army Corps of Engineers Engineer Research and Development Center: Vicksburg, MS, USA, 2016; 11p. Available online: http://hdl.handle.net/11681/20282 (accessed on 7 October 2022).
48. Larson, M.; Kraus, N.C. *Analysis of Cross-Shore Movement of Natural Longshore Bars and Material Placed to Create Longshore Bars*; Technical Rep. No. DRP-92-5; Department of the Army, Waterways Experiment Station, Corps of Engineers: Vicksburg, MS, USA, 1992; 124p. Available online: http://hdl.handle.net/11681/4617 (accessed on 7 October 2022).
49. McFall, B.C.; Brutsché, K.E. *User's Guide for the Sediment Mobility Tool Web Application*; ERDC/TN RSM-18-4; U.S. Army Corps of Engineers Engineer Research and Development Center: Vicksburg, MS, USA, 2018; 11p. [CrossRef]
50. Komar, P.D. *Beach Processes and Sedimentation*, 2nd ed; Prentice Hall: Upper Saddle River, NJ, USA, 1998.
51. Dean, R.G.; Dalrymple, R.A. *Coastal Processes with Engineering Applications*; Cambridge University Press: Cambridge, UK, 2002.
52. Dean, R.G. *Beach Nourishment: Theory and Practice*; Advanced Series on Ocean Engineering, vol. 18; World Scientific Publishing Company Inc.: Hackensack, NJ, USA, 2002.
53. Soulsby, R.L. *Dynamics of Marine Sands*; Thomas Telford Publications: London, UK, 1997.
54. Bain, R.; McFall, B.; Krafft, D.; Hudson, A. Evaluating Transport Formulations for Application to Nearshore Berms. *J. Waterw. Port Coast. Ocean Eng.* **2021**, *147*, 4021031. [CrossRef]
55. Shaeri, S.; Etemad-Shahidi, A.; Tomlinson, R. Revisiting Longshore Sediment Transport Formulas. *J. Waterw. Port Coast. Ocean Eng.* **2020**, *146*, 4020009. [CrossRef]
56. Dronkers, J. *Dynamics of Coastal Systems*, 2nd ed.; World Scientific Publishing: Hackensack, NJ, USA, 2016.
57. Hudson, A.; Moritz, H.R.; Norton, J. *Sediment Mobility, Closure Depth, and the Littoral System–Oregon and Washington Coast*; ERDC/TN RSM-22-7; U.S. Army Corps of Engineers Engineer Research and Development Center: Vicksburg, MS, USA, 2022; 10p. [CrossRef]
58. Otsu, N. A Threshold Selection Method from Gray-Level Histograms. *IEEE Trans. Syst. Man Cybern.* **1979**, *9*, 62–66. [CrossRef]
59. Grunnet, N.M.; Ruessink, B.G. Morphodynamic response of nearshore bars to a shoreface nourishment. *Coast. Eng.* **2005**, *52*, 119–137. [CrossRef]
60. Vos, K.; Harley, M.D.; Splinter, K.D.; Simmons, J.A.; Turner, I.L. Sub-annual to multi-decadal shoreline variability from publicly available satellite imagery. *Coast. Eng.* **2019**, *150*, 160–174. [CrossRef]

Journal of
Marine Science and Engineering

MDPI

Article

Natural and Anthropogenic Factors Shaping the Shoreline of Klaipėda, Lithuania

Vitalijus Kondrat *, Ilona Šakurova, Eglė Baltranaitė and Loreta Kelpšaitė-Rimkienė

Marine Research Institute, Klaipėda University, Universiteto Ave. 17, LT-92294 Klaipėda, Lithuania; ilona.sakurova@ku.lt (I.Š.); egle.baltranaite@ku.lt (E.B.); loreta.kelpsaite-rimkiene@ku.lt (L.K.-R.)
* Correspondence: vitalijus.kondrat@ku.lt

Abstract: Port of Klaipėda is situated in a complex hydrological system, between the Curonian Lagoon and the Baltic Sea, at the Klaipėda strait in the South-Eastern part of the Baltic Sea. It has almost 300 m of jetties separating the Curonian Spit and the mainland coast, interrupting the main path of sediment transport through the South-Eastern coast of the Baltic Sea. Due to the Port of Klaipėda reconstruction in 2002 and the beach nourishment project, which was started in 2014, the shoreline position change tendency was observed. Shoreline position measurements of various periods can be used to derive quantitative estimates of coastal process directions and intensities. These data can be used to further our understanding of the scale and timing of shoreline changes in a geological and socio-economic context. This study analyzes long- and short-term shoreline position changes before and after the Port of Klaipėda reconstruction in 2002. Positions of historical shorelines from various sources were used, and the rates (EPR, NSM, and SCE) of shoreline changes have been assessed using the Digital Shoreline Analysis System (DSAS). An extension of ArcGIS K-means clustering was applied for shoreline classification into different coastal dynamic stretches. Coastal development has changed in the long-term (1984–2019) perspective: the eroded coast length increased from 1.5 to 4.2 km in the last decades. Coastal accumulation processes have been restored by the Port of Klaipėda executing the coastal zone nourishment project in 2014.

Keywords: Baltic Sea; Port of Klaipėda; shoreline changes; DSAS; clusterization; regime shift detection; dredging; sand nourishment

Citation: Kondrat, V.; Šakurova, I.; Baltranaitė, E.; Kelpšaitė-Rimkienė, L. Natural and Anthropogenic Factors Shaping the Shoreline of Klaipėda, Lithuania. *J. Mar. Sci. Eng.* **2021**, *9*, 1456. https://doi.org/10.3390/jmse9121456

Academic Editor: Carlos Daniel Borges Coelho

Received: 18 November 2021
Accepted: 17 December 2021
Published: 20 December 2021

1. Introduction

Erosion is a significant problem affecting sandy beaches that will worsen with climate change and anthropogenic pressure. Sandy shorelines are highly dynamic due to altering wave conditions, sea levels and winds, geological factors, and human activity [1]. Therefore, identifying the most vulnerable areas to erosion is crucial for nearshore communities since it could significantly affect their socio-economic state through destruction of infrastructure, loss of land and property on the coast, and valuable beach areas used for recreation.

Shore regeneration is a slow process lasting for more than one year, while erosion usually occurs in a matter of a few days, making it difficult to detect visually. As short-term measurements do not reflect actual multi-annual dynamic trends, studies involving several shoreline decay and regeneration cycles are necessary to determine long-lasting changes in the shoreline dynamics. Typically, coastal research to assess and predict long-term shoreline dynamics and the erosion rates is based on the data covering up to 10 years (short-term), 10–60 years (medium-term), and more than 60 years (long-term) of shoreline position [2–4].

Shoreline dynamics depend on different causes, mainly on the sediments in the sea-land system [5–7]. Furthermore, the different coastal stretches have particular favorable hydrometeorological conditions for the accumulation or erosion processes. The rapid urbanization of the coastal zone has a significant impact on shoreline development [8–10]. Sustainable coastal development requires knowledge of the coastal processes combined with incorruptible urbanization and properly chosen shoreline erosion mitigation methods [10,11]. Often, an insufficient understanding of the coastal processes causes costly incidents.

A number of studies [8,12,13] show the impact of anthropogenic factors in particular port activities on shoreline positions. Erosion and accumulation are naturally occurring processes that often coincide in a dynamic equilibrium [14]. However, increasing anthropogenic pressure at the coast has disrupted the natural development of the coast, accelerating erosion processes in some places and causing accumulation in others [14]. Analysis of shoreline changes is a well-developed field that has progressed complex data processing and analytical protocols [15]. However, quantifying coastal development trends is only one aspect of the problem; it is necessary to understand the drivers of change and address local impacts in a broader regional context that is important from a decadal to a centennial timescale [15]. Understanding the causes of atypical coastal development is important to make sustainable coastal zone management plans. Such knowledge is crucial not only for the coastal dynamics experts, but also for the port managers, as it can serve as the basis for future decisions on how to reduce port damage to the coasts.

This paper analyses the shoreline dynamic in the context of climate change and increased anthropogenic pressure, focusing on identifying long- and short-term shoreline movement tendencies and identifying the direct impact zone of the Port of Klaipėda. As well as answering the question of whether and how shoreline evolution is affected by the artificial sand nourishment carried out in accordance with the Port of Klaipėda management plan.

2. Study Site

The Lithuanian coast of the Baltic Sea represents a generic type of almost straight, relatively high-energy, actively developing coasts that (i) contain a large amount of fine mobile sediment; (ii) are open to predominating wind and wave directions; and (iii) are exposed to waves from many directions [16]. The study area extends 10 km from Klaipėda seaport jetties to the north and 10 km to the south. This particular area was chosen based on the following aspects: (i) the broad demand spectrum of recreational uses [17]; (ii) the high risk of coastal erosion [18,19]; (iii) the possibility of direct and indirect anthropogenic impacts [20,21].

The South-Eastern coast of the Baltic Sea is formed by the presence of the Port of Klaipėda [21,22]. Historically, the Port of Klaipėda has been known from the 13th century when vessels of Lubeck and Bremen merchants used to moor in the small port neighboring the Klaipėda castle [23]. Port expansion to the Klaipėda strait started in 1745, and the chronicle of 1797 mentions that Port of Klaipėda consists of the Dane river port and a big water basin in the strait of the Curonian Lagoon. In the 19th century, wooden jetties were constructed [22]. 1924–1939 was a period when Klaipėda seaport was at its flourishing peak—new stony jetties and quays were assembled [24,25]. Since the occurrence of the first jetties, ongoing coastal engineering problems were encountered relating to wave exposure, siltation within the port, extensive dredging requirements, and seiching within the confines of the present harbor [22,26].

After the construction of the first port jetties, at the end of the 19th century, the shoreline moved seawards significantly on both sides of the jetties [20]. This insight raises doubts about the predominant sedimentary direction from south to north [6]. The dumping of the dredged sand can partly explain this accumulation tendency in the northern part of the jetties from the Klaipėda strait [22]. Up until the beginning of the 20th century, sand dredged from the port had been dumped at shallow depths north of the jetties, initiating coast accumulation [22].

After the prolongation and construction of new concrete jetties at the beginning of the 20th century (works finished till 1934) [21] alongside changed dredging policies [13], observations were made that sand dredged next to the port jetties returns into the inlet and continues dredging works to ensure the depth of the entrance channel.

Due to depth restrictions in the Danish Straits, vessels with a maximum draft of 16.5 m, and in some cases, vessels with a draft of up to 17 m can enter the Baltic Sea. Another limitation for ships entering the Baltic Sea is the bridge height to about 65 m entering the strait of the Great Belt, which connects the Danish islands of Zeeland and Funen. These restrictions prevent vessels with a draft greater than 16.5 m from entering the Baltic Sea from those of Class Panamax (Baltmax). The long-term competitiveness and sustainability of the Klaipėda seaport can be ensured only by increasing the technical capability of the port to receive and service ships of the maximum capacity [27].

Therefore, in 1999, the final design for reconstruction of the Port of Klaipėda jetties was established. The seaport jetties system was reconstructed by narrowing the entrance channel and changing the position of the northern pier. In 2002 the northern pier was extended by 205 m (up to 733 m) and the southern pier by 278 m (up to 1374 m) (Figure 1) [28]. At the same time, the entrance channel was dredged to a depth of 14.5 m. According to the recent port development plan, the entrance channel will be dredged up to 17 m by 2023.

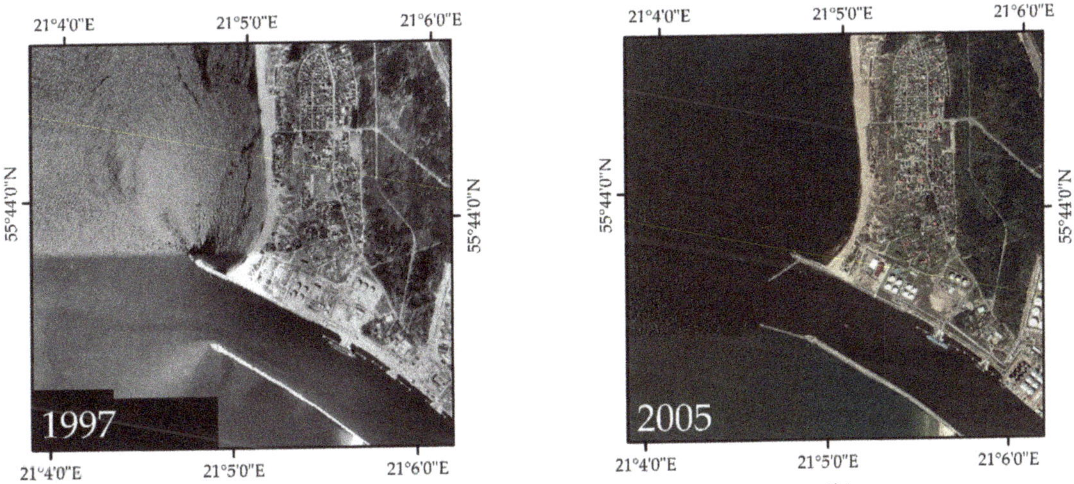

(a) (b)

Figure 1. The Klaipėda seaport jetties before and after the reconstruction of 2002, (**a**) 1997, and (**b**) 2005.

The Port of Klaipėda, located at the Klaipėda strait (Figure 2) (South-Eastern coast of the Baltic Sea), divides the Lithuanian coast into two geologically and geomorphologically different parts: southern—the coast of the Curonian spit, northern—mainland coast (Figure 3) [29]. Port jetties interrupt the main sediment transport path and significantly influence the Lithuanian coast's northern (38.49 km long) part [6,20]. Only Quaternary sediments are found on the Lithuanian coast of the Baltic Sea [6,30]. From the geological point of view, the mainland coast and the Curonian Spit coast are not homogenous (Figure 4). The geological structure of the mainland coast was mainly determined by the sediments formed during the last few glaciations. The sediments of the Curonian Spit coast were formed in the Baltic Sea basin—starting with the Baltic Ice Lake and ending with the modern Baltic Sea [6,30].

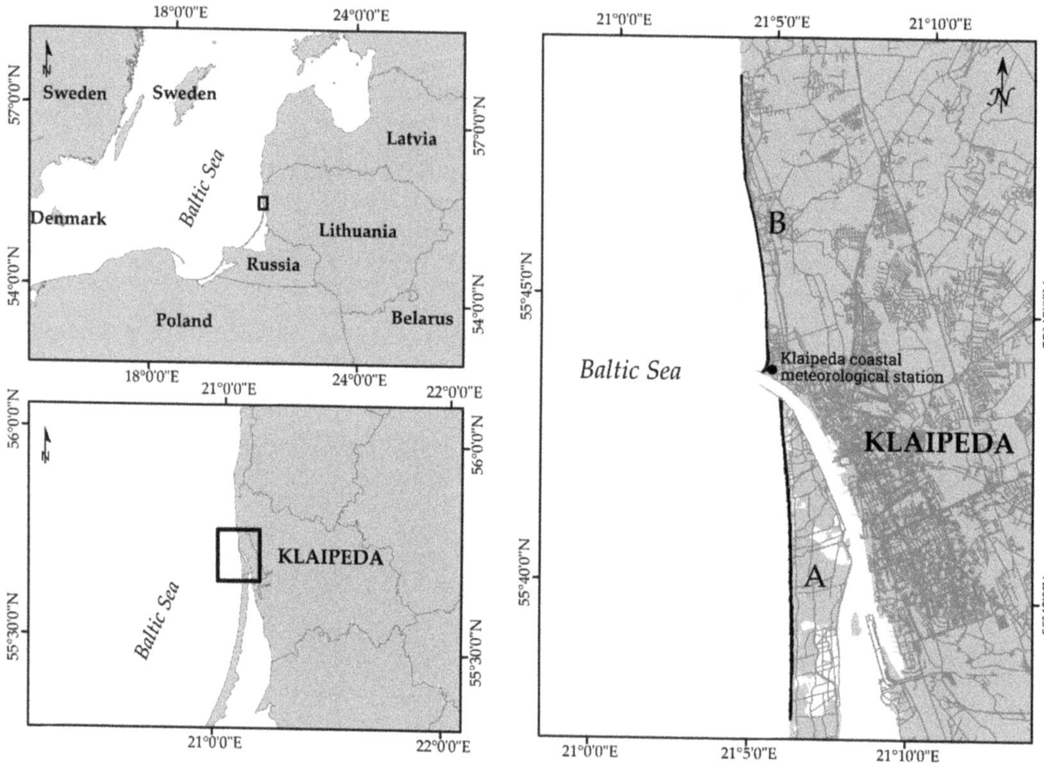

Figure 2. Location of the study site in the south-eastern Baltic Sea, A: the Curonian Spit coast, B: the mainland coast.

The sandy sediments form the part of the Curonian Spit coast: this Lithuanian coastal sector is characterized by accumulation relief [6]. The mainland coast of Lithuania is geologically heterogeneous: the northern part of the mainland coast is mainly formed of fine-grained sand (0.25–0.1 mm), while the southern and central parts of the mainland coast are formed by the medium-grained (0.5–0.25 mm) and coarse-grained (1–2.5 mm) sand [6,14]. A detailed description of the Lithuanian coast geomorphological and geological structure is provided by Bitinas et al. (2005)

According to the granulometric analysis of sediment samples from 2019 along the study area (Figure 4), on the Curonian Spit coast (A, a), very well and moderately sorted (σ = 1.21–1.47 mm) fine sand (Md = 0.20–0.37 mm) prevails, while on the mainland coast (B, b, c), the sorting of the sediments differs in a cross-shore profile. In profile *b*, moderately well-sorted (σ = 1.44 mm) medium sand (Md = 0.32 mm) prevails in a shoreline area, well-sorted (σ = 1.19 mm) slightly very fine gravelly medium sand (Md = 0.21 mm) prevails in a beach area, and moderately well-sorted (σ = 1.47 mm) sand prevails (Md = 0.36 mm) in a foredune area. In profile *c*, poorly sorted (σ = 3.84 mm) very fine gravelly fine sand (Md = 0.24 mm) prevails in a shoreline area, poorly sorted (σ = 10.69 mm) medium gravelly fine sand (Md = 0.23 mm) prevails in a beach area, and poorly sorted (σ = 18.21 mm) sandy very fine gravel prevails (Md = 1.12 mm) in a foredune area.

Figure 3. Study site shoreline features: (**a**) the Curonian Spit coast Smiltynė I beach (© I. Šakurova); (**b**) the Curonian Spit coast Smiltynė I beach (© I. Šakurova); (**c**) the mainland coast Giruliai beach (© L. Kelpšaitė-Rimkienė); (**d**) the mainland coast Melnragė I beach (© V. Kondrat).

During the dredging of the Klaipėda strait, the glaciogenic moraine deposits and alluvial sediments are mainly excavated—sand (0.002 mm 10–30%–2 mm 50%) and silt (0.002 mm 10–30%–2 mm 30–50%). All lithological sediment types are dumped in dumping area I (Figure 4) at a depth of 45–50 m. The II dumping area (Figure 4) is intended only for the dumping of sandy sediments—fine (0.25–0.1 mm > 50%) and aleuritic (<0.063 mm 10–30%) sand at a depth of 28–35 m. Since 2001, clean sand that meets sanitary–hygienic requirements excavated from the port entrance channel has been dumped in the dumping area III (Melnragė-Giruliai) at a 4–6 m depth. This area is intended to replenish the sediment balance and restore beach sand reserves [24].

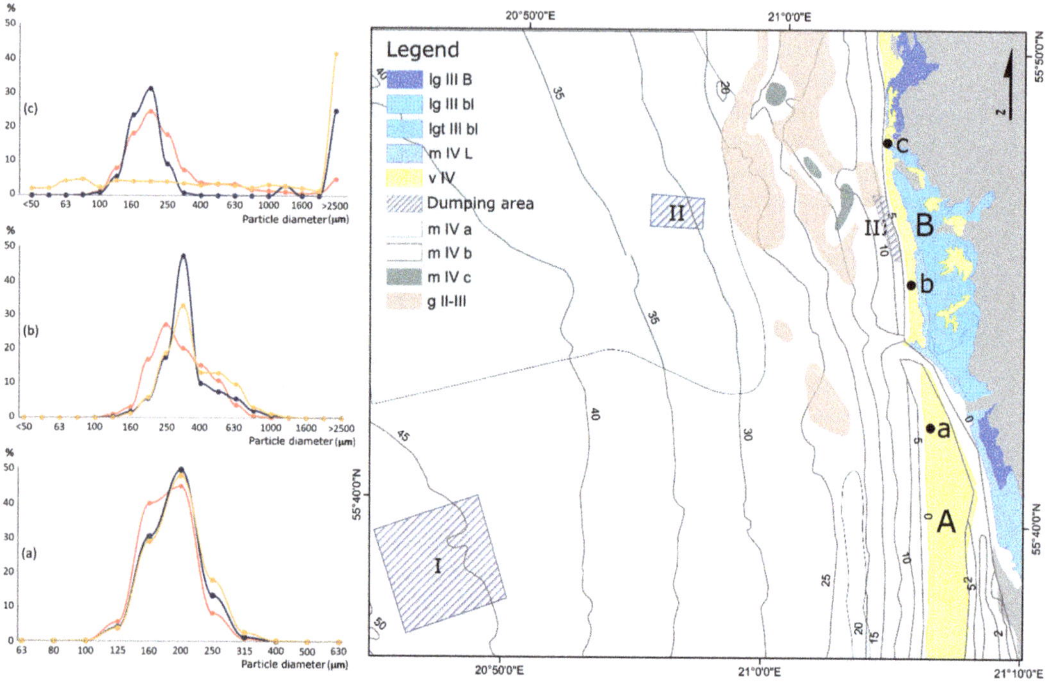

Figure 4. Map of Quaternary sediment type of coastal area and dumping zones of dredging material. Lg III B—glaciolacustrine sediments of the Baltic Ice Lake (fine sand); lg III bl—glaciolacustrine sediments (various sand); lgt III bl—marginal glaciolacustrine deposits (fine sand); m IV L—Litorina Sea sediments (fine sand); v IV—aeolian deposits (fine sand); m IV a—nearshore sediments (extra fine sand (0.05—0.1 mm)); m IV b—nearshore sediments (fine sand (0.1—0.25 mm)); m IV c—nearshore sediments (gravel with sand); gII-III—glacial deposits of Middle and Upper Pleistocene (unseparated), glacial loam, boulders and gravel (washed till). I—distant dumping area; II—near dumping area; III—nearshore dumping area (adapted from Bitinas et al., 2004). Grain-size composition of surface sediments at Smiltynė I (**a**), Melnragė I (**b**), and Karklė (**c**). Orange color line—western part of the dunes (foredune); blue line—the middle of the beach; red line—a dynamic shoreline in July of 2019.

3. Materials and Methods

3.1. Analysis of Cartographical Data

In this paper, we evaluate a period of 35 years of shoreline position variation tendencies for 1984–2019. All shoreline position changes were determined using the available high accuracy (1:10,000) cartographic data for the years: 1984, 1990, 1995, and 2005 (Table 1) obtained from Lithuania's National Land Service under the Ministry of Agriculture and GPS survey data for 2015 and 2019. The shoreline position was established at the middle of the swash zone by dual-band GPS receiver "Leica 900".

Shoreline position changes were analyzed with the ArcGIS extension DSAS v. 5.0 (Digital Shoreline Analysis System) package, developed by the United States Geological Survey (USGS) [31,32]. The DSAS is executed in five steps: (1) shorelines digitizing and uniforming to WGS-84 coordinate systems (UTM Zone 34); (2) computation of the uncertainties; (3) baseline creation and transects generation; (4) computation of distances between baseline and shorelines at each transect; and (5) computation of shoreline change statistics.

Three statistical parameters—net shoreline movement (NSM), end-point rate (EPR), and the shoreline change envelope (SCE)—were estimated and analyzed along with each transect every 25 m along the shoreline (796 transects in total). NSM values report the net change of the shoreline in the study period between the oldest and most recent shoreline.

EPR rate (m/yr) indicates change rates between the earliest and most recent shoreline positions. SCE capacity provides the envelope of shoreline variability, and it is the only measure of the total shoreline change among all the available shoreline positions [33].

Table 1. Shoreline positioning and detection errors. Ed—digitization error, Ep—pixel error, Es—sea-level fluctuation error, Ec—shoreline line detection or resolution errors, Etc—T-sheets plotting errors, Er—rectification error, Ut—shoreline capture error.

Data Source	Errors (m)						
	Ed	Ep	Es	Ec	Etc	Er	Ut
T-Sheets (1984)	2.961	0.987	0.680	3.948	7.500		9.058
T-Sheets (1990)	2.760	0.920	0.570	2.680	7.500		8.498
Orthophotos (1995)	2.500	0.506	0.490	2.024		0.500	3.331
Orthophotos (2005)	2.500	0.513	0.720	2.052		0.500	3.390
GPS (2010)			0.590	0.295			0.660
GPS (2015)			0.610	0.295			0.678
GPS (2019)			0.570	0.295			0.642

3.2. Data Reliability and Limits of Uncertainty

The shoreline position is highly variable in short time scales due to heavy storms, waves, and wind setup when extreme natural variations induce significant temporary shoreline retreat. Mapping the historical shorelines introduces additional uncertainties [34]. Although most researchers have similar techniques for estimating shoreline value changes, the methodology used to estimate changes varies considerably, significantly altering the accuracy and reliability of the data collected or determined. The dynamics of the shore itself may also cause certain differences and inaccuracies in shoreline surveys. Therefore, the values of the same shoreline determined by two independent scientists in the same field of science can vary considerably in their size and accuracy [35].

The most significant differences in the data occur during the digitization and processing of cartographic material. The differences in the values of shoreline changes may also occur due to the different statistical research methods used to determine the degree of shoreline change (shoreline change rate). The primary data and the analysis methods are the main factors defining the shoreline variations and accuracies. Therefore, prior to choosing a statistical research method, it is imperative to estimate the errors in determining the shoreline position in the cartographic material [36].

In this study, we determined three shoreline positioning and detection errors (Table 1) based on [14,36,37]:

The error in the position of the shoreline when determining in the T-Sheets:

$$Ut = \pm\,(Ed^2 + Ep^2 + Etc^2 + Es^2 + Ec^2)^{1/2} \tag{1}$$

The positioning error of the shoreline in orthophotos equals:

$$Ut = \pm\,(Er^2 + Ed^2 + Ep^2 + Es^2 + Ec^2)^{1/2} \tag{2}$$

GPS data error:

$$Ut = \pm\,(Es^2 + Ec^2)^{1/2} \tag{3}$$

Here: Ut—shoreline capture error, Er—rectification error, Ed—digitization error, Ep—pixel error, Ets—photo plan creation error, Ec—shoreline line detection or resolution errors, Eg—georeferencing error; Es—sea-level fluctuation error; Etc—T-sheets plotting errors.

The shoreline uncertainty limit for different periods is equal to the sum of the shoreline fixation errors for different periods:

$$\sum Ut = (Utn_1 + Utn_2 + Utn)^{1/2} \quad (4)$$

Here n_1, n_2,—shoreline detection errors for different periods.

The shoreline uncertainty threshold (minimum time criterion) in the statistical methods for deter-mining shore change (EPR) equals:

$$\sum Ut/n \quad (5)$$

Here n—research period.

3.3. Clusterization

K-Mean cluster analysis for the net shoreline movement (NSM) values was applied to identify shoreline zones with similar evolution tendencies [38]. The K-means algorithm is a simple and popular clustering approach used in various applications [39]. It is a point-based clustering approach that starts with cluster centers located initially in arbitrary locations and goes through each stage of the cluster center to reduce the cluster error [39–41].

$$E = \sum ||X_i - m_i||^2 \quad (6)$$

where E is the sum of squared errors for all objects in the data, X_i is the point in a cluster, and m_i is the mean of cluster k_i. The objective of K-means is to minimize the sum of squared errors over all k clusters. The algorithm first places k points in the space represented by the objects clustered as initial group centroids. The second step is to assign each object to the nearest cluster center. Then, the mean of each cluster is calculated to obtain a new centroid. These steps are repeated until the centroids do not change. The within-cluster sum of squares measures the variability of the observations within each cluster. In general, a cluster with a small sum of squares is more compact than a cluster with a large sum of squares [38,39]. Clusters with higher values exhibit more significant variability of the observations within the cluster [38,39]. The number of clusters is chosen based on the elbow method [38], whose main idea is to define groups such that the total intra-cluster variation (or the total sum of squares within clusters (WSS)) is minimized. In this case, the elbow of the curve is formed for the five clusters (Figure 5).

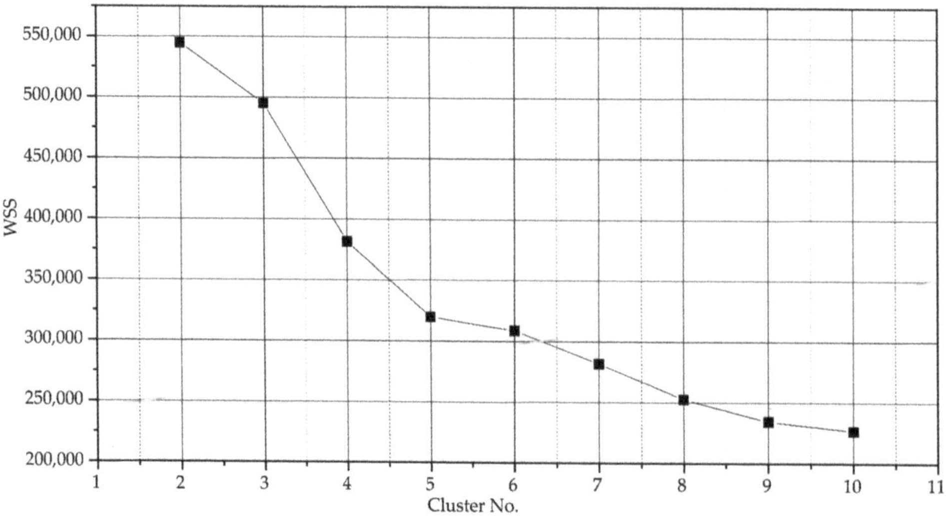

Figure 5. The number of clusters is chosen based on the within-cluster sum of squares parameter.

3.4. Analysis of Meteorological Data

The meteorological data (annual mean wind speed and direction) of the 1960–2019 time period were analyzed to detect the wind direction's regime shift. The meteorological data were acquired from the Marine Environment Assessment Division of the Environmental Protection Agency (EPA) and derived from the Klaipėda coastal meteorological station (Figure 2) under the Lithuanian Ministry of Environment's environmental monitoring program. The program has been prepared in line with the legislation of the European Union.

A STAR (Sequential T-test Analysis of Regime Shifts) algorithm was applied to determine regime shifts in the analyzed time series (https://www.beringclimate.noaa.gov/ (accessed on 10 October 2021)). The algorithm was built upon a sequential *t*-test that can signal the possibility of a real-time regime shift [42]. The algorithm can process the data regardless of whether it is presented in anomalies and/or absolute values or not. It can automatically calculate regime shifts in large sets of variables [42].

For this study, the following set of input parameters were used: cut-off length (I) was set to 10 years; Hubert's weight parameter (HWP) was set to 1. HWP determined the weight of outliers in the calculation of average values of the regime shift. The confidence level was set to 0.1.

4. Results

4.1. Long-Term Shoreline Changes

NSM for the entire study period **1984–2019** showed (Figure 6) that 60.43% of the shoreline was accumulative, 20.98% erosive, and 18.59% was stable or within the range of uncertainty ±9.08 m (Table 2). Generally, the studied coast can be described as accumulative with the average 14.46 ± 1.92 m shoreline movement offshore tendency; the average shoreline movement velocity was 0.42 ± 0.03 m/year.

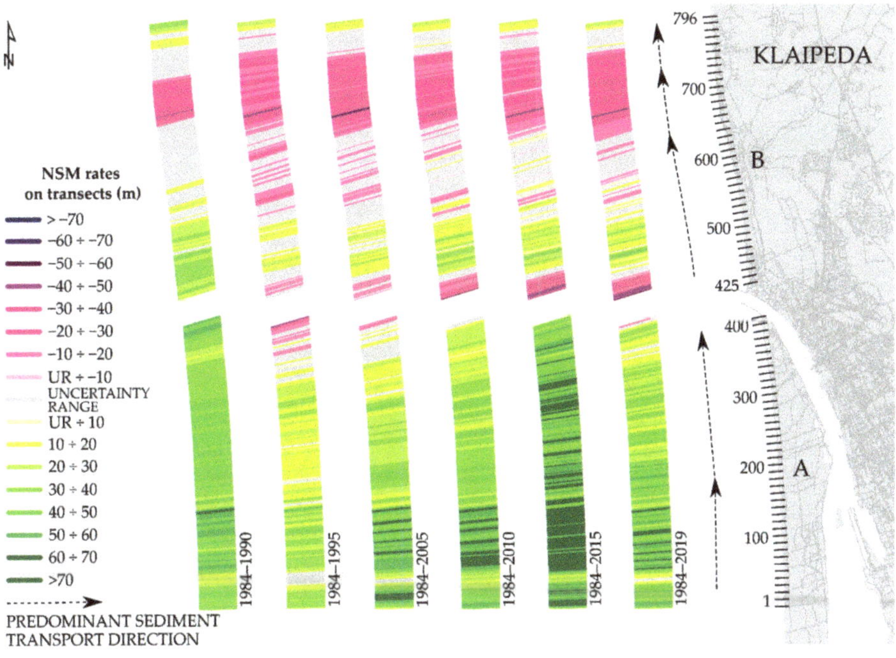

Figure 6. Net Shoreline Movement (NSM) rates 1984–2019 short-term vs long-term tendencies on the Curonian Spit coast (**A**) and the mainland coast (**B**). Annual shoreline change rates are shown on the transects graph. Purplish color tones on the transects indicate a trend of coastal erosion, while green tones indicate a trend of accretion, and grey color indicates shoreline variation values in its positioning and detection uncertainty range. Numbers and lines on the A and B coasts indicate transects distribution along the study site.

Table 2. Shoreline uncertainty range.

Years	±Uncertainty Range		Years	±Uncertainty Range	
	(m)	(m/yr)		(m)	(m/yr)
1984 * and 1990 *	±12.42	±2.07	1990 * and 1995 **	±9.13	±1.83
1984 * and 1995 **	±9.65	±0.88	1995 ** and 2005 **	±4.75	±0.48
1984 * and 2005 **	±9.67	±0.46	2005 ** and 2010 ***	±3.45	±0.69
1984 * and 2010 ***	±9.08	±0.35	2010 *** and 2015 ***	±0.95	±0.19
1984 * and 2015 ***	±9.08	±0.29	2015 *** and 2019 ***	±0.93	±0.23
1984 * and 2019 ***	±9.08	±0.26			

* T-Sheets; ** Orthophotos; *** GPS.

Comparing trends of shoreline changes in **1984–2019**, we found that the accumulation processes on the shores of the Curonian Spit accounted for 96.12% (396 out of 412) of transects. The shoreline moved towards the sea at an average speed of 1.01 ± 0.02 m/year (Figure 7), with the highest rates of the EPR 2.15 m/year. The NSM value was 35.97 ± 0.69 m, stable shoreline changes were found in 3.64% of transects and erosions in 0.24% of transects. The highest intensity of erosion processes at the Curonian Spit was recorded in 1984–1995. The negative shoreline shift towards the mainland was found in 6.07% (25 out of 412) of transects, where the average NSM value was −19.38 ± 2.50 m. Stable shoreline changes were found in 18.69% (77 of 412) of transects, and accumulation was detected in 75.24% (310 of 412) of transects with an accumulation rate of 2.17 ± 0.05 m/year, NSM value was 23.86 ± 0.52 m.

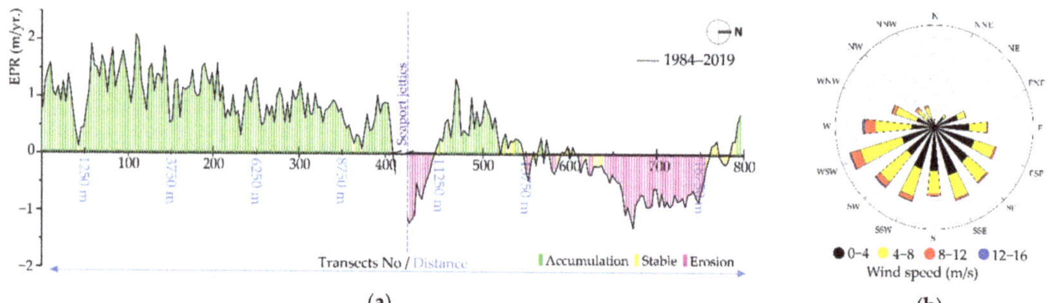

Figure 7. Graph showing the distribution of EPR (**a**) and wind rose (**b**) for 1984–2019.

In **1984–2019**, accumulation processes occurred in 22.14% (85 out of 384) of transects on the mainland coast. The shoreline shifted towards the sea within 20.30 ± 1.04 m, with an average speed of 0.57 ± 0.03 m/year (Figure 7). Erosion during this period accounted for 43.23% (166 out of 384) of transects, and the shoreline shifted towards the mainland at an average velocity of −0.70 ± 0.02 m/year; the NSM value was −24.84 ± 0.74 m. Stable shoreline was found in 34.64% (133 of 384) of transects. Significant coastal erosion extends at the northern pier of the Port of Klaipėda −56.9 m in transect 413 (Figure 6). Accumulation processes in the accesses of Port of Klaipėda piers changed to intensive erosion, which in 2019 covered 700 m (28 transects) of the coast; the total NSM in them was −28.28 m, the EPR value was −0.76 ± 0.04 m/year.

4.2. Short-Term Shoreline Changes

Comparison of the shoreline changes in 1984–1990 and 1984–2019 showed that the area of eroded coast increased 2.7 times, from 1.50 km (60 transects) to 4.15 km (166 transects).

The effect of accumulation processes in 1984–2019 was recorded in 85 transects instead of 145 transects in 1984–1990. The accumulation rate decreased from 4.33 ± 0.11 m/year to 0.57 ± 0.03 m/year. The area of stable shores decreased from 3.325 km (133 transects) to 4.475 km (179 transects).

During the **1984–1990** period (Figure 8), the overall shoreline change was positive—the coast moved seawards on average 23.95 ± 0.76 m. During this period, the predominant wind direction was W, WSW, and the average wind speed variated from 0 to 16 m/s.

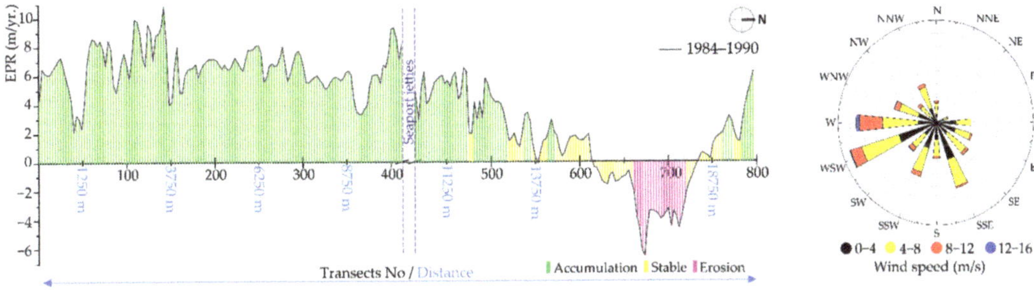

(a) (b)

Figure 8. Graph showing the distribution of EPR (**a**) and wind rose (**b**) for 1984–1990.

Accumulation was detected in all transects of the Curonian Spit coast, where the shoreline moved seawards by 12.99–65.08 m with an average velocity of 6.56 ± 0.08 m/year. On the mainland coast, the shoreline position changes were observed within the range of determination ± 12.42 m and can be considered as quasi-stable.

Coastal erosion was observed in a 1.5 km (60 transects) area to the north in the 6.2 km from the northern seaport jetty. The shoreline moved landward at an average velocity of −3.97 ± 0.13 m/year. The most significant negative change occurred in the 672nd transect and reached −41.58 m. Accumulation occurred in 37.8% of transects on the mainland coast, and here the shoreline moved seawards, with an average velocity of 4.33 ± 0.11 m/year.

In the **1990–1995** period (Figure 9), the coast has been intensively eroded, with the predominant 0–16 m/s W, WSW, SW wind direction. The shoreline moved landwards in 620 (77.9%) from 796 transects with an average of −22.85 ± 0.46 m. Significant changes in shoreline movement were observed in the immediate proximity of the seaport jetties. In the Curonian Spit coast, the maximum value of NSM was −100.85 m and was detected in the 412rd transect, next to the southern Klaipėda seaport jetty (Figure 8). The most significant shoreline movement landwards was observed in a 250 m (402–412 transects) coastal area to the south from the southern seaport jetty. Here the shoreline moved toward land on average 77.88 ± 1.11 m with an average velocity of (EPR) −15.58 ± 0.22 m/year.

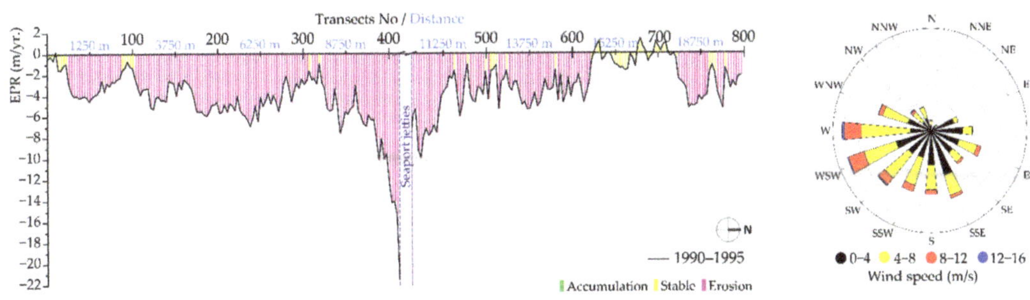

Figure 9. Graph showing the distribution of EPR (**a**) and wind rose (**b**) for 1990–1995.

65.9% of transects on the mainland coast can be described as erosive. The average change velocity reached −4.09 ± 0.10 m/year, and the shoreline moved landwards about −20.45 ± 0.51 m. The quasi-stable coast was observed in 131 transects (34.1%), and an average EPR value was 0.44 ± 0.08 m/year. The most significant shoreline movement >30 m was detected in the 419–443 transect. The maximum value was observed in the 424th transect and reached 49.61 m (EPR −9.92 m/year).

The following ten years, **1995–2005**, with the predominant SW, SSW, and WSW (0–16 m/s velocity) winds (Figure 10), had accumulative tendencies at the Curonian spit coast. The coast started recovery after the previous erosive period. Furthermore, hurricane Anatoly, which occurred in December 1999 [20], was not visible in the coastal evolution processes. It is evident that the quasi-stable part became erosive during the last five years at the mainland coast, and all other parts stayed accumulative.

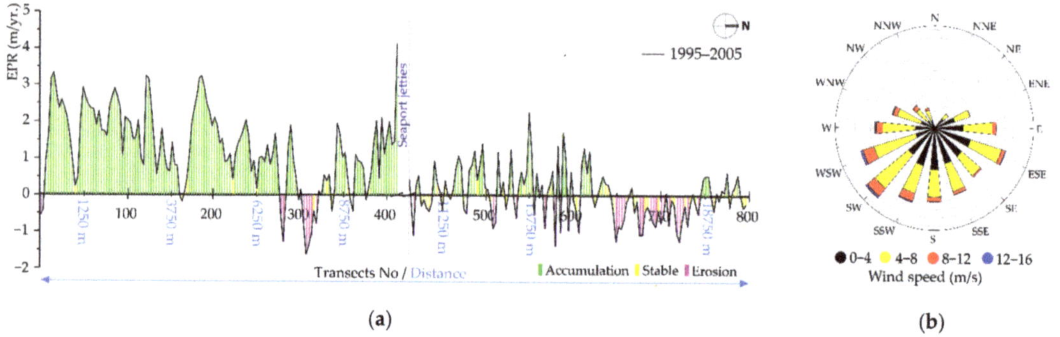

Figure 10. Graph showing the distribution of EPR (**a**) and wind rose (**b**) for 1995–2005.

The total change of the shoreline in the studied area in 1995–2005 was positive and amounted to 6.72 ± 0.39 m with an EPR value of 0.67 ± 0.04 m/yr. The Curonian Spit coast was characterized as accumulative. Here accumulation processes were observed in 320 transects from 412, and the accumulation rate was 1.70 ± 0.044 m/yr. Erosion was observed in 27 transects (650 m). From 304 to 320 the transect EPR value was −1.00 ± 0.03 m/yr. From 277 to 282, the EPR value reached −0.86 ± 0.10 m/yr. The significant accumulation rate of 4.15 m/yr. (NSM 41.52) was noted in the immediate proximity of the jetties.

In the next five years, **2005–2010** (Figure 11), wind accumulation processes prevailed, with the WSW, SW, S, SE (0–12 m/s). In 61.1% of transects, the shoreline moved seawards with an averaged velocity of 2.12 ± 0.05 m/yr., and NSM value reached 10.62 ± 0.25 m.

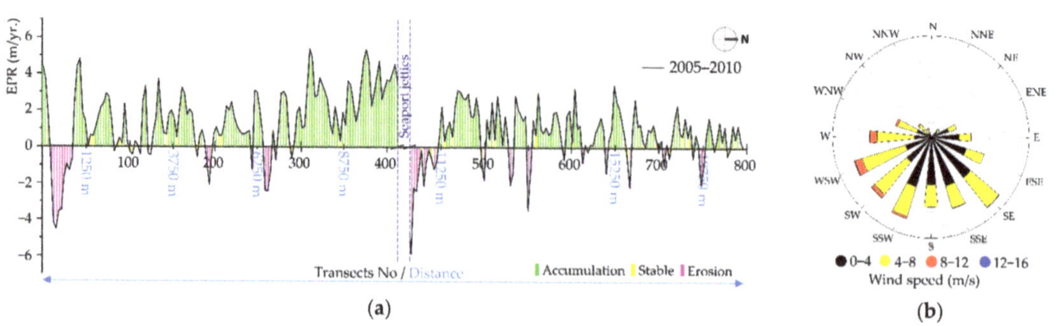

Figure 11. Graph showing the distribution of EPR (**a**) and wind rose (**b**) for 2005–2010.

Accumulation processes were more frequent on the Curonian Spit coast, which was observed in 67.7% of transects. The average velocity of shoreline movement seawards was +2.42 ± 0.07 m/yr. During 2005–2010 the shoreline erosion on the Curonian Spit coast occurred only in 10.40% of transects that amounted to 1075 m out of 10.3 km. The significant erosive coastal stretch was found in the southern part of the Curonian Spit between 10 and 34 transects. In the 625 m section, the shoreline moved landwards, on average −12.82 ± 0.29 m (EPR −2.56 ± 0.26 m/yr). The maximum value of NSM was noted in the 26th transect and reached −26.69 m.

On the mainland coast, accumulation was detected in 53.9% (270 out of 384) of transects, and the shoreline moved towards the sea by an average of 8.62 ± 0.28 m. The average EPR value was 1.73 ± 0.06 m/yr. Stable shoreline changes or changes in the shoreline determination uncertainty range within ±0.69 m/yr were detected at 119 or 31% of transects. Coastal erosion was recorded in 15.1% of transects (58 transects), in which the shoreline moved landwards at an average speed of −2.01 ± 0.19 m/yr. The most significant adverse changes in the shoreline position were found between 413 and 446 transects. In this 850 m-long coast stretch, the shoreline shifted to the mainland on average by −9.64 ± 0.28 m (EPR was −1.93 ± 0.06 m/yr).

During the **2010–2015** period (Figure 12), with the predominant WSW, SW, S, SE (0–12 m/s) winds, accumulation processes were noticed in 94.9% of transects (391 out of 412 transects) on the coast of the Curonian Spit, in which the shoreline moved seawards at an average speed of 3.40 ± 0.09 m/yr. In 50% of transects (206 out of 412 transects), the shoreline shifted from land to sea by an average of 27.82 ± 0.04 m (NSM). The maximum value of NSM reached 49.67 m in the 318th transect.

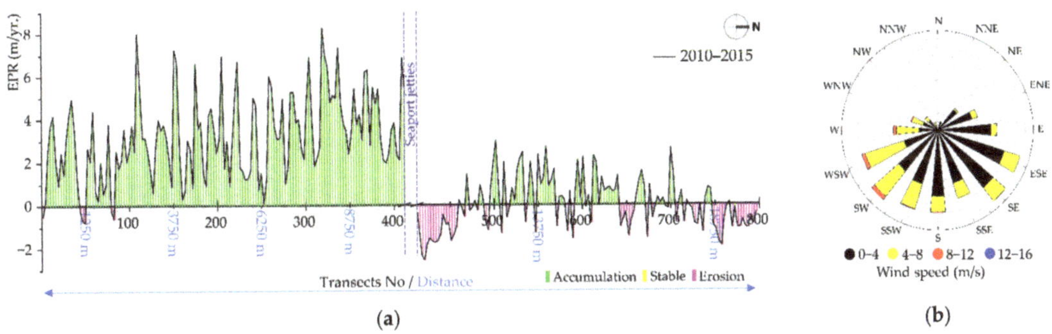

(a) (b)

Figure 12. Graph showing the distribution of EPR (**a**) and wind rose (**b**) for 2010–2015.

On the mainland coast, erosive processes were observed during 2010–2015. Negative tendencies of shoreline displacement landwards were recorded in 47.4% of transects (182 out of 384), in which the shoreline generally shifted at an average speed of −0.51 ± 0.07 m/yr. The significant shoreline movement towards land was recorded in the 1175 m shoreline section north of the northern seaport jetty (between tr. 413 and 459). The average EPR value was −1.49 ± 0.01 m/yr, and the average NSM value was −8.63 ± 0.07 m; the maximum value of EPR was −2.57 m/yr in 421 transects, and the maximum NSM value was −14.84 m. The section of the shore from 746 to 796 transects stands out. This shore of 1275 m in 2010–2015 moved towards the sea in total −5.10 ± 0.07 m, and the erosion rate reached −0.88 ± 0.01 m/yr. The central part of the mainland coast was mainly formed by accumulation processes, which accounted for 41.9% of all transects (182 out of 284). The average accumulation rate in these transects was 1.14 ± 0.06 m/yr, the value of NSM was 6.80 ± 0.34 m. Stable shoreline fluctuations of about ± 0.19 m/yr were recorded in the 41st transect.

During the last analyzed period **2015–2019** (Figure 13), the predominant wind direction was WSW, SW, SWS, S, SSE (0–12 m/s velocity) and all of the coast was erosive. Over these 4 years, the shoreline moved seawards in 80.9% of transects (644 out of 796) with the average EPR value −4.24 ± 0.12 m/yr., and NSM — −15.91 ± 0.46 m.

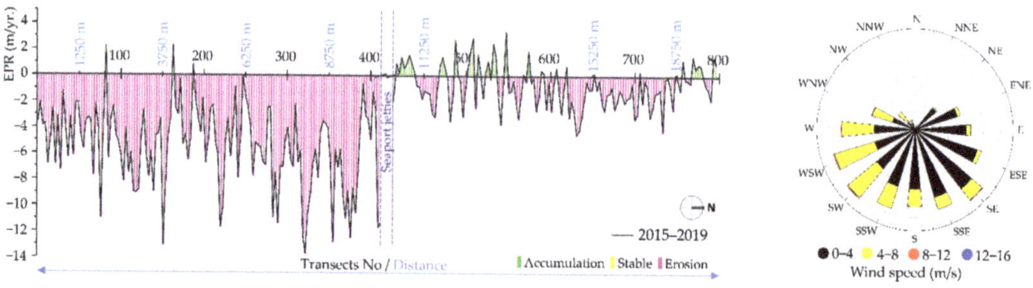

(a) (b)

Figure 13. Graph showing the distribution of EPR (**a**) and wind rose (**b**) for 2015–2019.

On the Curonian Spit coast, erosion processes were detected at 97.3% of transects (401 out of 412) and were 3 times more intense than on the mainland. Here the EPR value reached -5.72 ± 0.15 m/yr., and the NSM respectively was -1.80 ± 0.07 m/yr.

The mainland coast moved seawards in 63.3% of transects (243 out of 384). In the southern part of the mainland coast, 105 transects (27.3%) were accumulative with an average velocity of 1.30 ± 0.08 m/yr; here, the NSM value was 4.86 ± 0.29 m.

In 2015, the Klaipėda seaport authorities started a nearshore nourishment project in front of the mainland coast (Figure 2). As a result, the additional sediments in the longshore sediment transport system led to milder coastal erosion on the mainland coast.

4.3. Clusterization

K-Means cluster analysis was used to group the transects to identify stretches of shoreline with similar development tendencies. Net Shoreline Movement (NSM) values over the study period were grouped into five clusters (Figure 14). The NSM and SCE values and results of the cluster analysis distinguish different processes in different stretches of the Curonian Spit and the mainland coast and reflection of the influence of Klaipėda seaport piers on the morpho-lytodynamic processes of the coast.

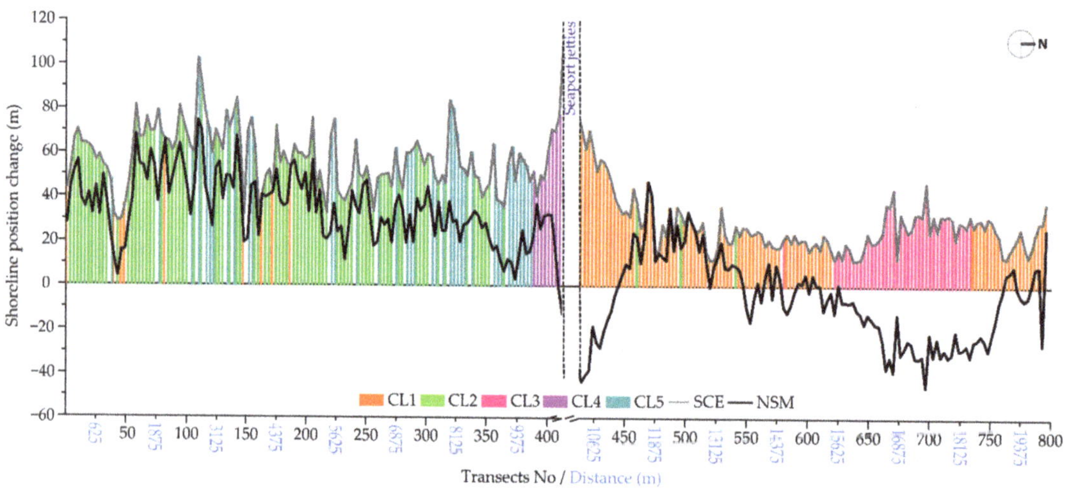

Figure 14. Graph showing the distribution of shoreline change envelope (SCE) (gray line) and net shoreline movement (NSM) (black line) along the study area for 1984–2019, and five clusters: cluster No. 1 (CL1), cluster No. 2 (CL2), cluster No. 3 (CL3), cluster No. 4 (CL4), cluster No. 5 (CL5).

The SCE corresponds closely with the NSM, implying that progressive and continuous change is more common than cyclical or reverse behavior in the spatial pattern of shoreline variability along the Curonian Spit. This stretch of coast connects Clusters No. 2 and No. 5, where the shoreline shifted towards the sea at an average of 38.93 ± 1.53 m and 27.66 ± 2.17 m, respectively (Table 1). Both clusters indicate accumulation processes on the coast. In cluster No. 2, the accumulation rate was 1.10 ± 0.04 m/yr., the SCE range was 65.14 m. In cluster No. 5, the shoreline moved towards the sea at an average velocity of 0.78 ± 0.06 m/yr. The SCE ranged between 38.01 m and 102.62 m (64.61 m). Moreover, on the coast of the Curonian Spit, Cluster No. 4 enters the southern port pier impact zone, which includes 27 transects (675 m long shoreline), where the shoreline may have different trends onshore dynamics at different times depending on hydrometeorological conditions. During the study period, the total change of the shoreline position in this cluster was positive and reached 20.74 ± 5.52 m, and the accumulation speed was 0.58 ± 0.16 m/yr. NSM values in this cluster ranged from −11.66 to 37.07 m.

The SCE closely corresponds with NSM along the mainland coast, except for the 445 and 547 transect section. The section of Cluster No. 1 is alternating, mainly due to anthropogenic activity, such as beach nourishment.

The majority (67.2%) of the mainland coast transects belong to cluster No. 1 (No. 2—3.1%, No. 3—29.7%). Four coast sections can be distinguished in this area, where the shoreline has different movement tendencies in the transects in the 675 m long section of the coast (from 415 to 442 tr.) North of the northern port jetty, erosion processes took place during the study period. The average erosion rate (EPR) was −0.64 ± 0.04 m/year, and the NSM value was −24.59 ± 1.31 m. The NSM range covered values from −4.19 m to −43.49 m, with a mean SCE of 56.74 ± 0.96 m. From 445 to 547 transects, the shoreline position changed at an average speed of 0.47 ± 0.01 m/year. The total NSM in transects was 16.67 ± 0.36 m. from −0.33 m to 47.25 m. SCE from 11.8 m to 47.25 m. In 2014–2018, by order of the Klaipėda seaport Authority, 237.78×10^3 m^3 of sand was dumped on the coast near the beaches of Melnragė-Giruliai (Figure 2).

Another group of transects from 519 to 619 in Cluster No. 1 showed slightly negative shoreline position changes, in which the shoreline moved towards the mainland during the study period by −0.05 ± 0.01 m/yr., the mean NSM value was −1.93 ± 0.30 m. SCE ranges from 15.78 m to 26.37 m, NSM from −16.07 to 10.73 m. In the northern part of cluster No. 1, from 736 to 796 transects, changes in the shoreline influenced by erosive processes were recorded. Here the shoreline changed at an average velocity of −0.20 ± 0.02 m/yr. NSM was −7.15 ± 0.72 m (from −29.23 to 25.7 m), SCE covered an overall change of 23.83 ± 0.32 m and ranged from 12.82 m to 37.52 m.

Cluster No. 3 covers the central part of the mainland coast and indicates transects in which negative trends in shoreline dynamics have occurred during the study period. The shoreline of the 117 transects of this cluster moved towards the mainland at an average velocity of −0.64 ± 0.05 m/yr. The overall change in NSM was −22.70 ± 1.74 m.

This indicates the accretion processes in the Curonian Spit coast. The clusterization approach also suggests the accretion processes on the Curonian Spit coast with positive values of SCE and NSM (Table 3).

Table 3. Net Shoreline Movement (NSM) values and Shoreline Change Envelope (SCE) values per cluster.

Clusters	Transects	SCE (m)			NSM (m)		
No.	No.	Mean	Min	Max	Mean	Min	Max
1	285	29.34 ± 1.38	11.8	70.36	4.07 ± 2.07	−43.49	65.62
2	255	55.74 ± 1.44	27.29	92.43	38.93 ± 1.53	4.3	69.97
3	117	25.41 ± 1.41	11.92	46.76	−22.70 ± 1.74	−45.53	0.7
4	27	64.25 ± 6.91	40.51	108.85	20.74 ± 5.52	−11.66	37.07
5	114	62.68 ± 2.18	38.01	102.62	27.66 ± 2.17	3.13	74.44

4.4. Meteorological Data Analysis

Changes in the wind direction are determined as the primary driver for sediment transport and drive coastal erosion [1,16,43,44]. The long-term wind direction and velocity at the studied area were analysed to indicate such changes.

The time series of yearly mean wind direction at Klaipėda is presented in Figure 15, and demonstrates changes in the regime of wind direction in the 1960–2019 period and suggests that at least two regime shifts have occurred during this period. The regime shift timings are found using a cut-off length of 10 years and Hubert's weight parameter of 1 [42]. This method detected that from 1960 till 1992, the wind direction on average was 216° (SW), then an average direction shifted to 188° (S), and the recent shift that occurred in 2011 was to 177° (S). The applied Rodionov regime shift method indicates that the average wind direction is shifting to the southern direction.

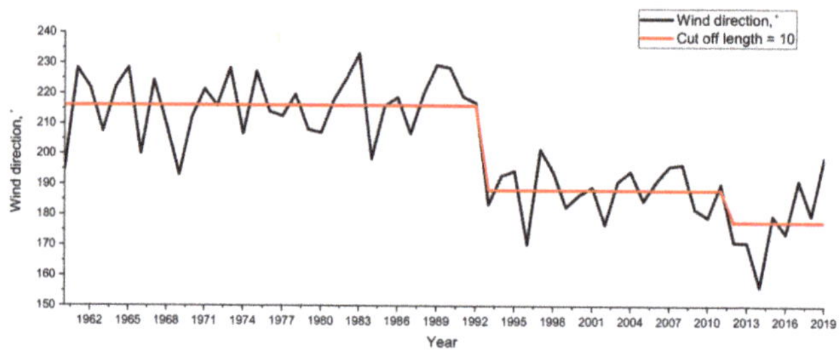

Figure 15. A shift in the annual average wind direction in Klaipėda in 1960–2019.

The first observed regime shift in the mean values of wind direction occurred in 1992 (Figure 15). At this point, we observed that the wind direction shifted to the west–south direction. This change in the regime coincides with the changes in the shoreline that occurred when erosion was observed both on the Curonian Spit and on the mainland coast. Another detected regime shift occurred in 2011 with the same shift to the southern direction. During this period on the mainland coast, erosion processes were observed, and accumulation prevailed on the Curonian Spit coast.

The frequency distribution (Figure 16) of the predominant wind direction at Klaipėda in the 1960–2019 period determines that the predominant wind, up to1995, was 270° (W). The applied Rodionov shift detection method (Figure 15) confirms that in 1995 the predominant wind direction shifted to 209° (SSW).

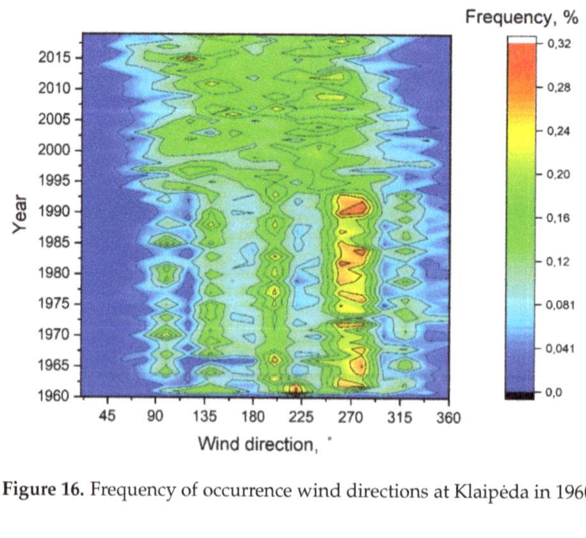

Figure 16. Frequency of occurrence wind directions at Klaipėda in 1960–2019.

5. Discussion

The Port of Klaipėda jetties location interrupts the natural longshore sediment transport path from the south to north at this point of the South-East Baltic Sea [6,23,45,46]. This should create favorable conditions for the two different processes: accumulation on the Curonian Spit south of the jetties and erosive—north of the jetties. Although the long-term analysis of shoreline changes in the whole study area indicates a total positive shoreline shift towards the sea, on the average velocity of 0.43 ± 0.03 m/yr, over the 35 years, the shoreline had different trends in both geomorphological and temporal changes. From the long-term perspective, the 10 km long Curonian Spit coast to the south of the southern Klaipėda seaport jetties is attributed to the accumulating coastal stretch. The mainland coast encompassing the northern part of the study site is affected by erosive processes.

The jettie's seaport systems on a straight sandy shore block the natural littoral drift [47,48], which determines the development of shoreline configurations. Typically, an up and down littoral drift is formed when hard breaking structures interrupt the predominant sediment transport direction. Due to the prevailing W and SW winds off the coast of Lithuania, sand transport is directed from south to north [49–52]. As a result, up-drift accretion occurs on the Curonian Spit coast on the south side of the jetties. Down-drift erosion occurs after losing its replenishment to maintain stability on the mainland coast (on the north side of the jetties).

The morphological changes of sandy beaches occur rapidly on a spatio-temporal scale as a response to natural (wind direction and speed, wave climate, sea-level fluctuations, etc.) processes [53]. Signs of climate change in the Baltic Sea can be more than just seawater level rise [54–56], increase in storminess [1], but also changes in the predominant wind and wave climate [43]. The climate change indicator in the wind regime is characterized as increasing in the wind velocity or intense wind events and changes in the predominant wind direction. This indicates changes in the cyclone patches over the Baltic Sea [57]. Changes in the wind direction and wave climate can alter longshore sediment transport magnitude and the dominant direction [58,59].

During this study, changes were observed in the predominant wind direction since 1992 (Figure 14), when the first regime shift occurred. The second shift in the wind direction regime was observed in 2012 (Figure 14). Significant changes in the predominant coastal evolution processes were observed after the wind direction shifts. Observed shifts of wind direction regime correspond with short-term changes of shoreline dynamics.

Shifts of wind direction regimes influenced intensified coastal erosion on both the Curonian Spit and the mainland coasts. In particular, the change in the wind direction regime influenced the short-term development of the Curonian Spit coast. In the periods of 1990–1995 and 2015–2019, the degree of erosion on this coast reached the respective levels of 4.57 ± 0.09 and 4.24 ± 0.12 m/year. The shoreline movement tendency of the 19th century was observed when the shoreline shifted towards the sea on both the Curonian Spit and the mainland coast [21]. This tendency reoccurred in the period of 2015–2019, on the usually accumulative Curonian Spit coast, which became erosive, while the average rate of erosion processes on the mainland coast decreased. In order to identify shoreline movement changes related to shifts in hydrometeorological conditions, a detailed investigation of wave climate (height, direction, period), sea-level fluctuations, and stormy events is required. Wave climate is driven by the wind climate [1,60] combined with the wind-driven coastal currents, and these are the major drivers for erosion and sedimentation, especially along the sandy sections of sandy beaches, dunes and soft moraine cliffs [2,61]. Future coastal process predictions are complicated as potential changes in the long-term mean and extreme wind speeds have a high uncertainty rate [1,62].

Moreover, significant changes in shoreline dynamics were observed in periods after the 2002 Klaipėda seaport reconstruction. Intensive erosion was observed on the mainland coast in the nearest proximity to the seaport jetties. Erosion after the reconstruction is acknowledged in other authors' [13,63] research. However, nowadays, as well as in the past, the main factor for the coastal erosion processes was attributed to dredging works in the Klaipėda seaport and especially in port jetty area [22,64].

Dredging works are carried out to maintain proper water levels in fairways, waterways, and ports. Work related to the extraction of bottom sediments includes various areas of activity related to their extraction, transportation, storage, cleaning, and practical use. Dredging works disturb the natural integrity of bottom sediments (benthos) and directly and indirectly impact all marine environment elements [65,66]. Sediments excavated from the Baltic Sea coast are stored in designated areas at sea or on land. Such sites are usually located near port areas for economic motives [65]. Current environmental trends encourage the recycling or practical use of excavated sediments. One of the essential practical advantages is the beach nourishment with extracted sand if it meets the established physical and chemical properties. Artificial sand nourishment can be used as a coastal erosion mitigating measure by adding sediments directly to the coast or supplementing the natural longshore sediment transport budget.

In 2014–2018, by order of the Klaipėda seaport Authority, 237.78 $\times$ 10^3 m^3 of fine sand was dumped on the nearshore beaches of Melnragė-Giruliai at 4–6 m depth (Figure 4). The extracted sediments from the Klaipėda strait were used to restore the mainland sediment budget and replenish the coast. Beach sand nourishment is a widely known method to widen and restore the subaerial beach and decrease coastal erosion [67–69]. The nourishment material redistribution is driven by local hydrodynamic conditions (waves and currents). The predominant longshore current is directed from south to north along the Lithuanian coast [49,51]. Therefore, to mitigate the disrupted natural sediment transport by Klaipėda seaport jetties, the sediment dumping areas are located north of Klaipėda seaport jetties (Figure 4). The grain size distribution of the sand is dominated by grains with a size of 0.1–0.25 mm, representing 70–96% of grains with an Md between 0.14 mm and 0.22 mm, which corresponds precisely to the composition of the beach sand. Such sand composition detected on the mainland coast indicates that the nourishment material is transported in a predominant longshore direction and significantly influences cross-shore profile evolution.

Understanding the short- and long-term variability of the shoreline changes could help design shore nourishment in such a way that anthropogenic activity would be carried out in coherence with natural processes rather than in conflict [70,71]. Usually, shoreline change rates are best suited for the quasi-linear trend analysis. However, values of the shoreline variation are often non-linear and have different trend reversals. It is possible to single out the behaviors of certain groups that have the same or similar tendencies of change when using a joint shoreline change rates trend and cluster-based segmentation analysis.

According to K-means clustering of long-term changes in five different short-term periods in 796 transects, 369 transects covering clusters No. 2 and No. 5 are essentially distributed at the Curonian Spit and indicate accumulation processes. The positive dynamic characteristics of this coastal stretch are essentially in line with the multi-year shoreline changes in this coast type. Moreover, they reflect the main geomorphological and sedimentary conditions of the Curonian Spit.

The Klaipėda seaport impact zone was reflected in clusters No. 1 and No. 5. Still, cluster No. 1 identifies significant anthropogenic activities or impacts on the mainland coastal stretch due to shore replenishment. At the same time, on the mainland coast further from the direct port jetties impact area [20,28], Cluster No. 3 shows the presence of other factors with a more significant impact on the shoreline evolution. The trend in the SCE indicator also distinguishes the accumulative stretch of shore from 445 to 550 transects, which proves the impact of damping of the dredged sand from the Klaipėda strait.

6. Conclusions

Forecasting and continuous estimation of the intensity of the sandy South Eastern Baltic Sea coast dynamics are essential to customizing coastal development management methods and techniques that affect the nature and economics of the coastal environment. The analysis of long- and short-term shoreline changes should provide the required knowledge for reducing the extent of the anthropogenic intervention factors into the natural coastal system with long-lasting consequences.

This study aims to qualitatively and quantitatively identify the sandy South-Eastern Baltic Sea coast shoreline evolution tendencies. The reconstruction of Klaipėda jetties disrupted the settled equilibrium stage, interrupted the longshore sediment transport, and activated erosion processes. As a result, in the long-term (1984–2019) perspective, the northern part of the coast became abrasive, eroded coast length increased three times, from 1.5 to 4.2 km.

Assessment of short-term shoreline changes combined with K-means cluster analysis has helped identify the direct impact zone of the Port of Klaipėda. In this study, short-term shoreline changes correspond with shifts in wind direction and reflect the effect of the dredging works in the Klaipėda strait. The research helped identify the part of the mainland coast (transects from 445 to 550) that acquires other dynamic properties of the shore—accumulation. Although according to the hydrometeorological and litho-geomorphological characteristics and the impact of the port, erosion processes should prevail. It occurs due to coastal zone nourishment works. Therefore, this site needs continuous research because it is sensitive to anthropogenic and meteorological conditions. It also requires regular monitoring of the coast nourishment, as the development of coastal infrastructure, coastal use for recreational purposes, and planning of coastal protection measures depend on it.

Author Contributions: Conceptualization, V.K.; methodology, V.K., L.K.-R. and I.Š.; software, V.K.; validation, V.K., I.Š. and E.B.; formal analysis, V.K.; investigation, I.Š.; data curation, V.K., E.B. and I.Š.; writing—original draft preparation, V.K.; writing—review and editing, I.Š. and E.B.; visualization, V.K. and I.Š.; supervision, L.K.-R. All authors have read and agreed to the published version of the manuscript.

Funding: The APC was funded by Lithuanian science foundation project "Development of doctoral studies" Nr. 09.3.3-ESFA-V-711-01-0001. Kelpšaitė-Rimkienė was also supported by the Baltic Research Programme (EEA Financial Mechanisms 2014–2021) project "Solutions to current and future problems on natural and constructed shorelines, eastern Baltic Sea" (EMP480).

Institutional Review Board Statement: Not applicable.

Informed Consent Statement: Not applicable.

Data Availability Statement: The data presented in this study are available on request from the corresponding author.

Acknowledgments: We would like to thank the Klaipėda State Seaport Authority for supporting this research and providing data.

Conflicts of Interest: The authors declare no conflict of interest.

References

1. Weisse, R.; Dailidiene, I.; Hünicke, B.; Kahma, K.; Madsen, K.; Omstedt, A.; Parnell, K.; Schöne, T.; Soomere, T.; Zhang, W.; et al. Sea Level Dynamics and Coastal Erosion in the Baltic Sea Region. *Earth Syst. Dyn. Discuss.* **2021**, *12*, 871–898. [CrossRef]
2. Zhang, W.; Schneider, R.; Kolb, J.; Teichmann, T.; Dudzinska-Nowak, J.; Harff, J.; Hanebuth, T.J.J. Land-sea interaction and morphogenesis of coastal foredunes—A modeling case study from the southern Baltic Sea coast. *Coast. Eng.* **2015**, *99*, 148–166. [CrossRef]
3. Montaño, J.; Coco, G.; Cagigal, L.; Mendez, F.; Rueda, A.; Bryan, K.R.; Harley, M.D. A Multiscale Approach to Shoreline Prediction. *Geophys. Res. Lett.* **2021**, *48*. [CrossRef]
4. Davidson, M.A.; Splinter, K.D.; Turner, I.L. A simple equilibrium model for predicting shoreline change. *Coast. Eng.* **2013**, *73*, 191–202. [CrossRef]
5. Phillips, B.T.; Brown, J.M.; Bidlot, J.R.; Plater, A.J. Role of Beach Morphology in Wave Overtopping Hazard Assessment. *J. Mar. Sci. Eng.* **2017**, *5*, 1. [CrossRef]
6. Viška, M.; Soomere, P.D.T. *Sediment Transport Patterns along the Eastern Coasts of the Baltic Sea*; Tallin University of Technology: Tallin, Estonia, 2014.
7. Soomere, T.; Viška, M.; Lapinskis, J.; Räämet, A. Linking wave loads with the intensity of erosion along the coasts of Latvia. *Est. J. Eng.* **2011**, *17*, 359–374. [CrossRef]

8. Bulleri, F.; Chapman, M.G. The introduction of coastal infrastructure as a driver of change in marine environments. *J. Appl. Ecol.* **2010**, *47*, 26–35. [CrossRef]
9. Schlacher, T.A.; Dugan, J.; Schoeman, D.S.; Lastra, M.; Jones, A.; Scapini, F.; McLachlan, A.; Defeo, O. Sandy beaches at the brink. *Divers. Distrib.* **2007**, *13*, 556–560. [CrossRef]
10. Hegde, A.V. Coastal erosion and mitigation methods-Global state of art. *Indian J. Geo-Mar. Sci.* **2010**, *39*, 521–530.
11. Rashidi, A.H.M.; Jamal, M.H.; Hassan, M.Z.; Sendek, S.S.M.; Sopie, S.L.M.; Hamid, M.R.A. Coastal Structures as Beach Erosion Control and Sea Level Rise Adaptation in Malaysia: A Review. *Water* **2021**, *13*, 1741. [CrossRef]
12. Bezerra, M.O.; Pinheiro, L.; Morais, J.O. Shoreline Change of the Mucuripe Harbour Zones (Fortaleza-Ceará, Northeast of Brazil) 1972–2003 on JSTOR. Available online: https://www.jstor.org/stable/26481755 (accessed on 15 December 2021).
13. Bagdanavičiūtė, I.; Kelpšaitė-Rimkienė, L.; Galinienė, J.; Soomere, T. Index based multi-criteria approach to coastal risk assesment. *J. Coast. Conserv.* **2019**, *23*, 785–800. [CrossRef]
14. Bagdanavičiūtė, I.; Kelpšaitė, L.; Daunys, D. Assessment of shoreline changes along the Lithuanian Baltic Sea coast during the period 1947–2010. *Baltica* **2012**, *25*, 171–184. [CrossRef]
15. Burningham, H.; French, J. Understanding coastal change using shoreline trend analysis supported by cluster-based segmentation. *Geomorphology* **2017**, *282*, 131–149. [CrossRef]
16. Kelpšaite, L.; Dailidiene, I. Influence of wind wave climate change on coastal processes in the eastern Baltic Sea. *J. Coast. Res.* **2011**, *27*, 220–224.
17. Baltranaitė, E.; Kelpšaitė-rimkienė, L.; Povilanskas, R.; Šakurova, I.; Kondrat, V. Measuring the impact of physical geographical factors on the use of coastal zones based on bayesian networks. *Sustainability* **2021**, *13*, 7173. [CrossRef]
18. Bagdanavičiūtė, I.; Umgiesser, G.; Vaičiūtė, D.; Bresciani, M.; Kozlov, I.; Zaiko, A. GIS-based multi-criteria site selection for zebra mussel cultivation: Addressing end-of-pipe remediation of a eutrophic coastal lagoon ecosystem. *Sci. Total Environ.* **2018**, *634*, 990–1003. [CrossRef]
19. Žilinskas, G.; Pupienis, D.; Jarmalavičius, D. Possibilities of regeneration of palanga coastal zone. *J. Environ. Eng. Landsc. Manag.* **2010**, *18*, 92–101. [CrossRef]
20. Jarmalavičius, D.; Žilinskas, G.; Pupienis, D. Impact of Klaipda port jetties reconstruction on adjacent sea coast dynamics. *J. Environ. Eng. Landsc. Manag.* **2012**, *20*, 240–247. [CrossRef]
21. Žaromskis, R.P. *Baltijos Jūros Uostai: Monografija*; Vilnius University: Vilnius, Lithuania, 2008; ISBN 9789955332510.
22. Žilinskas, G.; Janušaitė, R.; Jarmalavičius, D.; Pupienis, D. The impact of Klaipėda Port entrance channel dredging on the dynamics of coastal zone, Lithuania. *Oceanologia* **2020**, *62*, 489–500. [CrossRef]
23. Demereckas, K. *Klaipėdos Uostas = Port of Klaipeda*; Libra Mamelensis: Klaipėda, Lithuania, 2007.
24. History. Available online: https://www.portofklaipeda.lt/history (accessed on 14 November 2021).
25. Vareikis, V.; Bareiša, E. *Technika ir Gamta: Klaipėdos Uostas XIX a. Pabaigoje—XX a. Pirmojoje Pusėje = Technology and Nature: The Port of Klaipėda in the End of the 19th and the First Half of the 20th Century*; Klaipedos Apskrities Archyvas: Klaipėda, Lithuania, 2014; ISBN 9789955188148.
26. Kelpšaite-Rimkiene, L.; Soomere, T.; Bagdanavičiute, I.; Nesteckite, L.; Žalys, M. Measurements of Long Waves in Port of Klaipeda, Lithuania. *J. Coast. Res.* **2018**, *85*, 761–765. [CrossRef]
27. Ministry of Transport and Communications of the Republic of Lithuania. *The Master Plan of the Port of Klaipeda (Land, Internal Water Area, External Raid, and Related Infrastructure) N. 15088*; Ministry of Transport and Communications of the Republic of Lithuania: Vilnius, Lithuania, 2019.
28. Pupienis, D.; Jonuškaitė, S.; Jarmalavičius, D.; Žilinskas, G. Klaipėda port jetties impact on the Baltic Sea shoreline dynamics, Lithuania. *J. Coast. Res.* **2013**, *165*, 2167–2172. [CrossRef]
29. Bitinas, A.; Žaromskis, R.; Gulbinskas, S.; Damušyte, A.; Žilinskas, G.; Jarmalavičius, D. The results of integrated investigations of the Lithuanian coast of the Baltic Sea: Geology, geomorphology, dynamics and human impact. *Geol. Q.* **2005**, *49*, 355–362.
30. Bitinas, A.; Aleksa, P.; Damušytė, A.; Gulbinskas, S.; Jarmalavičius, D.; Kuzavinas, M.; Minkevičius, V.; Pupienis, D.; Trimonis, E.; Šečkus, R.; et al. *Baltijos Jūros Lietuvos Krantų Geologinis Atlasas*; Geological Survey of Lithuania: Vilnius, Lithuania, 2004.
31. Thieler, E.R.; Himmelstoss, E.A.; Zichichi, J.L.; Ergul, A. *The Digital Shoreline Analysis System (DSAS) Version 4.0—An ArcGIS Extension for Calculating Shoreline Change*; Open-File Report; U.S. Geological Survey: Reston, VA, USA, 2009. [CrossRef]
32. Himmelstoss, E.A.; Henderson, R.E.; Kratzmann, M.G.; Farris, A.S. *Digital Shoreline Analysis System (DSAS) Version 5.0 User Guide*; Open-File Report 2018–1179; U.S. Geological Survey: Reston, VA, USA, 2018; Volume 126.
33. Oyedotun, T.D.T. Shoreline Geometry: DSAS as a Tool for Historical Trend Analysis. *Geomorphol. Tech.* **2014**, *2*, 1–12.
34. Byrnes, M.R.; Anders, F.J. Accuracy of Shoreline Change Rates as Determined From Maps and Aerial Photographs. *Shore Beach Obs.* **2016**, *58*, 30.
35. Dolan, R.; Fenster, M.S.; Holme, S.J. Temporal analysis of shoreline recession and accretion. *J. Coast. Res.* **1991**, *7*, 723–744.
36. Fletcher, C.; Rooney, J.; Barbee, M.; Lim, S.; Beach, W.P.; Fletchert, C.; Rooneyt, J.; Barbeef, M.; Limf, S.; Richmond, B. Mapping Shoreline Change Using Digital Orthophotogrammetry on Maui, Hawaii Stable URL Linked References are Available on JSTOR for This Article: Mapping Shoreline Change Using Digital Orthophotogrammetry on Maui. *J. Coast. Res.* **2003**, *18*, 106–124.
37. Crowell, M.; Leatherman, S.P.; Buckley, M.K. Shoreline Change Rate Analysis: Long Term Versus Short Term Data. *Shore Beach* **1993**, *61*, 13–20.

38. Laccetti, G.; Lapegna, M.; Mele, V.; Romano, D.; Szustak, L. Performance enhancement of a dynamic K-means algorithm through a parallel adaptive strategy on multicore CPUs. *J. Parallel Distrib. Comput.* **2020**, *145*, 34–41. [CrossRef]
39. Kanungo, T.; Mount, D.M.; Netanyahu, N.S.; Piatko, C.; Silverman, R.; Wu, A.Y. An Efficient *k*-Means Clustering Algorithm: Analysis and Implementation 1 Introduction. In Proceedings of the 16th Annual Symposium on Computational Geometry, New York, NY, USA, 12–14 June 2000; pp. 1–21.
40. Kelpšaitė-Rimkienė, L.; Parnell, K.E.; Žaromskis, R.; Kondrat, V. Cross-shore profile evolution after an extreme erosion event—Palanga, Lithuania. *J. Mar. Sci. Eng.* **2021**, *9*, 38. [CrossRef]
41. Likas, A.; Vlassis, N.; Verbeek, J. The global k-means clustering algorithm. *Pattern Recognit.* **2003**, *36*, 451–461. [CrossRef]
42. Rodionov, S.N. A sequential algorithm for testing climate regime shifts. *Geophys. Res. Lett.* **2004**, *31*, 2–5. [CrossRef]
43. Soomere, T.; Pindsoo, K. Spatial variability in the trends in extreme storm surges and weekly-scale high water levels in the eastern Baltic Sea. *Cont. Shelf Res.* **2016**, *115*, 53–64. [CrossRef]
44. Bagdanavičiute, I.; Kelpšaite, L.; Soomere, T. Multi-criteria evaluation approach to coastal vulnerability index development in micro-tidal low-lying areas. *Ocean Coast. Manag.* **2015**, *104*, 124–135. [CrossRef]
45. Gudelis, V. *Lietuvos Įjūris ir Pajūris = The Lithuanian offshore and Coast of the Baltic Sea: Monograph*; Science and Arts of Lithuania: Vilnius, Lithuania, 1998; ISSN 132-4044.
46. Dean, R.G.; Dalrymple, R.A. *Coastal Processes with Engineering Applications*; Cambridge University Press: Cambridge, UK, 2001; ISBN 9780521495356.
47. De Boer, W.; Mao, Y.; Hagenaars, G.; de Vries, S.; Slinger, J.; Vellinga, T. Mapping the sandy beach evolution around seaports at the scale of the African continent. *J. Mar. Sci. Eng.* **2019**, *7*, 151. [CrossRef]
48. Bruun, P. The Development of Downdrift Erosion Author (s): Per Bruun Stable URL. The Development of Downdrift Erosion. *J. Coast. Res.* **1995**, *11*, 1242–1257.
49. Viška, M.; Soomere, T. Simulated and observed reversals of wave-driven alongshore sediment transport at the eastern baltic sea coast. *Baltica* **2013**, *26*, 145–156. [CrossRef]
50. Krek, A.; Stont, Z.; Ulyanova, M. Alongshore bed load transport in the southeastern part of the Baltic Sea under changing hydrometeorological conditions: Recent decadal data. *Reg. Stud. Mar. Sci.* **2016**, *7*, 81–87. [CrossRef]
51. Pupienis, D.; Buynevich, I.; Ryabchuk, D.; Jarmalavičius, D.; Žilinskas, G.; Fedorovič, J.; Kovaleva, O.; Sergeev, A.; Cichoń-Pupienis, A. Spatial patterns in heavy-mineral concentrations along the Curonian Spit coast, southeastern Baltic Sea. *Estuar. Coast. Shelf Sci.* **2017**, *195*, 41–50. [CrossRef]
52. Žilinskas, G.; Jarmalavičius, D.; Pupienis, D. The influence of natural and anthropogenic factors on grain size distribution along the southeastern Baltic spits. *Geol. Q.* **2018**, *62*, 375–384. [CrossRef]
53. Benkhattab, F.Z.; Hakkou, M.; Bagdanavičiūtė, I.; El Mrini, A.; Zagaoui, H.; Rhinane, H.; Maanan, M. Spatial–temporal analysis of the shoreline change rate using automatic computation and geospatial tools along the Tetouan coast in Morocco. *Nat. Hazards* **2020**, *104*, 519–536. [CrossRef]
54. Chechko, V.A.; Chubarenko, B.V.; Boldyrev, V.L.; Bobykina, V.P.; Kurchenko, V.Y.; Domnin, D.A. Dynamics of the marine coastal zone of the sea near the entrance moles of the Kaliningrad Seaway Channel. *Water Resour.* **2008**, *35*, 652–661. [CrossRef]
55. Soomere, T.; Viška, M. Simulated wave-driven sediment transport along the eastern coast of the Baltic Sea. *J. Mar. Syst.* **2014**, *129*, 96–105. [CrossRef]
56. Dailidienė, I.; Davulienė, L.; Tilickis, B.; Stankevičius, A.; Myrberg, K. Sea level variability at the Lithuanian coast of the Baltic Sea. *Boreal Environ. Res.* **2006**, *11*, 109–121.
57. Lehmann, A.; Höflich, K.; Post, P.; Myrberg, K. Pathways of deep cyclones associated with large volume changes (LVCs) and major Baltic inflows (MBIs). *J. Mar. Syst.* **2017**, *167*, 11–18. [CrossRef]
58. Dada, O.A.; Li, G.; Qiao, L.; Ma, Y.; Ding, D.; Xu, J.; Li, P.; Yang, J. Response of waves and coastline evolution to climate variability off the Niger Delta coast during the past 110 years. *J. Mar. Syst.* **2016**, *160*, 64–80. [CrossRef]
59. Chowdhury, P.; Behera, M.R. Effect of long-term wave climate variability on longshore sediment transport along regional coastlines. *Prog. Oceanogr.* **2017**, *156*, 145–153. [CrossRef]
60. Meier, H.E.M.; Kniebusch, M.; Dieterich, C.; Gröger, M.; Zorita, E.; Elmgren, R.; Myrberg, K.; Ahola, M.; Bartosova, A.; Bonsdorff, E.; et al. Climate Change in the Baltic Sea Region: A Summary. *Earth Syst. Dyn. Discuss.* 2021; in review. [CrossRef]
61. Harff, J.; Furmańczyk, K.; von Storch, H. *Coastline Changes of the Baltic Sea from South to East*; Springer: Berlin/Heidelberg, Germany, 2017; Volume 19, p. 386.
62. Räisänen, J. Future Climate Change in the Baltic Sea Region and Environmental Impacts. *Oxf. Res. Encycl. Clim. Sci.* **2017**, *1*, 1–39.
63. Jarmalavičius, D.; Pupienis, D.; Žilinskas, G.; Janušaite, R.; Karaliunas, V. Beach-foredune sediment budget response to sea level fluctuation. Curonian Spit, Lithuania. *Water* **2020**, *12*, 583. [CrossRef]
64. Žilinskas, G. Kranto Linijos Dinamikos Ypatumai Klaipėdos Uosto Poveikio Zonoje. *Geogr. Metraštis* **1998**, *31*, 99–109.
65. Staniszewska, M.; Boniecka, H. Managing dredged material in the coastal zone of the Baltic Sea. *Environ. Monit. Assess.* **2017**, *189*, 46. [CrossRef]
66. Rangel-Buitrago, N.; de Jonge, V.N.; Neal, W. How to make Integrated Coastal Erosion Management a reality. *Ocean Coast. Manag.* **2018**, *156*, 290–299. [CrossRef]
67. Ludka, B.C.; Guza, R.T.; O'Reilly, W.C. Nourishment evolution and impacts at four southern California beaches: A sand volume analysis. *Coast. Eng.* **2018**, *136*, 96–105. [CrossRef]

68. De Schipper, M.A.; Ludka, B.C.; Raubenheimer, B.; Luijendijk, A.P.; Schlacher, T.A. Beach nourishment has complex implications for the future of sandy shores. *Nat. Rev. Earth Environ.* **2021**, *2*, 70–84. [CrossRef]
69. Guillén, J.; Hoekstra, P. Sediment distribution in the nearshore zone: Grain size evolution in response to shoreface nourishment (Island of Terschelling, The Netherlands). *Estuar. Coast. Shelf Sci.* **1997**, *45*, 639–652. [CrossRef]
70. Hamm, L.; Capobianco, M.; Dette, H.H.; Lechuga, A.; Spanhoff, R.; Stive, M.J.F. A summary of European experience with shore nourishment. *Coast. Eng.* **2002**, *47*, 237–264. [CrossRef]
71. Pinto, C.A.; Silveira, T.M.; Teixeira, S.B. Beach nourishment practice in mainland Portugal (1950–2017): Overview and retrospective. *Ocean Coast. Manag.* **2020**, *192*, 105211. [CrossRef]

Journal of
Marine Science and Engineering

MDPI

Article

The Making of a Gravel Beach (Cavo, Elba Island, Italy)

Irene Cinelli [1,2], Giorgio Anfuso [3], Enrico Bartoletti [2,4], Lorenzo Rossi [2,5] and Enzo Pranzini [1,3,*]

1 Dipartimento di Scienze della Terra, Università di Firenze, Via Micheli 6, 50121 Firenze, Italy; irene.cinelli@unifi.it
2 Gruppo Nazionale per la Ricerca sull'Ambiente Costiero, 16132 Genova, Italy; enrico.bartoletti@regione.toscana.it (E.B.); lrossi@geocoste.com (L.R.)
3 Department of Earth Sciences, Faculty of Marine and Environmental Sciences, University of Cádiz, Polígono del Río San Pedro s/n, 11510 Puerto Real, Spain; giorgio.anfuso@uca.es
4 Regione Toscana, Settore Idrologico e Geologico Regionale, Lungarno Pacinotti no. 49, 56100 Pisa, Italy
5 GeoCoste s.n.c., 50141 Firenze, Italy
* Correspondence: enzo.pranzini@unifi.it

Abstract: This paper presents the history and evolution of the different projects carried out from 1999 to 2008 at Cavo beach in the Elba Island, Italy. The village of Cavo almost completely lost its beach in the 1970s due to the reduction of sedimentary input, and the backing coastal road was defended by a revetment and two detached breakwaters. Such severe erosion processes continued in the following years and impeded any possibility of beach tourist development. In 1999, a project based on the removal of existing breakwaters and beach nourishment works based on the use of gravel as borrow sediment and the construction of two short groins to maintain nourished sediment, raised environmental concern and did not find the approval of the stakeholders. They were worried about the characteristics of the sediments, i.e., waste materials from iron mining rich in red silt and clay. Such sediment fractions made the sea red during the nourishment and deposited on the *Posidonia oceanica* meadow in front of the beach, with a potential environmental impact. Furthermore, they cemented the gravel fraction forming a beach rock. Between 2006 and 2008, these materials were covered with better quality gravel, extending and raising the beach profile, which required the elevation and lengthening of the two existing groins. Beach evolution monitoring following the second project, based on morphological and sedimentological data acquired before, during and after the works, demonstrated the great stability of the newly created beach. The wider beach has allowed the construction of a promenade and the positioning, in summer, of small structures useful for seaside tourism, increasing the appeal of this village. Data presented in this paper shows an interesting study case, since few examples exist in international literature regarding gravel nourishment projects monitoring and evolution.

Keywords: beach nourishment; coastal erosion; gravel beaches; sediment budget; shore protection structures

Citation: Cinelli, I.; Anfuso, G.; Bartoletti, E.; Rossi, L.; Pranzini, E. The Making of a Gravel Beach (Cavo, Elba Island, Italy). *J. Mar. Sci. Eng.* **2021**, *9*, 1148. https://doi.org/10.3390/jmse9101148

Academic Editor: Carlos Daniel Borges Coelho

Received: 15 September 2021
Accepted: 11 October 2021
Published: 19 October 2021

1. Introduction

In small islands, pocket beaches quite often represent one of the most important tourist assets [1], and this is even more true for those in the Mediterranean Sea [2,3], where a strong transition from traditional activities (agriculture and fishing) to tertiary activities (almost exclusively tourism) occurred in the 20th century [4,5]. Furthermore, in small islands, pocket beaches with a limited sediment stock are extremely vulnerable to sedimentary input reduction [6], which is frequently produced by the abandonment of cultivated lands [7]. In those sites where attractive landscape values support beach tourism, shore protection projects based on the emplacement of hard structures should be limited and artificial nourishment preferred, at best associated with small containment structures. This is the most sustainable option to preserve the natural scenic beach value and all beach-related activities [8–10].

Beach nourishment projects on small islands are quite complex because their limited surface and complex orography often prevent finding suitable natural borrow materials in land deposits, and the importing of sediments from the mainland is extremely expensive [11]. In addition, environmental constraints do not allow shelf sediment dredging and, in any case, mob-demob cost for deep operating dredges is not justified by the small volumes of sediments usually required. Riverbed quarrying, which however is forbidden in Italy, reduces sediment input to the coast and cannot be a solution to contrast coastal erosion.

Therefore, one option is rock crushing to produce gravel to create coarse sediment beaches, whose pros are:

Higher stability than sand beaches [12];
Higher dry beach expansion at a given fill volume [13];
Clearer water, since there are no fine sediments that can be suspended [14];
No wind erosion [15];
No sticking on the beachgoers' skin [16].

However, some cons must be considered, e.g.,:

Less ease in walking and lying on the beach and entering the sea [17];
Steeper swash zone, which is an obstacle for elderly, children and disabled people [17];
Reduced play possibilities for children and limitations to beach games and sports [18,19].

In Italy, gravel and grains have been used both to build a new beach where it was completely lost, e.g., at Cala Gonone [20], Marina di Pisa [21], or to expand an eroding sand beach, e.g., at Massa [22] and Terracina [23]. The case considered in this paper refers to a pocket beach in front of a small village named Cavo, in the eastern coast of the Elba Island, in the Thyrrhenian Sea, Italy (Figure 1).

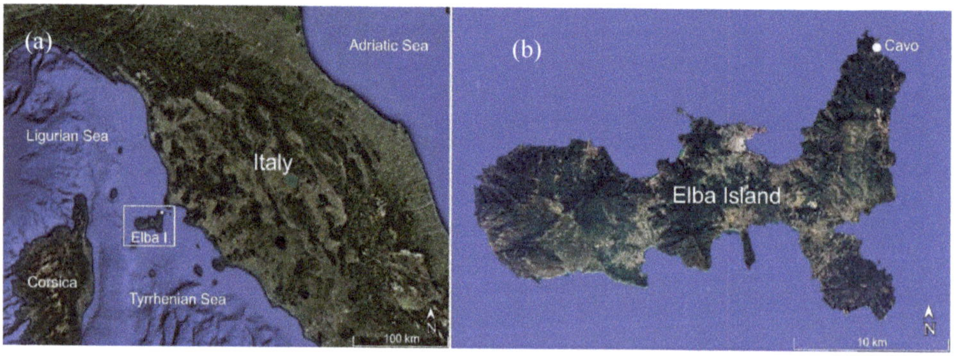

Figure 1. Location map of central-northern part of Italy (**a**) and Elba Island (**b**) (Google Earth).

At such beach, in the 1970s, sand was almost completely lost, and a revetment was positioned adjacently to the coastal road and two detached breakwaters were constructed, thus preventing access to the sea and creating very dangerous nearshore conditions (Figure 2). After an attempt to restore the original mixed sediment sand and gravel beach carried out in 1999 (Figure 3), which created a hard surface disliked by tourists, and environmental concern for the water turbidity on the *Posidonia oceanica* meadow, a pure gravel beach was constructed in 2006–2008 (Figure 3) with the satisfaction of beachgoers and local stakeholders.

Figure 2. The central part of the beach at Cavo in June 1995.

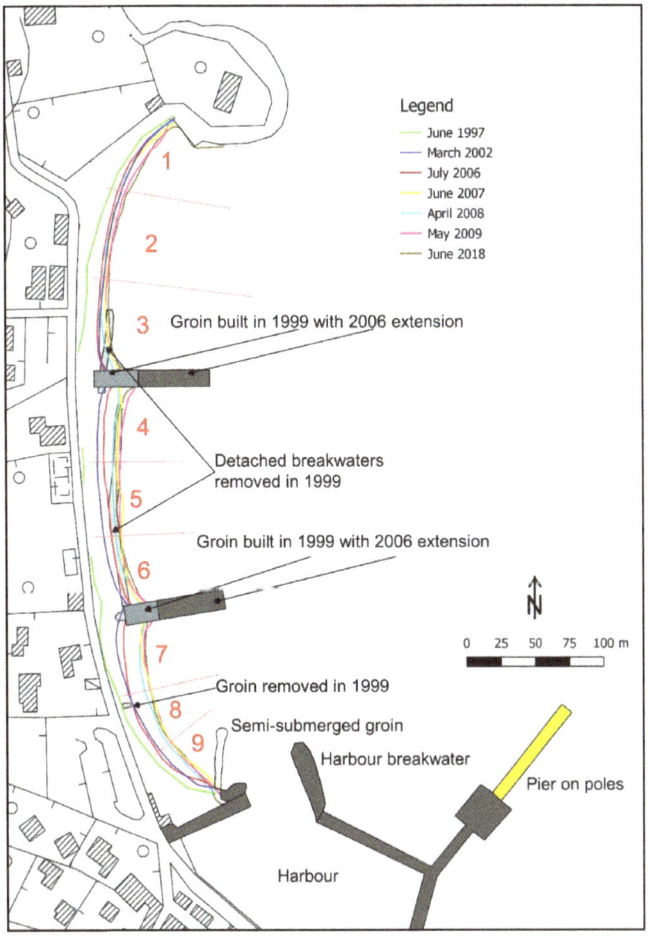

Figure 3. Shoreline (zero isobath) evolution at Cavo from 1997 (before the first nourishment) to 2018 (ten years after the end of the second nourishment). In red sectors number.

The artificial fill was not accompanied by a specific monitoring, as frequently happens for coastal defence projects in Italy. However, for this case, a three-year monitoring was commissioned to the University of Florence (Italy) by the works contracting entity (the Provincia di Livorno, i.e., the provincial administration), and results of such studies appeared only in grey literature and essentially concerned administrative and project management issues. Ten years later, a survey of Cavo beach was commissioned by the Regione Toscana (i.e., the regional administration) within the framework of a wide regional project focused on the monitoring of all regional beaches. Despite the nonhomogeneous temporal spacing of acquired data regarding Cavo beach evolution, this paper describes the different coastal projects actuations and assesses the effectiveness of the nourishment project carried out in 2006–2008, giving only some indications on a previous project performed in 1999 for which data are extremely poor. Results, which demonstrate the fill stability and the approval of the type of material by beachgoers, are of interest to coastal planners and can be useful for designing new gravel beaches or to expand eroding ones in similar environments.

2. The Beach at Cavo (Elba Island, Italy)

The beach of Cavo (Figure 1) is located inside a bay facing NNE (fetch width from 32° N to 107° N) sheltered by the Central Tuscany coast. Fetch length at the extremes of the angle is 6.3 and 17.0 nautical miles (nmi), respectively (13.4 nmi along the bisector of the opening angle 69.5° N). No physical or virtual wave gouges are present in this sea sector, and data on wave climate can be obtained from the "Wind and Wave Atlas of the Mediterranean" [24], where the nearest point is 22 nmi south of Cavo at 25 nmi from the coast, thus has a longer fetch than the real one observed at Cavo. At such point, waves approach from the 30–120° N sector and significant wave height (H_s) > 2.0 m approach from 30° N and represent 0.2% of records. Tidal range is 36 cm at the Livorno Gouge, on the continental coast [25].

2.1. The Loss of the Beach

Cavo is a little village located on the eastern side of Elba Island (Figure 1) where, since Roman times, the main traditional activity has been iron mining, flanked by some agriculture consisting mainly of vine cultivation, the main occupation in the rest of the Island.

After WWII, all the island recorded a transformation of the economic activity, from agriculture to tourism, but such a shift took place a bit later on the eastern side, since mining activity lasted, although with reduced production, until 1981 [26]. All the beaches at Elba Island are eroding because sediment input was reduced when the crops were abandoned, and the forest grew [27–29]. On the eastern side, additional sediment input to beaches was linked to mining activity because waste materials from excavations were abandoned on the slopes of mountains and thus easily transported by run-off processes to the coast [30]. Therefore, the beach of Cavo and others on the eastern coast, formed thanks to land erosion of cultivated areas and erosion of quarry waste deposits, but when both activities were interrupted, coastal retreat also interested this part of the island. In addition, a small pier with impermeable root was created on the southern side of the bay as a docking structure for ferries connecting the island to the continent and a small marina added at its northern side. These structures interrupted the limited longshore sediment transport and contributed to the urban beach disappearance, a loss only apparently compensated by the expansion of the beach placed updrift, the latter being in a marginal area of lesser tourist value.

At the same time, this village also started to look for new opportunities in 3S (Sun, Sea and Sand) tourism, with few hotels and some second houses, but the beach was almost inexistent and could not support this activity.

2.2. The Making of the Gravel Beach

At the end of the 1990s, the beach in front of the village of Cavo was constituted by only two narrow strips of sand, one close to the northern headland and one leaned on the marina downcoast dock. In addition, the coastal road was defended by two detached breakwaters and by a revetment (before 1954) making difficult and extremely dangerous the use the beach for bathing activity (Figure 2).

The wide beach, which is now in front of the village, is the result of a complex and controversial story that presents technical and legal aspects, both of which were important for the achievement of the design solution that led to the current configuration of the coastline. This story can be synthetized in two projects: the first carried out in 1999 and the second between 2006 and 2008; the latter being the main topic of this paper.

2.2.1. The First Project (1999)

In 1999 a project was carried out to create a 10-meter-large beach in front of the street wall. The previous detached breakwaters and the revetment were removed, and two short groins constructed to divide in three parts the 497 m long coastal sector delimited by the headland to the north and the marina to the south (Figure 3).

Due to the lack of suitable natural aggregate deposits on the Island, wastes of the old iron mines were used as borrow material. They consisted of unsorted sand and gravel, with a high percentage of fines (silt and clay = 13%) formed by yellow-red iron oxides [29]. The presence of heavy minerals in excess respect to environmental regulations was later assessed as well.

The nourishment was carried out just before the tourist season, but results were not those expected: the sea water acquired a red colour that persisted during all the summer and several tourists cancelled their hotel reservations. The *Posidonia oceanica* meadow, present in the nearshore, was covered by a thin layer of clay and the risk of some permanent ravages was raised. Local stakeholders claimed environmental and economic damages, and the case arrived at the court. Monitoring of the Posidonia proved that no long–lasting injury was done, and economic loss not motivated. As far as heavy minerals are concerned, further analyses showed that all the beaches present along the eastern side of the Island have similar concentrations, mostly deriving from more than two thousand years of mining activity.

In the months after the nourishment, the borrow material compacted and became impermeable because something similar to a beach rock was formed (Figure 4) so that, during the following winter storms, run up water was not able to infiltrate into beach sediments and reached the coastal road [31]. Permeability measurements were performed by the University of Florence in three points of the beach (north, centre, south), giving a permeability coefficient (K_s) between 5.0×10^{-6} and 1.3×10^{-7} m/s (typical of silty to silty-clay sediments).

Figure 4. The beach at Cavo in January 2002, after the nourishment performed with iron rich materials with puddles showing the low permeability of the sediments (**a**); outcropping of the consolidated and impermeable fill material layer (**b**).

2.2.2. The Second Project (2006–2008)

In the following years, the eventuality of removing all the material still present ashore was considered but the risk of favouring further fines and heavy minerals offshore dispersion discouraged such a solution. This led to the design a new gravel beach, large and high enough to allow run-up water percolation even during extreme storms. This solution forced a modification of the groin configuration by elevating the crest and extending it offshore to host a higher and wider berm (Figure 3). Nourishment comprised approx. 30,000 m^3 (*ca.* 80 m^3/m for the project sector) of gravel 4.0–4.5 phi (16–24 mm) in mean size (Figure 5). To not increase water turbidity, local authorities asked to maintain the quantity of fines (<0.063 mm) lower than 2%, as occurred for many projects carried out in Italy in sensible sites (e.g., at Cala Gonone, Sardinia [20]).

Figure 5. Fill materials used at Cavo: on the left side the iron rich mining waste, on the right side the new carbonate sandstone of the second nourishment. One euro coin for reference at the centre of the photo.

As in the 1999 project, aggregates were transported by truck and downloaded directly on the beach and later distributed with a bulldozer, pushing them even into the water.

Work initiated in January 2006 with the groins modification and ended in May 2008, with interruptions during the summer tourism seasons. Circa 25,000 m^3 of gravel were deposited. In addition, a small (unknown) volume of very fine sand dredged at the harbour entrance was discharged in the sectors near the marina, but being a volume moved inside the area, it did not change the overall sedimentary budget.

Dry beach expansion, on the 497-meter-long coast, was 3777 m^2. Mean shoreline progradation was 7.48 m, with important differences from the southern and central sectors (10 to 14 m) and the northern one (less than 3 m) where the original beach was not nourished with iron-rich materials and did not necessitate additional protection. In spring 2008, a volume of 3000 m^3 was placed to complete the project [31].

Part of the gravel was deposited in front of the swash zone, as feeding groins [16] that theoretically allow to have a more natural beach profile since wave action should move grains onshore, after a first phase in which grains are in situ cleaned and rounded.

An extensive beach scraping was performed in March 2009 to redistribute sediments accumulated in December 2008 in front of the promenade wall during a severe storm, but similar works are frequently carried out at the beginning of the summer season to flatten the storm berm crest that constitutes an obstacle to dive for children and elderly or disabled people. This activity does not influence the fill stability assessment, since beach volume is not modified, and the profile soon adapts to the autumn–winter storms.

The overall beach expansion and stabilization allowed to provide the coastal road with a large sidewalk with a balustrade and trees, which transformed it into a promenade. In this way, part of the fill material is delimitated by the promenade wall and is no more part of the beach sediment stock. This partially explains why the volume of sediments added resulting from the comparison of the two first surveys (pre- and post-works) is lower than what was actually deposited on the beach. At the root of the two groins, where the beach was expected to be wider and stable, the promenade was expanded into semi-circular exedras (Figure 6). All this transformed the coastal landscape and triggered a revival of the tourist activity of this location.

Figure 6. One of the two exedras constructed along the promenade (Google Earth).

3. Materials and Methods

As explained in the Introduction, different topographic, bathymetric and sedimentological surveys (not homogeneously spaced in time, Table 1) are available to assess the evolution of nourishment projects realised in 1999 and in 2006–2008 at Cavo. They were performed within (i) a scientific research agreement between the Province of Livorno and the Earth Science Department of the University of Florence and (ii) the surveys commissioned in 2018 by the Regione Toscana to monitor several beaches at Elba Island, including the one at Cavo.

Table 1. Available data concerning Cavo beach area.

	June 1997	March 2002	July 2006	June 2007	April 2008	May 2009	June 2018
Shoreline	•	•	•	•	•	•	•
Bathymetry		•		•		•	•
Sedimentology		•		•		•	

Shoreline position was acquired by means of GPS surveys (LEICA system RX 900, Leica Geosystems, Heerbrugg, Swiss), in June 1997 and RTK-GPS (GPS NRTK 1250, Leica

Geosystems, Heerbrugg, Swiss) in the following surveys: June 1997, March 2002, July 2006, June 2007, April 2008, May 2009 and June 2018 (Figure 3, Table 1). Bathymetric surveys, consisting of 34 single beam (Hydrotrac, Teledyne Odom, Slangerup, Denmark) profiles each, were performed in March 2002, June 2007 and May 2009, whereas a multibeam survey (Seabat 7k, Teledyne Reason, Slangerup, Denmark) was performed in 2018. Together with the 2002, 2007 and 2009 surveys, sediment samples (N = 91, 42 and 35) were collected with a Van Veen grab along 6 profiles from +2 m to −5 m, and grain size analysis via dry sieving was performed to obtain Folk and Ward (1957) textural parameters [32].

Shoreline evolution was analysed dividing the coast in 9 sectors, each approximately 55 m long, 3 in each cell in which the beach is divided by the two groins. Mean shoreline displacement was computed for each sector (Table 2) using the Surface Based Analysis (SBA), since the traditional Profile Based Analysis (PBA) was not considered reliable due to the nonlinear shape of the shoreline given by the groins [33]. Surface measurements were performed with QGIS Rel. 3.0.

Beach morphology evolution was studied by comparing pairs of surveys with Surfer Rel. 14 and producing vertical changes maps for the time intervals 2002–2007, 2007–2009 and 2009–2018. Although surveys were performed with standard calibrations (check bar, tide and draft, pitch and roll) and linked to geodetic points, further corrections were done on sea-true points located on some rocky shoals at the border of the bay. Despite all the above, the accuracy of vertical changes in bathymetric maps was approximately 20 cm [34] and, therefore, such maps were only used for a semi-quantitative assessment. Sedimentological maps were drawn to represent Mean size (Mz) and Sorting (σ_I) parameters, and a Mean size *vs* Depth graph was plotted.

No measured wave data are available for this bay, and reference can be done only with two buoys operated by the Tuscany Region, one near Gorgona Island, 75 km NNE of Cavo, the other near Giannutri Island, 90 km SE of Cavo, but considering that this coast is exposed to the East, towards the nearby continent (approx. 10 km), wave energy is significantly lower. However, after the last survey, on 31 December 2018a storm with significant wave height (H_s) of 5.40 m at Gorgona and 6.50 m at Giannutri was registered, the highest since the buoys were installed (2008).

4. Results

4.1. General Considerations on the First Nourishment Evolution

As previously stated, no specific monitoring was performed on this project, and beach transformation could be evaluated only by the comparison between two surveys done by the University of Florence in 1997 (two years before the fill), within a regional study on Elba Island beaches erosion, and one in 2002 as a basis of a second project. A further survey, performed in June 2006 (before the second nourishment), allows to assess fill evolution in the following four years.

The first nourishment induced a notable dry beach expansion, with the March 2002 beach 9.8 m wider on average respect to the June 1997 one (Figures 3 and 7; Table 2).

Obviously, fines were lost only from the upper part of the deposits, which strongly modified optical properties of the nearshore water that acquired a deep red colour induced by the fact that the grainsize fraction was mostly composed by clay (and not silt). Actually, a very limited fill volume was dispersed.

From March 2002 to July 2006 the beach was as a whole stable, with only some sediment shift from the side sectors to the central ones (no. 4 to no. 7; Figure 3; Table 1), favoured by the low groins height.

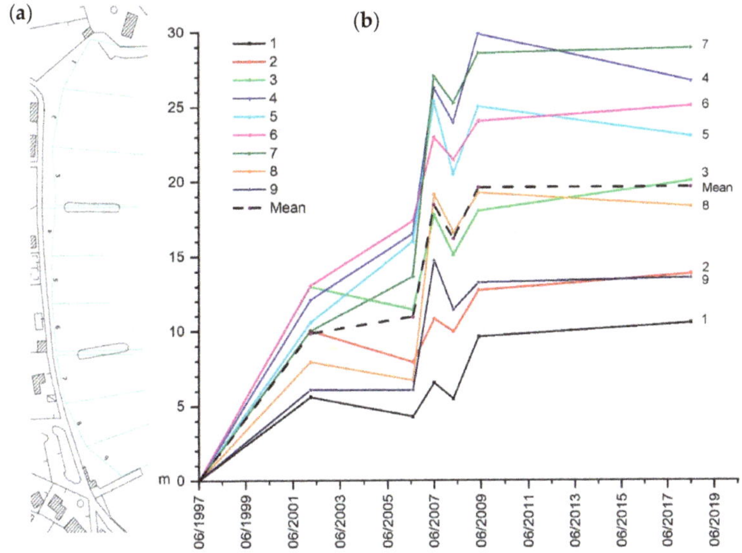

Figure 7. Location of sectors (**a**) and sectors and total mean beach width evolution (**b**) from 1997 to 2018. First nourishment 1999; second nourishment 2006–2008.

Table 2. Mean shoreline displacement (m) in the different periods from June 1997 to June 2018 in the 9 considered beach sectors and in the entire beach.

Sect. no.	June 1997 March 2002	March 2002 July 2006	July 2006 June 2007	June 2007 April 2008	April 2008 May 2009	May 2009 June 2018
1	5.60	−1.32	2.24	−1.05	4.13	0.94
2	10.03	−2.11	2.89	−0.86	2.78	1.10
3	12.97	−1.53	6.33	−2.68	2.94	2.03
4	12.10	4.43	9.72	−2.29	5.92	−3.16
5	10.58	5.41	9.40	−4.89	4.52	−1.99
6	13.05	4.33	5.58	−1.48	2.55	1.02
7	10.02	3.63	13.39	−1.79	3.32	0.37
8	7.95	−1.25	12.45	−2.56	2.64	−0.89
9	6.09	−0.03	8.67	−3.31	1.82	0.31
Entire beach	9.85	1.09	7.48	−2.26	3.42	0.07

4.2. Second Nourishment Evolution Assessment

Shoreline Displacement

The survey performed in June 2007 shows the beach as it was immediately after the end of the main works (i.e., without the 3000 m^3 of the 2008 additional filling) and its comparison with the July 2006 allows to quantify the enlargement artificially obtained.

This beach nourishment was not homogeneous along the coast but concentrated where the beach was narrower. The different dry beach expansion is visible in Figures 3 and 7 and in Table 2: external sectors 1, 2 and 9 received a limited or null volume of sediments, whereas the central ones obtained more (Table 2). Such distribution of nourished sediments was essentially aimed to cover the previously deposited iron-rich material (1999 nourishment), more than to expand the beach.

Mean shoreline progradation was 7.48 m, ranging from 2.24 (sector 1) to 13.39 m (sector 7). Adding these values to what was obtained with the 1999 nourishment, the beach was 18.42 m wider on average than it was in 1997. However, part of this increment will be later used to enlarge the coastal road adding a wide footstep and two exedras.

From June 2007 to April 2008, i.e., in the first winter after the nourishment, all the sectors of the beach recorded moderate erosion (Figure 3): overall beach surface reduction was approximately 1189 m^2, with a mean beach retreat of 2.26 m.

Concerning beach morphology, the beach profile constructed during the nourishment works was flat and wide; in the following months, after the impacts of winter waves, storm berm crests were observed. This involved the migration of material from the swash zone to a more internal position, with a reduction of the beach surface, but without changes in gravel volume, as proved by elevation change maps (Figure 8).

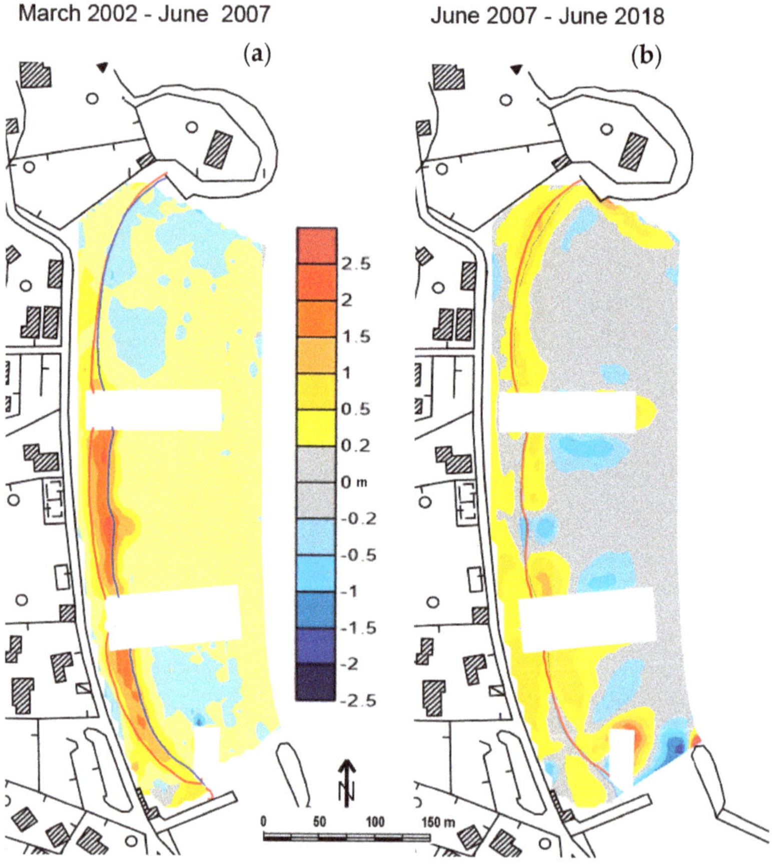

Figure 8. Beach vertical changes (m) at Cavo considering the 2002–2007 (**a**) and 2007–2018 (**b**) surveys.

In the following year (April 2008–May 2009) advancement is recorded in all the sectors with an average dry beach expansion of 3.42 m, resulting in a beach wider than the one obtained immediately after the works.

When comparing the evolution of the different sectors (Figure 3), a very homogeneous behaviour is observed and characterised by synchronous accretion and erosion in the period 2006–2009, showing that no significant beach rotation occurred in each cell.

It is only in the nine following years (May 2009–June 2018) that the three cells show a slightly different behaviour. In the northern one (sectors 1–3), sector no. 1 accretes more than the others, which results in a slight clockwise rotation, although within a general beach progradation. On the contrary, the central cell (sectors 4–6) behaves in an opposite way, with the first sectors eroding and the latter accreting, with an anticlockwise rotation and limited beach erosion. In the southern cell (sectors 7–9) a widespread but limited beach accretion is measured, especially at the central segment (no. 8, Figure 3).

All those mentioned changes represent small variations (between +4 m and −2 m) for a long period, with negative values in the central cell only, which is the most exposed to the incoming waves. All these data show that waves strong enough to move these coarse sediments approach the coast almost orthogonally.

Unfortunately, no high temporal resolution data are available for these nine years, but several inspections and photographs show that the beach was almost stable, so much that the municipality built public toilets at the base of the promenade and gave some surfaces in concession to carry out commercial activities (e.g., beach bars).

A stable and attractive beach was the goal of the project, and collected data, although incomplete, show that this was achieved. Its expansion, necessary to cover the nourishment of 1999, was a welcome side effect of the project.

4.3. Nearshore Morphology Evolution

To assess the morphological evolution of the beach, and the sediment budget of the emerged and the submerged parts, 3D topo-bathymetric models produced in each survey were compared limiting the offshore part to that in which full overlapping was possible, i.e., down to ca. 5 m water depth (Figure 8).

From May 2002 to June 2007 the direct effect of the nourishment is visible with beach surface rising both on the dry beach and on the submerged profile, especially in the central cell, where most of the volume was deposited. At the centre of this cell a lobe is evident, where the main feeding groin was positioned (Figure 9).

Figure 9. Northward (**a**) and southward (**b**) views of the remains of the filling groins on 18 June 2008.

In the following eleven years (2007–2018), limited morphological changes occurred, except the disappearance of what remained of the filling groin that experienced a concentrated lowering of ca. 1.5 m (evident the blue area in the central cell in Figure 7). However, it must be considered that small morphological changes and volumetric variations at the beginning of the tourist season are associated with artificial beach smoothing carried out to eliminate storm berms crest (Figure 10), as a result the emerged profile is expanded and amounts of sediments are moved to the nearshore. Therefore, beach scraping makes the interpretation of the post-nourishment beach evolution more complex, but it is a procedure frequently carried out also in sand beaches with relevant tourist use.

Figure 10. Storm berm crest on the Northern Cell, October 2018 (Google Earth).

Beach volume increase on the dry beach and decrease on the wet one in the 2007–2018 period can be explained with the onshore sand moving, very likely induced by the higher permeability and porosity of the gravel fill, as observed in other gravel nourishments [21]. Similar to what was observed by previous authors, sand patches are seldom present in the swash zone and sand saturates the pores of the gravel as evident in some trenches excavated on the berm.

Tentative sediment budget estimation was performed and presented in Table 3. Data concerning the dry beach volumes are reliable since topographic measurements have the accuracy of few centimetres (i.e., gravel size) but, unfortunately, it is not the case for the nearshore. As previously said, accuracy of bathymetric data is approximately ±10 cm, which corresponds to changes in height of 20 cm between surveys. Since the study area surface is approximately 40,000 m^2, a volume change of ca. 8000 m^3 is within the accuracy of the methodology used.

Table 3. Beach volumetric changes during the studied period (m^3).

	Dry Beach	From 0 to −5 m	Total
March 2002–June 2007	3730	696	4426
June 2007–June 2018	4017	−97	3920

However, according to our data, the artificial input of ca. 30,000 m^3 of gravel did not produce an equivalent increase in the beach sediment budget. One reason could be that part of the gravel forms now, together with stones, asphalt and different aggregates, the promenade embankment, which is out of the area considered in this study. Analysis of dry beach evolution confirms the stability of nourished sediments, which is not demonstrated when the sedimentary balance is calculated up to a depth of 5 m.

4.4. Sedimentological Evolution

A first sedimentological study was performed in June 2007 (Figure 11a), i.e., at the end of the second nourishment work, when sediments have not yet been sorted by wave action and still have their original size (between −4 and −3 phi; 16–8 mm). A very thin strip of coarse to medium sand (0.0–2.0 phi) runs on the seaside of the gravel, but its bi-modal distribution shows that it is a mixture of the fill gravel and the native sand previously present on the nearshore and in the two lateral segment of the bay. Pure sand is present only offshore of the investigated depth.

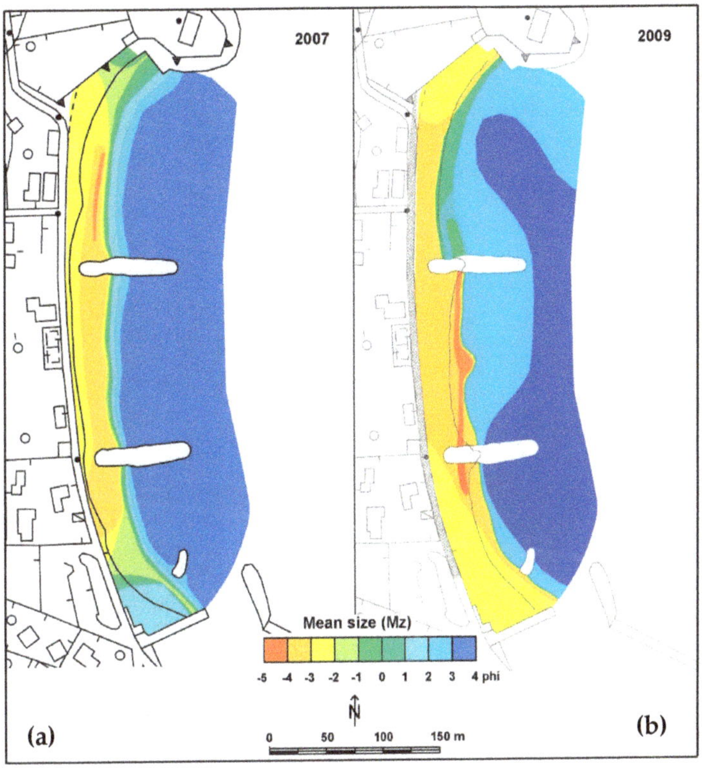

Figure 11. Sediment mean size (M_Z) in June 2007 (**a**) and May 2009 (**b**).

A second sedimentological survey was performed in May 2009, before beach raking, to assess whether offshore gravel displacement occurred or not. Results (Figure 11b) show that all the gravel remained in the dry beach or in the nearshore close to the shoreline, with no sediments coarser than 2.0 phi (0.250 mm) offshore of the 2 m isobath, confirming the cross-shore sorting of sediments observed in other natural [34–37] and artificially nourished beaches [21,37]. The coarsest grains of the nourishment, between −5 and −4 phi (16 and 32 mm) were found in the central cell, all along the beach step and, between −1.5 m and −2.0 m, in the point where the filling groin was located. There, coarse and very well sorted lag deposits are present. In addition to this, evident is a lobe of 2–3 phi (0.250–0.125 mm) sediments in the central area, possibly due to a limited migration of the fine grains present in the fill material.

Both 2007 and 2009 grain size data show that below the 2 m isobath, where wave energy is lower, no gravel is present (Figure 12).

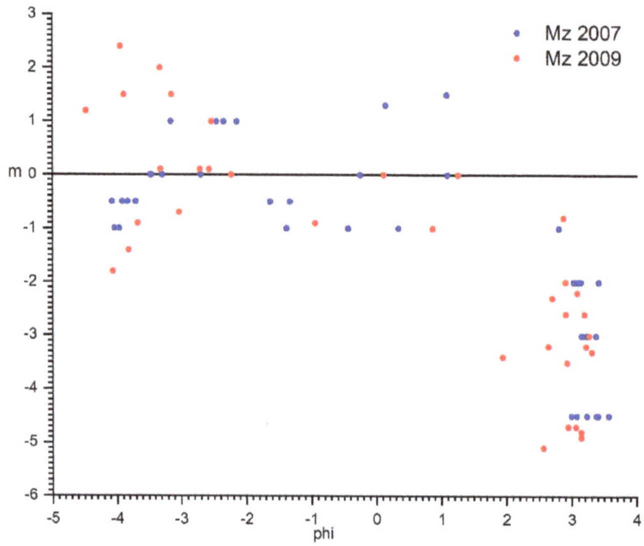

Figure 12. Mean size (phi) vs. Depth (m) for the sediments at Cavo in June 2007 and May 2009.

5. Discussion

The discontinuity of the data, with a long temporal gap from 2009 to 2018, the known but not quantified reshaping of the beach carried out to better satisfy tourism purposes, and the absence of in situ wave climate data, make the study of the evolution of this gravel nourishment very difficult.

Nevertheless, it is important to share this experience within the scientific and technical community, since this kind of nourishment is more and more frequently performed, being cost effective in terms of shore protection and sustainability.

It is sure that the fill proved to be very stable since, after twelve years, almost all the deposited volume is still on the beach, and the beach width is approximately the same as it was after the nourishment.

The behaviour of gravel to gather on the dry part of the profile, as occurred in other gravel nourishments, is confirmed together with the fact that in the case of highly permeable beach sediments, sand can approach the coast and fill the intergranular space [20,21].

Interviews done to beach goers [38] show that this kind of sediments are strongly appreciated, and traditional frequenters of this beach recognize that the coastal environment has improved, for water transparency, beach width (Figure 13) (that gave the possibility to expand the promenade and provide beach services, Figure 14), and for the fact that grains do not stick to the skin. As grain sharpness is concerned, only 30% of the interviewed claimed it as a negative element.

Figure 13. The beach at Cavo before (**a**) and after (**b**) the works (Courtesy Elbaworld).

Figure 14. A small bar (**a**) and toilets (**b**) on the new gravel beach (April, 2017).

6. Conclusions

The beach at Cavo was almost completely lost in the 1970s and a revetment and two detached breakwaters were built to protect the coastal road. However, it made this coastal segment very dangerous for diving and unsuitable for the tourist industry. An attempt to give the site a beach again was carried out with iron ore waste, which was rejected by the stakeholders and posed several environmental issues.

The gravel beach, created to cover this ugly and potentially dangerous material, was effective at this end and proved to be very stable and appreciated by the beachgoers.

The wide beach, the transparent coastal water, the new promenade (Figure 13), and beach services (Figure 14), were all made possible thanks to the choice to build a gravel beach, demonstrating that coarse sediment nourishment can be a viable solution to beach erosion, especially in areas where sand is not available, or its use should be accompanied with harder shore protection structure.

7. Update

The 31 October 2018, storm severely hit the entire Tuscany coast, so much that the region implemented an emergency plan to allow municipalities to dredge sediment in the nearshore to feed the beach in order to allow tourist activity in the following summers. Approximately 11 million euros were spent at that end for 17 small projects.

The beach of Cavo was very little affected by the storm, with a shoreline retreat of approximately 2 m (data from Regione Toscana), and the coastal road was not reached by the waves; therefore the gravel beach proved to be not only suitable for tourism, but also an effective shore protection structure.

Author Contributions: Conceptualization, E.P. and E.B.; methodology, E.P., L.R. and I.C.; software, I.C.; validation, G.A., E.B. and L.R.; formal analysis, I.C.; writing—original draft preparation, I.C.; writing—review and editing, G.A., L.R., E.P.; supervision, E.P. All authors have read and agreed to the published version of the manuscript.

Funding: This research received no external funding.

Institutional Review Board Statement: Not applicable.

Informed Consent Statement: Not applicable. The study did not involve humans or animals.

Data Availability Statement: Data supporting reported results can be asked to the first author.

Acknowledgments: This research is a contribution to the Ibero-American Beach Management and Certification Network—PROPLAYAS and to the Andalusia PAI Research Group 'RNM-328'.

Conflicts of Interest: The authors declare no conflict of interest.

References

1. Peña-Alonso, C.; Ariza, E.; Hernández-Calvento, L.; Pérez-Chacón, E. Exploring multi-dimensional recreational quality of beach socio-ecological systems in the Canary Islands (Spain). *Tour. Manag.* **2018**, *64*, 303–313. [CrossRef]
2. Andreadis, O.; Hasiotis, T.; Psarros, F.; Chatzipavlis, A.; Velegrakis, A. Erosion and sediment transport processes along Eresos coastal zone (Lesvos, Greece). In Proceedings of the 15th International Conference on Environmental Science and Technology, Rhodes, Greece, 31 August–2 September 2017.
3. Gómez-Pujol, L.; Orfila, A.; Morales-Márquez, V.; Compa, M.; Pereda, L.; Fornós, J.J.; Tintoré, J. Beach Systems of Balearic Islands: Nature, Distribution and Processes. In *The Spanish Coastal Systems*; Springer: Berlin/Heidelberg, Germany, 2018; pp. 269–287.
4. Tsartas, P. Tourism Development in Greek Insular and Coastal Areas: Sociocultural Changes and Crucial Policy Issues. *J. Sustain. Tour.* **2003**, *11*, 116–132. [CrossRef]
5. Taboada, T.; Carta, A.; Müller, J.V. Diachronic analysis using aerial photographs across fifty years reveals significant land use and vegetation changes on a Mediterranean island. *Appl. Geogr.* **2018**, *98*, 78–86.
6. Nobre Silva, A.; Rui Taborda, R.; Andrade, C.; Ribeiro, M. The future of insular beaches: Insights from a past-to-future sediment budget approach. *Sci. Total. Environ.* **2019**, *676*, 692–705. [CrossRef] [PubMed]
7. Cipriani, L.E.; Pranzini, E.; Rosas, V.; Wetzel, L. Landuse changes and erosion of pocket beaches in Elba Island (Tuscany, Italy). *J. Coast. Res.* **2011**, *64*, 1774–1778.
8. Anfuso, G.; Williams, A.T.; Casas, G.M.; Botero, C.M.; Cabrera, J.A.; Pranzini, E. Evaluation of the scenic value of 100 beaches in Cuba: Implications for coastal tourism management. *Ocean. Coast. Manag.* **2017**, *142*, 173–185. [CrossRef]
9. Mir Gual, M.; Pons, G.X.; Martín Prieto, J.A.; Rodríguez Perea, A. A critical view of the Blue Flag beaches in Spain using environmental variables. *Ocean. Coast. Manag.* **2015**, *105*, 106–115. [CrossRef]
10. Semeoshenkova, V.; Newton, A. Overview of erosion and beach quality issues in three Southern European countries: Portugal, Spain and Italy. *Ocean. Coast. Manag.* **2016**, *118*, 12–21. [CrossRef]
11. Ebejer, J. Creating a sandy beach in St. George's Bay. A new experience for Malta. In Proceedings of the 1st International Conference on the Nanagement of Coastal Resources, La Valletta, Malta, 20–23 October 2004; pp. 161–167.
12. Pranzini, E.; Anfuso, G.; Munoz-Perez, J.J. A probabilistic approach to borrow sediment selection in beach nourishment projects. *Coast. Eng.* **2018**, *139*, 32–35. [CrossRef]
13. Bitan, M.; Zviely, D. Sand beach nourishment: Experience from the mediterranean coast of Israel. *J. Mar. Sci. Eng.* **2020**, *8*, 273. [CrossRef]
14. Peterson, C.H.; Bishop, M.J. Assessing the Environmental Impacts of Beach Nourishment. *BioScience* **2005**, *55*, 887–896. [CrossRef]
15. Sherman, D.J.; Hotta, S. Aeolian sediment transport: Theory and measurement. In *Coastal Dunes: Form and Process*; Nordstrom, K.F., Psuty, N., Carter, B., Eds.; Edward Arnold: London, UK, 1990; pp. 17–37.
16. Pranzini, E.; Anfuso, G.; Botero Saltaren, C. Nourishing tourist beaches. In *Beach Management Tools—Concepts, Methodologies and Case Studies*; Botero, C.M., Cervantes, O.D., Finkl, C.W., Eds.; Springer: Cham, Switzerland, 2018; pp. 293–318.
17. Nordstrom, K.C.; Pranzini, E.; Jackson, N.; Coli, M. The Marble beaches of Tuscany. *Geogr. Rev.* **2008**, *98*, 280–300. [CrossRef]
18. Obrador, P. Building Castles in the Sand: Repositioning Touch on the Beach. *Senses Soc.* **2009**, *4*, 195–210. [CrossRef]
19. Rainis, M. French beach sports culture in the twentieth century. *Int. J. Hist. Sport* **2000**, *17*, 144–158. [CrossRef]
20. Pacini, M.; Pranzini, E.; Sirito, G. Beach nourishment with angular gravel at Cala Gonone (Eastern Sardinia, Italy). In Proceedings of the MEDCOAST, Qawra, Malta, 11–14 November 1997; pp. 1043–1058.
21. Cammelli, C.; Jackson, N.L.; Nordstrom, K.F.; Pranzini, E. Assessment of a gravel-nourishment project fronting a seawall at Marina di Pisa, Italy. *J. Coast. Res.* **2006**, *39*, 770–775.

22. Aminti, P.; Pelliccia, F.; Pranzini, E. Evoluzione del profilo di spiaggia a seguito di un ripascimento artificiale in ghiaia su di una spiaggia altamente protetta. *Studi Costieri* **2002**, *5*, 47–57.
23. Medatlas Group. Wind and Wave Atlas of the Mediterranean Sea, W.E.A.O. *Res. Cell* **2004**. Available online: https://core.ac.uk/download/pdf/52757034.pdf (accessed on 10 October 2021).
24. Istituto Idrografico della Marina. Tavole di Marea, Mediterraneo e Mar Rosso, Genova, II-3133.
25. Berriolo, G. Interventi di riequilibrio delle spiagge della provincia di Latina. In *La Difesa dei Litorali in Italia*; Aminti, L., Pranzini, E., Eds.; Edizioni delle Autonomie: Roma, Italy, 1993; pp. 153–173.
26. Tanelli, G.; Benvenuti, M.; Costagliola, P.; Dini, A.; Lattanzi, P.; Maineri, C.; Mascaro, I.; Ruggieri, G. The iron mineral deposits of Elba Island: State of the art. *Ofioliti* **2001**, *26*, 239–248.
27. Mannori, S.; Pranzini, E. From agriculture to tourism: A cause of beach erosion. In Proceedings of the 1st International Conference on the Management of Coastal Recreational Resources, La Valletta, Malta, 20–23 October 2004; pp. 79–85.
28. Pranzini, E.; Rosas, V. Pocket beach response to high energy—Low frequency floods (Elba Island, Italy). *J. Coast. Res.* **2007**, *50*, 969–977.
29. Pranzini, E.; Rosas, V.; Jackson, N.L.; Nordstrom, K.F. Beach changes due to sediment delivered by streams to pocket beaches during a major flood. *Geomorphology* **2013**, *199*, 36–47. [CrossRef]
30. Nordstrom, K.F.; Jackson, N.L.; Pranzini, E. Beach sediment alteration by natural processes and human action: Elba Island, Italy. *Ann. Ass. Am. Geogr.* **2004**, *94*, 794–806.
31. Bartoletti, E.; Bini, A.; Marasco, I.; D'Oriano, V.; Mercantelli, A.; Sani, L. La Spiaggia Ritrovata. *ACER* **2008**, *6*, 53–55.
32. Folk, R.; Ward, W.C. Brazos river bar: A study in the significance of grain size parameters. *J. Sediment. Petrol.* **1957**, *27*, 3–26. [CrossRef]
33. Anfuso, G.; Bowman, D.; Danese, C.; Pranzini, E. Transect Based Analysis versus Area Based Analysis to quantify shoreline displacement: Spatial resolution issues. *Environ. Monit. Assess.* **2016**, *188*, 568. [CrossRef] [PubMed]
34. Pranzini, E. *Misurare la Spiaggia per Gestire la Costa*; Edifir: Firenze, Italy, 2019; 128p.
35. King, C.A.M. *Beches and Coasts*, 2nd ed.; Edward Arnold: London, UK, 1972; 570p.
36. Pontee, N.I.; Pye, K.; Blott, S.J. Morphodynamic Behaviour and Sedimentary Variation of Mixed Sand and Gravel Beaches, Suffolk, UK. *J. Coast. Res.* **2004**, *20*, 256–276. [CrossRef]
37. Horn, D.P.; Walton, S.M. Spatial and temporal variations of sediment size on a mixed sand and gravel beach. *Sediment. Geol.* **2007**, *202*, 509–528. [CrossRef]
38. Saitta, M. *Valutazione Dell'efficacia di un Ripascimento in Ghiaia nel Golfo di Follonica (GR)*; University of Florence: Florence, Italy, 2021; Unpublished thesis.

Journal of *Marine Science and Engineering*

MDPI

Article

Artificial Nourishments Effects on Longshore Sediments Transport

Ana Margarida Ferreira and Carlos Coelho *

RISCO & Civil Engineering Department, University of Aveiro, Campus Universitário de Santiago, 3810-193 Aveiro, Portugal; margarida.ferreira@ua.pt
* Correspondence: ccoelho@ua.pt

Abstract: Serious erosion problems related to significant negative sediments budgets in the coastal systems are referred worldwide. Artificial nourishments are a coastal erosion mitigation strategy that allow for a decrease in those negative budgets by adding sediment to the coastal system. Thus, it is essential to understand and adequately model the shoreline response after a nourishment operation, in order to support the definition of the best intervention scenarios. The main goal of this work was to study the artificial nourishment effects on the longshore sediment transport and consequently on the morphological evolution at the intervention site and nearby areas, in a time horizon of 5 years. The longshore transport of the nourished sediments was evaluated, aiming to contribute to the evaluation of the sediment's permanence at the deposition site and the frequency required for new nourishments. The shoreline evolution numerical long-term configuration (LTC) model was applied in order to evaluate the spatial and temporal distribution of the nourished sediments along the coast, considering different types of beaches and incident wave climates. The adopted approach is generic and supported by simple numerical models, which can be useful for preliminary site-specific evaluations. The results show that the nourishment impact is mainly observed nearby the intervention site. It is highlighted that higher longshore sediment transport rates are associated with more energetic wave climates, but not necessarily with incident waves more oblique to the shoreline.

Keywords: morphology; LTC numerical model; erosion; littoral drift; coastlines

Citation: Ferreira, A.M.; Coelho, C. Artificial Nourishments Effects on Longshore Sediments Transport. *J. Mar. Sci. Eng.* **2021**, *9*, 240. https://doi.org/10.3390/jmse9030240

Academic Editor: Luis Moreno

Received: 5 February 2021
Accepted: 19 February 2021
Published: 24 February 2021

Publisher's Note: MDPI stays neutral with regard to jurisdictional claims in published maps and institutional affiliations.

1. Introduction

Large volumes of sand are transported along the coast (littoral drift) due to natural actions of waves, currents, and winds. Littoral morphology is dependent of the dynamic relationship between the wave climate, considered the main factor responsible for the sediments transport capacity, and the characteristics and availability of existing sediments [1]. If the quantity of sediments leaving a coastal stretch is balanced by the same quantity of sediments coming in, the system is in a dynamic equilibrium, not resulting in erosion or accretion. But if the sediment balance is not maintained, erosion or accretion occurs, representing, respectively, the retreat or advance of the shoreline position [2,3].

Due to the need to protect coastal areas from erosion problems, it is common to perform coastal defense interventions. Artificial sediment nourishment is an intervention strategy that has been adopted by several countries all over the world [4,5]. The nourishment refers to the process of adding sediments to the littoral drift, beach, and/or dune system in order to strengthen the cross-shore profiles of the coastal stretch [6]. In general, the most important benefits of artificial nourishments are the mitigation of coastal erosion, preventing flooding, and enabling the maintenance of wide beaches that also serve recreational purposes [7–9]. However, artificial nourishments interfere with the natural sediment dynamics and are not a permanent solution. The removal of sediments that occur naturally from the deposition site require re-nourishment over time, to maintain their design function [10]. Periodic re-nourishment intervals range on average from 2 to 10 years and depend on several factors related to initial design, wave climate, sand used, and the number and type of storms [11–13]. Therefore, although artificial nourishments are widely applied in different sites as a soft coastal intervention, questions related to the fill material fate, lifetime, and long-term impacts (to downdrift areas) still need to be answered with more confidence.

The main objective of this study was to present contributions to anticipate how sediments are spatially and temporally distributed after artificial nourishment interventions. The adopted approach was intended to be generic and supported by simple numerical theories/models, which could be useful for preliminary site-specific evaluations. The nourishment performance depends on multiple variables related to the sediment's dynamics and the artificial sand nourishment process itself. Thus, the nourishment impact was evaluated for different types of beaches (defined in the numerical model by different boundary conditions) and wave climates (wave heights and directions). Additionally, the longshore transport patterns of the nourished sediments were evaluated, aiming to anticipate their permanence at the deposition site and the frequency required for new nourishments.

2. Nourished Sediments Transport Processes

The concept of artificial nourishment is based on an attempt to replicate natural processes: Sand is deposited in coastal systems (beaches, dunes, or littoral drift) and nature, with the actions of waves, currents, and winds taking care of its distribution [8,14].

The artificially filled material is usually distributed in a limited length along the coastline and defines a cross-shore profile that is different from the stable profile at the deposition site. After sand deposition, the sediments spread out and tend toward a straight or slowly curving shoreline (Figure 1). In the direction perpendicular to the shoreline, the nourishment develops a stable configuration, in equilibrium with the relationship between the sediments' grain size and the incident wave climate. In the longshore direction, the center of the mass of the nourished sediments moves in the direction of the net sediment transport [14].

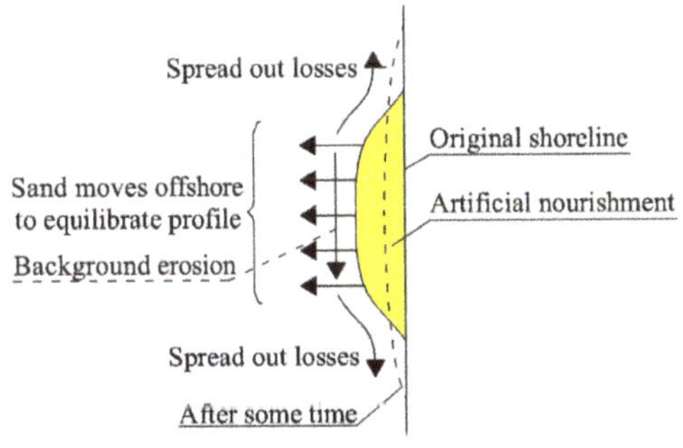

Figure 1. Artificial nourishment shoreline evolution and plan view of the erosion processes associated with the nourishment (based on [14,15]).

The sediments from artificial nourishment mix with native sediments and are distributed transversally and alongshore by wave actions and currents. The mixture of sediments occur by sediment transport processes related with local sediment dynamics together with losses that have occurred during the nourishment operations [15]. Depending on local conditions and intervention characteristics, an artificially nourished beach profile may take months or years to reach an equilibrium condition [16].

Due to the complexity of the coastal processes and taking into account the incompatibility of their time scales of interest, it is common to divide the assessment of the littoral sediments transport in two components [17,18]: Longitudinal transport, where sediments are transported in a direction parallel to the coastline; and cross-shore transport, where sediments are transported in a direction perpendicular to the coastline [19]. The cross-shore sediment transport reflects the short-term morphological evolution (storm to seasonal behavior) and the longshore sediment transport is associated with long-term shoreline changes, related to years or decades [3,14,19].

2.1. Theorical Approach

Several formulas to estimate longshore sediment transport rates are available in the literature. For example, Shaeri et al. [20], in their literature review about longshore sediment transport rate calculation, list the following equations: CERC (Coastal Engineering Research Center) [19], Kamphuis [21], Bayram et al. [22], van Rijn [23], and Tomasicchio et al. [24]. However, due to the coastal processes' complexity, longshore sediment transport evaluation is still challenging. Despite CERC formulation being introduced several decades ago, the CERC formula [19] is still widely used in both practice and fundamental research [20]. This formula [19] allows the longshore sediment transport to be computed based essentiality on the parameters related to waves at breaking: Significant wave breaking height and wave angle at breaking (Equation (1)). When offshore wave characteristics are known, breaking wave characteristics can be predicted, considering wave propagation phenomena, namely shoaling and refraction. Assuming a linear theory of wave propagation and that bottom contours are parallel to the shoreline, the wave angle at breaking (θ) depends on shoreline orientation. If waves reach the shore obliquely, breaking wave energy suspends and transports sediments alongshore [25].

$$Q = k\left(\frac{\rho\sqrt{g}}{16\gamma_b^{\frac{1}{2}}(\rho_s-\rho)(1-n)}\right)H_{sb}^{5/2}\sin(2\theta_b), \quad (1)$$

Since sand deposition during an artificial nourishment intervention changes the local bathymetry, refraction effects can be particularly important in a nourishment intervention with consequences on alongshore sediment flux (Figure 2).

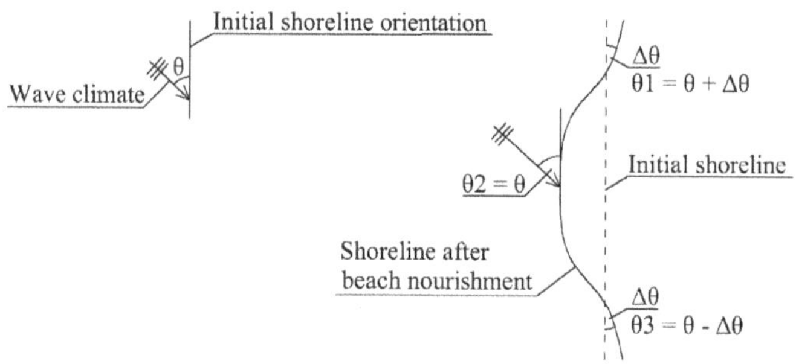

Figure 2. Wave angles before (θ) and after sand nourishment (θ1, θ2, and θ3, respectively, at updrift, middle, and downdrift of the intervention limits).

Murray and Asthon [26] evaluated the alongshore sediment flux for different wave angles and concluded that coastline segments with different orientations experience different alongshore sediment fluxes. The wave angle that leads to the highest value of sediment transport is not necessarily the most oblique wave in shallow water. Their results were produced based on a semi-empirical equation for alongshore sediment flux as a function of breaking-wave quantities, transforming it into a relationship involving offshore wave characteristics [25,26].

Considering the CERC formula and manipulating the equation in order to involve offshore wave characteristics, considering the linear wave theory, it is concluded that the longshore sediment transport rate increases with wave height and reaches the maximum for an offshore wave direction (θ) of around 48° (Figure 3a). However, if the nourishment changes the shoreline and wave breaking depth contours and, consequently, the wave breaking orientation, the maximum longshore sediment transport rates occur for different wave angles offshore (Figure 3b).

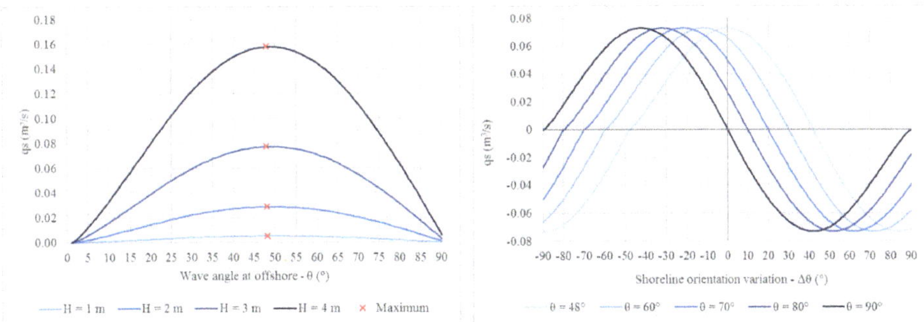

(a) (b)

Figure 3. Longshore sediment transport rate according to CERC formula: Sediment transport for different wave heights, depending on the wave angle offshore (**a**); sediment transport for different wave angles offshore (H = 3 m), depending on the shoreline orientation variation due to the nourishment—Δθ positive angles are in a clockwise direction (**b**).

For example, considering wave angles at offshore (θ) of 60° and 80°, it is possible to observe two main results for the longshore sediment transport rate at the nourishment updrift and downdrift boundaries, considering a shoreline rotation (Δθ) of 25° due to the intervention: (i) There is a change in the longshore sediment transport direction at updrift, for the wave with θ = 80°. This wave also presents higher sediment transport capacity to updrift than the 60° wave to downdrift (Figure 4a); (ii) the less oblique wave at offshore (80°) presents higher sediment transport capacity at the downdrift limit of the nourishment than the 60° wave, which reverses the behavior observed before the nourishment (Figure 4b).

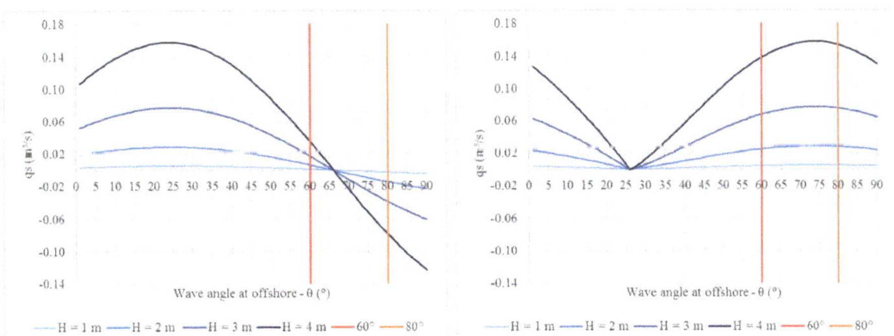

(a) (b)

Figure 4. Longshore sediment transport rate according to CERC formula for different wave heights in function of the wave angle at offshore: Δθ = +25° (**a**); Δθ = −25° (**b**).

2.2. Numerical Modeling Approach

Numerical models can be very important and useful tools to anticipate the performance of nourishment operations [17,27]. However, due to the complexity of the physical processes involved and the high computational effort required, each model focuses on a limited range of processes, acting over a certain scale, and usually not merging cross-shore and longshore processes at the same time [28–30].

In nourishment modeling applications, where both short- and long-term effects must be simulated, 3D models that simultaneously include cross-shore and longshore processes are still lacking to a large degree. At present, the profile change and contour line models are the ones that are regularly employed in engineering projects for forecasting coastal evolution. Modeling coastal evolution in a long-term perspective is also linked to the study of the sand nourishments, although some probabilistic studies related to modeling of both longshore and cross-shore processes in a medium-term perspective can be referred, e.g., Dong and Chen [31], Wang and Reeve [32], Callaghan et al. [33], and Reeve et al. [34].

Burcharth et al. [35] present some examples of coastal zone evolution studies through 3D models. However, because fully 3D beach change models are used in attempt to simulate local characteristics of waves, currents, and sediment transport, they require extensive verification and sensitivity analyses. Thus, these models are time demanding, due to both field data collection and computational time [35,36]. According to Burcharth et al. [35], 3D models are most suitable for medium-term morphological investigation over a limited coastal area with dimensions of about 10 km in the alongshore direction and 2 km in the offshore direction. Larson et al. [30] highlight the need for long-term morphological modeling, emphasizing the importance of reducing the computational time required, the model reliability and robustness being key properties to achieve useful simulation results.

3. Numerical Modeling

The alongshore distribution of the sediment volumes added by the artificial nourishments was evaluated by numerical modeling. For this purpose, the numerical long-term configuration (LTC) model was applied, which combines a simple classical one-line model with a rule-based model for erosion/accretion volume distribution along the beach profile [37–39]. The model was applied to a generic study area with regular and parallel bathymetry (modeled domain), allowing us to evaluate the effects of the wave climate and the domain boundary conditions in the nourishment performance.

3.1. LTC Numerical Model

The numerical LTC model, developed by Coelho [3], was specifically designed for sandy beaches, where the main cause of medium-term shoreline evolution is the longshore sediment transport gradients. The latter mainly depends on the wave climate and the domain boundary conditions. The model inputs are the water level and the bathymetry/topography of the modeled area, which is modified during the simulation [3,29,38].

The LTC numerical model assumes that each wave acts individually during a certain period of time, the computational time step. For each wave, the computational structure of the model performs three main steps: (1) Wave propagation; (2) calculation of the longshore sediment transport volume; and (3) update beach morphology, depending on longshore sediment transport gradients.

The model considers the refraction and shoaling phenomena during wave propagation and diffraction, close to coastal structures. The wave breaking occurs for the depth $H_b \approx 0.78$ Hb (Hb is the wave height at breaking and hb is the local depth at breaking).

The LTC numerical model is based on a one-line theory and simulates medium- to long-term coastal configuration evolution, considering the sediment continuity equation (Equation (2)), where V corresponds to the volume of sediments in a section of infinitesimal width ∂y, Q is the longitudinal sediments transport rate, q_x are any external supplies of sediments (artificial nourishments, for instance) per unit of the beach width, and t is the time [3].

$$\frac{\partial V}{\partial t} = \left(\frac{\partial Q}{\partial y} + q_x\right) \tag{2}$$

According to Coelho [3], the discretization of Equation (2) in time intervals equal to Δt and the analysis along the shoreline for stretches of length Δy allows the volume variation in each stretch with the variation in time of the longshore sediment transport to be related:

$$\Delta V = (Q_i - Q_{i-1} + Q_{ext})\Delta t = (\Delta Q + Q_{ext})\Delta t \quad (3)$$

The longshore sediment transport variation in the length Δy results from the difference between the incoming (Q_i) and leaving (Q_{i-1}) sediment volumes in the stretch of Δy length, for each time interval Δt. The longshore sediment transport volumes are computed for each coastal stretch Δy, estimated considering the angle of the shoreline to oncoming breaking waves [19,21], the wave breaking height, beach slope, and sediment grain size [21,36].

The model considers that $Q_{ext} = q_{ext}\Delta y$, corresponding to the volume variation of sediments (ΔV) in the length Δy, is uniformly distributed over the active cross-shore profile, representing a variation Δz in the elevation (Equation (4)), in every time step, as shown in Figure 5 [3]. The active profile is adjusted with the adjacent zones; therefore, the shoreline position, besides depending on Δz, also depends on the bathymetry and topography associated with each cross-shore profile (Figure 5). Thus, LTC computes a new shape of the cross-shore profile in each time step [3,40].

$$\Delta z = \frac{\Delta V}{(width\ of\ the\ active\ profile)\Delta y} \quad (4)$$

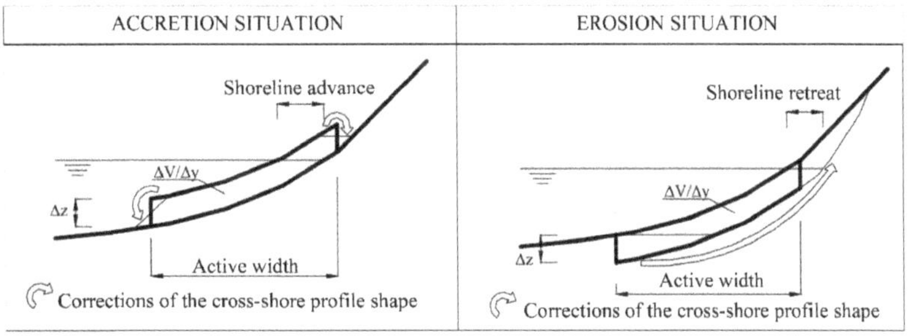

Figure 5. Schematic representation of long-term configuration (LTC) assumptions for cross-shore profiles updated during computational calculations (adapted from [3]).

Through LTC, it is possible to evaluate the shoreline evolution considering different coastal defense interventions, namely artificial nourishments. The user needs to define the parameters related to the total nourishment volume, the area covered by the intervention, and the time duration to perform the nourishment (starting and ending instants of the intervention). Since the model computes the wave-breaking characteristics, the location of the artificial nourishment should include the breaking depth in order to be adequately modeled and distributed in the cross-shore profile.

3.2. Modeled Scenarios

The artificial nourishment performance (sediment distribution along time and space) was evaluated for a generic situation, to better control the response to different scenarios of wave height and direction, plus the boundary conditions of the modeled domain, representing three different types of coasts: (i) Open coast; (ii) confined beach; and (iii) beach with a fixed position at the updrift side.

The domain area is 5000 × 20,000 m^2, with regular and parallel bathymetry, represented by a grid of points spaced 20 m in the cross-shore direction and 50 m in the longshore direction. The longshore direction was divided into 10 coastal stretches of 2 km in length (Figure 6). In each stretch, a representative cross-shore profile was selected for results control, located in the middle of the stretch length. Figure 6 also shows the initial cross-shore profile geometry of the entire modeled area.

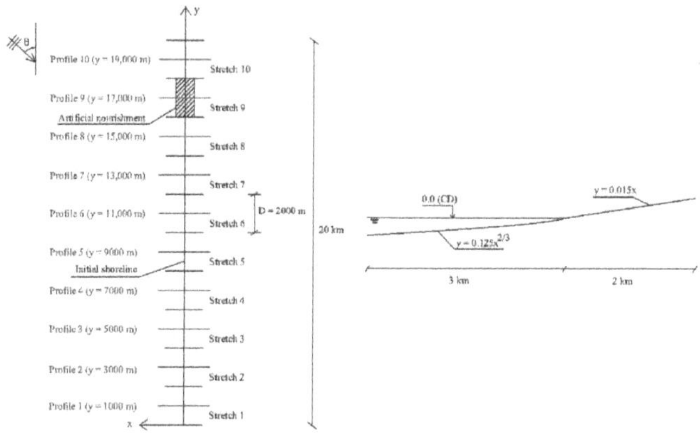

Figure 6. Study area: Plan view (**left**) and cross-shore profile (**right**) (based on [41,42]).

The nourishment location was defined near the updrift end of the modeled grid, Stretch 9, filling the entire active width of the beach profile, which makes up an area of 1640×2000 m^2 covered by the nourishment. The nourishment is performed a single time at the beginning of the simulation, with a volume of 5×106 m^3, taking ten months (7200 h) to deposit the sediments.

Twenty-four wave climate scenarios were generated, considering constant throughout the five years simulation period. These scenarios combined four values of wave heights (H = 1, 2, 3, and 4 m) with 6 different wave directions (θ = 60°, 65°, 70°, 75°, 80°, and 85°).

Additionally, aiming to understand the influence of beach exposure and sedimentary balance inputs in the coastal cell, three different situations were considered for the modeled domain boundary conditions. If there were no sedimentary exchanges with the adjacent cell, such as fluvial sources or updrift inputs on coastal balance, the cell was considered with the boundaries closed. Otherwise, it was considered that the domain may receive sediments from the exterior. The first boundary scenario considered no sediments going in or out, at both boundaries of the domain. This scenario was named Clo/Clo (representative of an embedded beach-closed area). The second boundary scenario extrapolates similar longshore sediments transport conditions in the neighboring areas of both boundaries of the domain, giving continuity to the longshore sediment transport rates along the coast, typical of open coasts or when fluvial sources are present just updrift the domain (Ext/Ext scenario). In the third scenario, no sediments were considered at updrift and extrapolation of neighboring conditions were adopted at downdrift (Clo/Ext). Combining each wave climate scenario (24) with the described boundary conditions (3) made up 72 different scenarios. Each scenario was modeled twice to evaluate the performance with and without the artificial nourishment.

In all the scenarios, a constant cross-shore active profile was considered, with a closure depth and runup limit of −15 (CD) and +5 m (CD), respectively. The CERC formula was adopted to estimate the longshore sediment transport. The parameters related to the water level, sediment's characteristics and cross-shore profiles evolution were considered the same in all the tested scenarios.

3.3. Methodology

The numerical modeling tests were carried out for a total period of 5 years (43,800 h of simulation), obtaining annual graphical outputs to analyze the nourished sediments distribution along the coast. The nourishment impact was assessed in each coastal stretch, by quantifying the sediments volume induced by the nourishment and the gain or loss of area due to the shoreline evolution (accretion or erosion). This was achieved by subtracting the results of the scenarios without nourishment to the ones in which the artificial nourishment was considered.

4. Results

The results of artificial nourishment impacts, evaluated in terms of the sediment volume distribution in each coastal stretch and related gains or losses of territory due to shoreline evolution, are presented in this section. The results were analyzed under the numerical model assumptions that shoreline position evolution depends on longshore sediments transport rate and neglects the coastal processes related to cross-shore sediment transport. Comparisons for different wave height and wave direction scenarios, in combination with the different type of beach exposures, are presented.

4.1. Wave Height

The nourished sediments' distribution along the coast shows that for the same wave direction and independently of the boundary conditions, the higher the wave, the earlier the downdrift stretches presented benefits from the nourishment operation. These benefits were observed in an increase of the robustness of the cross-shore profiles (increasing of volumes in the coastal stretches) and gain of territory (both accretion area and/or non-eroded area in comparison with the equivalent non-nourished scenario). As the wave height increased, the dispersion of the sediments became greater along the coast, causing the benefits to diminish in some coastal stretches close to the intervention site, where the beneficial effects of the nourishment tended to disappear after 5 years of simulation (Figure 7).

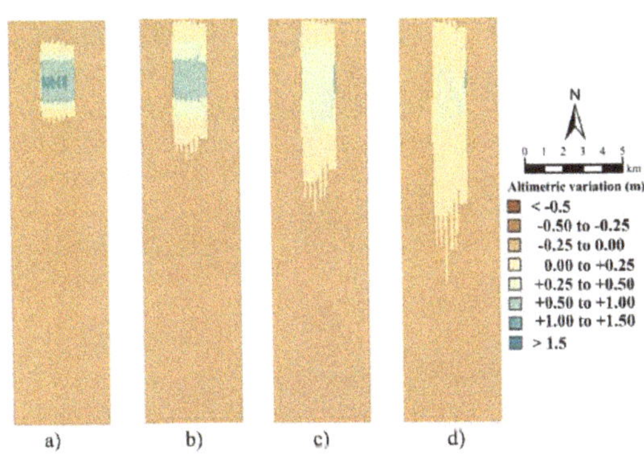

Figure 7. Beach nourishment impact on morphology after 5 years of simulation ($\theta = 60°$; Ext/Ext): (**a**) H= 1 m; (**b**) H = 2 m; (**c**) H = 3 m; (**d**) H = 4 m (based on [41,42]).

4.2. Wave Direction

The wave direction results indicated that the stretches located downdrift began to show benefits of nourishment sooner in time and in greater magnitude, the closer to perpendicular to the shoreline the wave direction was (Figure 8). These results are in agreement with the three types of boundary conditions tested. Although this result may seem counterintuitive, since it could be expected than the most oblique wave presents more longshore sediment transport capacity, the behavior can be attributed to wave refraction effects in the boundaries of the nourishment area, as previously discussed (see Section 2.1). Sediment deposition changed the local bathymetry and the resulting longshore sediment flux, which is dependent on the new bathymetry at each time step.

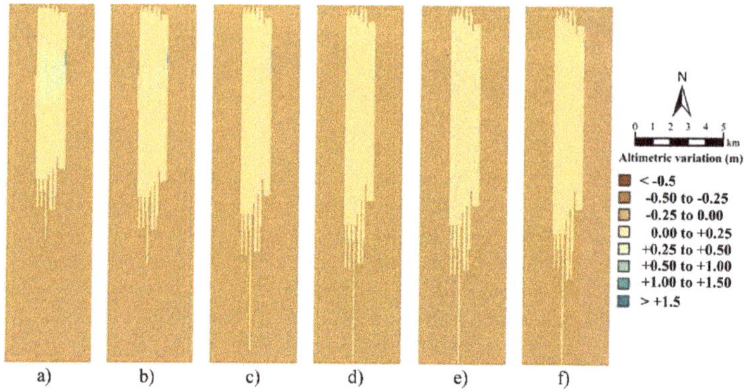

Figure 8. Beach nourishment impact on morphology after 5 years of simulation (H = 4 m; Ext/Ext): (**a**) θ = 60°; (**b**) θ = 65°; (**c**) θ = 70°; (**d**) θ = 75°; (**e**) θ = 80°; (**f**) θ = 85°.

4.3. Domain Boundary Conditions

Artificial nourishment effects are conditioned by the type of beach, which is here represented by the boundary conditions of the modeled domain. Two types of main effects were observed. In the scenarios represented by extrapolation of nearby sediment transport capacity at the updrift boundary, the impact of the artificial nourishment, although not reflected so significantly in the gained area, contributed to an increase of the robustness of the cross-shore profiles. When the scenarios corresponded to the closed updrift boundary condition (no sediments going in), the nourishment represented an important gain of territory, also causing an increase of sediment volume in the cross-shore profile. Both effects were greater in coastal stretches located nearby the fill placement (Figures 9 and 10).

Considering the boundary conditions, the difference of area gained observed at the updrift stretches was related to the absence of sediments going in, when the boundary was closed. In these scenarios, the shoreline presented a high retreat trend, in opposition to the extrapolation boundary, when the shoreline presented a stable configuration (Figure 11). The artificial nourishment effects decreased with the distance to the fill placement and the magnitude of the results at the most downdrift coastal stretches were similar for the three boundary scenarios tested.

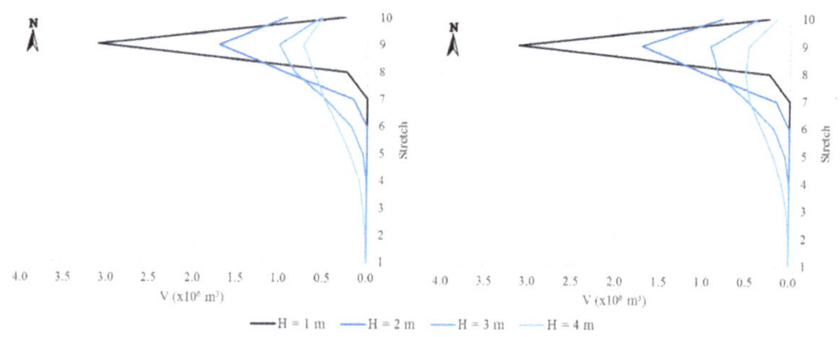

Figure 9. Sediment volumes due to artificial nourishment after the fifth year of simulation (θ = 85°): Clo/Clo (**left**); Ext/Ext (**right**) (based on [41,42]).

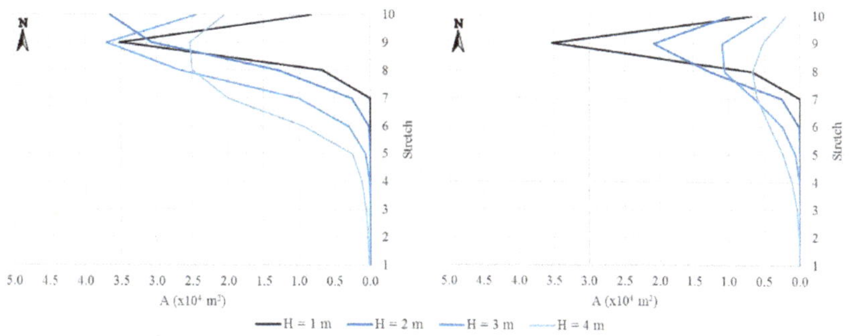

Figure 10. Area gained due to artificial nourishment after the fifth year of simulation (θ = 85°): Clo/Clo (**left**); Ext/Ext (**right**) (based on [41,42]).

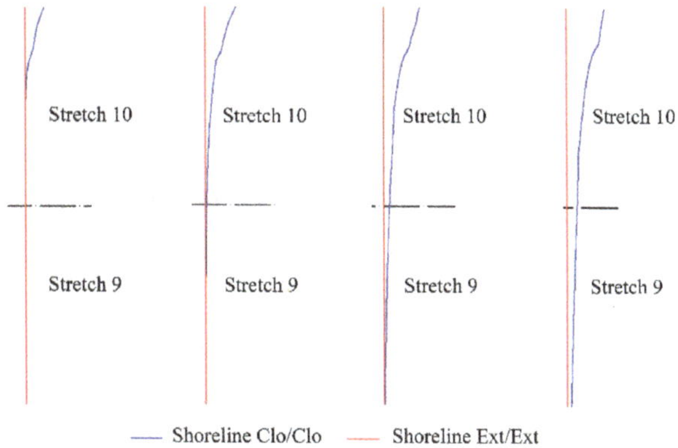

Figure 11. Shoreline position without nourishment after the fifth year of simulation (θ = 60° and from left to the right, H = 1 to H = 4 m).

4.4. Evolution along Time

Two types of behavior were mainly observed in time. On the one hand, there were coastal stretches that successively increased the benefits of nourishment in the 5-year simulation period. On the other hand, some stretches reached a maximum benefit and then they registered a diminishing trend of the benefits over time. In general, the stretches located further downdrift successively increased the beneficial effect over the 5 years of simulation. The loss of benefit behavior of the nourishment at the end of some time occurred in the most updrift stretches and was mainly observed for the more energetic wave climates (Figure 12).

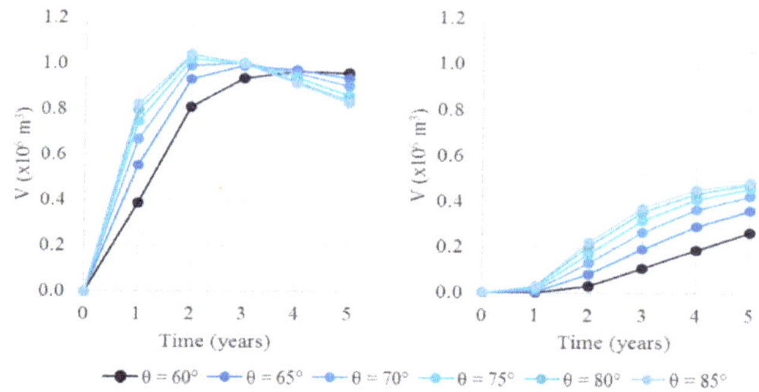

Figure 12. Nourishment volumes' distribution in time (H = 3 m; Ext/Ext): Stretch 8 (**left**) and stretch 7 (**right**).

5. Discussion

The evaluation of the longshore transport of the nourished sediments in 72 scenarios aimed to contribute to the understanding of sediments distribution in the coastal zone. In general, the artificial nourishment had a beneficial impact, both in terms of area gained and cross-shore profile robustness, represented by the gain in sediment volume. The nourishments presented significantly lower impacts at the most downdrift stretches, when compared to the stretches located closer to the intervention site. Sediments were mainly transported to the downdrift direction, but there were also some benefits at the region located updrift of the nourishment (Stretch 10). This conclusion was aligned with the results obtained by Beachler et al. [43] through monitoring studies in the City of Delray Beach (Florida) from 1973 to 1994. During the monitoring period in Delray Beach, about 3.5 million cubic meters of sediments were placed, and, according to Beachler et al. [43], in this same period a loss of sediments at the deposition site was observed, but those sediments were located on adjacent beaches, 3 km to the north and south of the fill area. Another example of where the updrift benefits of the nourishment are referred to is in the work of Slott et al. [44].

The position of the artificial nourishment volumes was mainly located close to the intervention site, near the updrift end of the study area (stretches 8 and 9), along the 5 years of simulation. As the wave height increased, the sediments were distributed to more downdrift locations. Some transport of the nourished sediments to updrift could be observed in the scenarios with smaller θ angles, due to sediment transport processes related with wave refraction effects (Section 2.1). Benedet et al. [45] highlighted that the changes in the shoreline orientation and morphology induced by nourishment interventions play an important role in the hydrodynamics of the nourished area and consequently, on the sediment transport processes.

The longshore nourished sediments' transport was a function of the wave climate and the boundary conditions of the modeled domain. It was difficult to predict the behavior trends of the nourished sediments. For the highest wave heights, higher nourished longshore sediments transport rates occurred. However, by setting a wave height, the trends due to wave direction were not completely clear and standard behaviors in the longshore transport of the nourished sediments were not identified. The boundary conditions of the modeled domain also affected the way the sediments were distributed, with impacts on the nourished longshore sediments transport rates. Campbell et al. [46] presented an overview of the most significant nourishment programs across the U.S., concluding that the nourishment magnitudes across the country varies, decreasing from north to south on the Atlantic Coast, and from the Atlantic to the Gulf Coast. This variation detaches the influence of beach morphology and wave climate in the evolution of beach nourishment interventions.

6. Conclusions

The present study allowed us to identify the impact of the artificial nourishments on coastal morphology through the numerical LTC model. Knowledge about the morphology evolution of the coastal zones after artificial nourishments is fundamental, making it possible to evaluate the intervention performance and to support the decision process in the choice of the most adequate solution. The performance and longevity of artificial nourishments were dependent on several factors related to the design of the intervention, as well the physical sediment transport processes that occurred at the deposition site. The sediments were distributed by the wave action, currents, and winds. However, the characteristics of the incident wave action were considered the main modeling agent and the most important in evaluating sediment transport capacity.

From the present study, it was concluded that nourishment interventions present a positive impact on morphology evolution, reducing negative effects of shoreline retreat against the existing sediment deficit, as well as increasing the volume of sediments in the cross-shore profile, the performance and longevity of the intervention being a function of the wave climate and the boundary conditions of the modeled area.

Generally, it was verified that the sediments' dispersion along the coastal zone was higher for the higher waves, which presented greater longshore sediment transport capacity. Regarding wave direction, it was observed that sediment distribution from updrift to downdrift was higher for waves propagating from directions close to perpendicular to the shoreline. This result can be attributed to refraction effects in the nearby boundaries of the nourished area, since the sediment deposition changed the local bathymetry and different coastline orientations led to different longshore sediment fluxes. The simulation of artificial nourishments in different types of beach, distinguished in this study by the boundary conditions of the modeled area, showed that the existence of sedimentary inputs on the coastal cell led to different impacts of the artificial nourishment. In the updrift area, when there was no sediment coming in (closed boundary), the artificial nourishment produced important benefits, verifying that the accomplishment of the artificial nourishment contradicts the retreat of the shoreline and increases the robustness of the cross-shore profile. In open beaches, assuming that erosion was not occurring, the artificial nourishment mainly increased the robustness of the cross-shore profiles.

In the present study, the artificial nourishment was evaluated for only one possible solution, with a specific filling volume, on a specific deposition location, and considering just one time interval for the intervention. However, a first step was achieved, giving important insights on modeling artificial nourishment behaviors and sediments' permanence time in downdrift locations alongshore. Only higher waves justified the need for new nourishment after the 5-year simulated scenarios, due to the decrease of the nourishment effects on the intervention site.

The globally obtained results were in line with the monitoring works referred. Thus, in a very simple approach, the performed work allowed us to generically anticipate how sediments are spatially and temporally distributed after artificial nourishments interventions, depending on the type of beach and wave climate (wave heights and directions). Currently, LTC is being applied to real nourishment scenarios (over more complex and realistic beach configurations), in order to study the performance and longevity of artificial nourishments. For that, real bathymetry and topography, a recorded series of wave climate and existing coastal works on the study site, as groins and longitudinal revetments, are being considered. Simplifications are required to represent the real conditions in the model, but it is considered that a step forward on understanding artificial nourishments' performance along time and along the shoreline has beem given by the model results presented in this study.

Author Contributions: Conceptualization, C.C.; methodology, C.C.; software, A.M.F.; validation, A.M.F. and C.C.; formal analysis, A.M.F.; investigation, A.M.F.; resources, C.C.; data curation, A.M.F. and C.C.; writing—original draft preparation, A.M.F.; writing—review and editing, C.C.;

visualization, A.M.F.; supervision, C.C.; project administration, C.C.; funding acquisition, C.C. Both authors have read and agreed to the published version of the manuscript.

Funding: This research received no external funding.

Institutional Review Board Statement: Not applicable.

Informed Consent Statement: Not applicable.

Data Availability Statement: Publicly available datasets were analyzed in this study. This data can be found here: http://hdl.handle.net/10773/17377, accessed on 4 February 2021.

Acknowledgments: This work was financially supported by the project "Sandtrack-Beach nourishment: An integrated methodology for coastal management support", POCI-01-0145-FEDER-031779, funded by FEDER, through COMPETE2020-POCI, and by national funds (OE), through FCT/MCTES.

Conflicts of Interest: The authors declare no conflict of interest.

References

1. Antunes-Carmo, J. *Processos Físicos e Modelos Computacionais em Engenharia Costeira*; Imprensa da Universidade de Coimbra: Coimbra, Portugal, 2016; 452p, ISBN 978-989-26-1152-5. (In Portuguese)
2. Housley, J. Justification for beach nourishment. In Proceedings of the 25th International Conference on Coastal Engineering, Orlando, FL, USA, 29 January 1996.
3. Coelho, C. Riscos de Exposição de Frentes Urbanas para Diferentes Intervenções de Defesa Costeira. Ph.D. Thesis, University of Aveiro, Aveiro, Portugal, 2005; 404p. (In Portuguese)
4. Hanson, H.; Brampton, A.; Capobianco, M.; Dette, H.H.; Hamm, L.; Laustrup, C.; Lechuga, A.; Spanhoff, R. Beach nourishment projects, practices, and objectives—A European overview. *Coast. Eng.* **2002**, *47*, 81–111. [CrossRef]
5. Coelho, C.; Silva, R.; Veloso Gomes, F.; Rodrigues, L. Artificial nourishment and sand by-passing in the Aveiro inlet, PortugalNumerical Studies. In Proceedings of the 32nd International Conference on Coastal Engineering, Shangai, China, 21 January 2010.
6. Van Heuvel, T. *Climate of Coastal Cooperation, Part III—3-3-2: Sand Nourishment: A Flexible and Resilient, Adaptive Coastal Defense Measure*; Coastal & Marine Union: Leiden, The Netherlands, 2011; pp. 164–166.
7. Verhagen, H. Method for artificial beach nourishment. In Proceedings of the 23rd International Conference on Coastal Engineering, Venice, Italy, 29 January 1992.
8. Tondello, M.; Ruol, P.; Sclavo, M.; Capobianco, M. Model tests for evaluating beach nourishment performance. In Proceedings of the 26th International Conference on Coastal Engineering, Copenhagen, Denmark, 29 January 1998.
9. Marinho, B.; Coelho, C.; Larson, M.; Hanson, H. Short- and Long-Term Responses of Nourishments: Barra-Vagueira Coastal Stretch, Portugal. *J. Coast. Conserv.* **2018**, *22*, 475–489. [CrossRef]
10. Clark, J. *Coastal Zone Management Handbook*, 1st ed.; Lewis Publishers: Boca Raton, FL, USA, 1995; 720p, ISBN 1-56670-092-2.
11. NRC. *Beach Nourishment and Protection*; Nacional Research Council; The National Academies Press: Washington, DC, USA, 1995; 352p, ISBN 978-0-309-05290-0.
12. ASBPA. *Shore Protection Assessment: How Beach Nourishment Projects Work*; Shore and Beach Preservation Association, US Army Corps of Engineers: Vicksburg, MS, USA, 2007.
13. Schipper, M.; Vries, S.; Ruessink, G.; Zeeuw, R.; Rutten, J.; Gelder-Maas, G.; Stieve, M. Initial Spreading of a Mega Feeder Nourishment: Observations of the Sand Engine Pilot Project. *J. Coast. Eng.* **2016**, *111*, 23–28. [CrossRef]
14. Kamphuis, J.W. *Introduction to Coastal Engineering and Management*; Advanced Series on Ocean Engineering; World Scientific: Singapore, 2000; Volume 16, 437p.
15. Bodegon, M.J. Beach Nourishment: An Evaluation of Equilibration Design Methods. Cancún Beach Rehabilitation Project. Master's Thesis, Delft University of Technology, Delft, The Netherlands, 2004; 149p.
16. Hearon, G.; Ledersdorf, C.; Gadd, P. *California Beach Restoration Study, Part II: Beach Nourishment*; Department of Boating and Waterways and State Coastal Conservancy: Sacramento, CA, USA, 2002; 7p.
17. Galofré, J.; Montoya, F.; Medina, R. Study of the Evolution of a Beach Nourishment Based on Computer Models. *Trans. Built Environ.* **1995**, *9*, 250–253.
18. Dean, R. *Beach Nourishment Theory and Practice*; Advanced Series on Ocean Engineering; World Scientific: Singapore, 2002; Volume 18, 403p.
19. CERC. *Shore Protection Manual*; U.S. Army Corps of Engineers, Coastal Engineering and Research Center, U.S. Government Printing Office: Washington, DC, USA, 1984; Volume 1.
20. Shaeri, S.; Etemad-Shahidi, A.; Tomlinson, R. Revisiting longshore sediments transport formulas. *J. Waterw. Port Coast. Ocean Eng.* **2020**, *146*, 04020009. [CrossRef]
21. Kamphuis, W. Alongshore sediment transport rate. *J. Waterw. Port Coast. Ocean Eng.* **1991**, *117*, 624–640. [CrossRef]
22. Bayram, A.; Larson, M.; Hanson, H. A new formula for the total longshore sediment transport rate. *Coast. Eng.* **2007**, *54*, 700–710. [CrossRef]

23. van Rijn, L.C. A simple general expression for longshore transport of sand, gravel and shingle. *Coast. Eng.* **2014**, *90*, 23–29. [CrossRef]
24. Tomasicchio, G.R.; D'Alessandro, F.; Barbaro, G.; Musci, E.; De Giosa, T.M. Longshore transport at shingle beaches: An independent verification of the general model. *Coast. Eng.* **2015**, *104*, 69–75. [CrossRef]
25. Murray, A. Reducing model complexity for explanation and prediction. *Geomorphology* **2007**, *90*, 178–191. [CrossRef]
26. Murray, A.; Asthon, A. Instability and finite-amplitude self-organization of large-scale coastline shapes. *Phil. Trans. R. Soc. A* **2013**, *371*, 20120363. [CrossRef]
27. Lima, M. Ferramenta Numérica de Análise do Impacto de Intervenções de Defesa Costeira na Evolução da Linha de Costa: Custos e Benefícios. Ph.D. Thesis, University of Aveiro, Aveiro, Portugal, 2018; 294p. (In Portuguese)
28. Hanson, H.; Aarninkhof, S.; Capobianco, M.; Jiménez, J.; Larson, M.; Nicholls, R.; Plant, N.; Steetzel, H.; Stive, M.; Vriend, H. Modelling of coastal evolution on early to decadal time scales. *J. Coast. Res.* **2003**, *19*, 790–811.
29. Coelho, C.; Silva, R.; Veloso-Gomes, F.; Taveira-Pinto, F. Potential impacts of climate change on NW Portuguese coastal zones. *ICES J. Mar. Sci.* **2009**, *66*, 1497–1507. [CrossRef]
30. Larson, M.; Palalane, J.; Fredriksson, C.; Hanson, H. Simulating cross-shore material exchange at decadal scale. Theory and model component validation. *Coast. Eng.* **2016**, *116*, 57–66. [CrossRef]
31. Dong, P.; Chen, H. A probability method for predicting the time-dependent long term beach erosion. *Coast. Eng.* **1999**, *36*, 243–261. [CrossRef]
32. Wang, B.; Reeve, D.E. Probabilistic modelling of long-term beach evolution near segmented shore-parallel breakwaters. *Coast. Eng.* **2010**, *57*, 732–744. [CrossRef]
33. Callaghan, D.P.; Ranasinghe, R.; Roelvink, D. Probabilistic estimation of storm erosion using analytical, semi-empirical, and process based storm erosion models. *Coast. Eng.* **2013**, *82*, 64–75. [CrossRef]
34. Reeve, D.; Karunarathna, H.; Pan, S.; Horrilo-Caraballo, J.; Różyński, G.; Ranasinghe, R. Data-driven and hybrid coastal morphological preditiction methods for mesoscale forecasting. *Geomorphology* **2016**, *256*, 49–67. [CrossRef]
35. Burcharth, H.; Hawkins, S.; Zanuttigh, B.; Lamberti, E. *Environmental Design Guidelines for Low Crested Coastal Structures*; Elsevier Science: Oxford, UK, 2007; 448p.
36. Hanson, H.; Kraus, N.C. *GENESIS: Generalized Model for Simulating Shoreline Chang—Report 1 Technical Reference, Technical Report CERC-89-19*; Coastal Engineering Research Center, US Army Engineer Waterways Experiment Station: Vicksburg, MS, USA, 1989; 247p.
37. Coelho, C.; Veloso-Gomes, F.; Silva, R. Shoreline coastal evolution model: Two Portuguese case studies. In Proceedings of the 30th International Conference, San Diego, CA, USA, 19 November 2006; pp. 3430–3771.
38. Coelho, C.; Lima, M.; Veloso-Gomes, F. Relationship between cross-shore active profile and one-line shoreline evolution models performance. *J. Coast. Res.* **2013**, *65*, 2107–2112. [CrossRef]
39. Coelho, C.; Narra, P.; Marinho, B.; Lima, M. Coastal management software to support the decision-makers to mitigate coastal erosion. *J. Mar. Sci. Eng.* **2020**, *8*, 37. [CrossRef]
40. Silva, R. Avaliação Experimental e Numérica de Parâmetros Associados a Modelos de Evolução da Linha de Costa. Ph.D. Thesis, Faculdade de Engenharia da Universidade do Porto, Porto, Portugal, 2010; 488p. (In Portuguese)
41. Ferreira, A.M. Evolução Morfológica do Litoral Após Alimentações Artificiais de Sedimentos. Master's Thesis, University of Aveiro, Aveiro, Portugal, 2016; 126p. (In Portuguese)
42. Coelho, C.; Ferreira, M.; Marinho, B. Numerical modelling of artificial sediment nourishment impacts. *J. Coast. Res.* **2020**, *95*, 209–213. [CrossRef]
43. Beachler, P.; Mann, D. Long range positive effects of the Delray beach nourishment program. *Coast. Eng.* **1993**, *359*, 4613–4620.
44. Slott, J.; Smith, M.; Murray, B. Synergies between adjacent beach-nourishing communities in a morpho-economic coupled coastline model. *Coast. Manag.* **2008**, *36*, 374–391. [CrossRef]
45. Benedet, L.; Finkl, C.; Hartog, W. Processes controlling development of erosional hot spots on a beach nourishment project. *J. Coast. Res.* **2007**, *23*, 33–48. [CrossRef]
46. Campbell, T.; Benedet, L. Beach nourishment magnitudes and trends in the U.S. *J. Coast. Res.* **2006**, *39*, 57–64.

Journal of
Marine Science and Engineering

Article

Evaluation of Coastal Protection Strategies at Costa da Caparica (Portugal): Nourishments and Structural Interventions

Francisco Sancho

Laboratório Nacional de Engenharia Civil, 1700-066 Lisboa, Portugal; fsancho@lnec.pt; Tel.: +351-218443664

Abstract: Costa da Caparica beach, in Portugal, has suffered from chronic erosion for the last 50 years, a phenomenon that has been countered by various management interventions. This study aims at comparing sixteen possible interventions, thus identifying the most effective one(s) in terms of reducing beach erosion or even promoting beach accretion. This exercise is achieved using a one-line shoreline evolution model, calibrated with in situ field data, forced by local wave conditions. The target management period is 25 years. In the calibration phase, it is found that the annual mean alongshore net sediment transport along the 24 km sandy coast is variable in direction and magnitude, but it is mostly smaller than $\pm 50 \times 10^3$ m^3/year. This net transport results from the imbalance of northward/southward-directed bulk transports of circa tenfold-larger magnitudes. This affects the overall sediment balance at the urban beaches, as well as the effectiveness of the intervention strategies. The results show that the present management strategy is effective in holding the shoreline position, although deploying the same nourishment volume but over a shorter area could lead to better results. The best solutions, which are capable of promoting beach accretion, implicate the lengthening of the terminal groin at the northern extremity of the beach. The results from this study can support decision makers in identifying the most appropriate management action, not just locally but also at other coastal regions where similar problems persist and the same methodology could be applied.

Keywords: one-line model; coastal erosion; sediment transport; coastal management; groin field

Citation: Sancho, F. Evaluation of Coastal Protection Strategies at Costa da Caparica (Portugal): Nourishments and Structural Interventions. *J. Mar. Sci. Eng.* **2023**, *11*, 1159. https://doi.org/10.3390/jmse11061159

Academic Editor: Yuanzhi Zhang

Received: 17 April 2023
Revised: 22 May 2023
Accepted: 26 May 2023
Published: 31 May 2023

1. Introduction

Over the last century, coastal management has been addressed and confronted with varying strategies and responses, with many successful and failure stories from which we can learn and achieve a more efficient integration of the socio-economic systems with natural coastal ecosystems [1]. Sandy beaches can enter into sediment unbalance and suffer from erosion problems, affecting the socio-economic value of the area and, conceivably, endangering people and goods [2–4].

Fifty years ago, coastal management responses were typically reactive, based mainly on structural engineering measures that aimed to provide protection against flooding, safety and space for beach recreation and, occasionally, natural conservation. Currently, preventive strategies with risk-based analysis frameworks are often employed, which combine structural and non-structural measures such as hard and soft protective measures, land use planning and ecosystems services, early warning/evacuation strategies, education and insurance schemes [5–7].

While the protection of coasts is becoming a global priority, there is no consensus on the actual meaning of "coastal protection" [7]. To some, this may mean resistance management, to halt coastal erosion and protect property; to others, it means pro-adaptive management, allowing coastal ecosystems to function naturally [6] and, in many cases, retreat. The outcomes of these different perceptions are contrasting. This paper analyses different management policies aligned with the first option—halting coastal erosion and protecting property.

At Costa da Caparica town (near Lisbon, Portugal, see location in Figure 1), human encroachment is responsible for degradation and loss of coastal ecosystems over recent decades [8]. The beach suffers from chronic erosion [9,10] and has been subject to several coastal protection measures since the early 1970s, involving both hard and soft protection measures. In general, the hard measures built to prevent erosion (such as seawalls or groins) further limited the coast from adjusting to the changing conditions. The naturally existing foredunes could no longer translate upwards and landwards as the sea level rose and the coast eroded, because the infrastructure occupied the required space. Consequently, the resilience and resistance of the coast and its ecosystem services are being lost, diminished, or put at risk. In addition, this coastal town is exposed to rising sea levels, erosion, flooding, tsunamis and extreme hydro-meteorological events (e.g., [11]). Problems remain; in particular, the narrow beach width is unable to withstand the beach-use summer crowds and economic losses due to coastal overtopping (e.g., [12]); hence, future actions and strategies need to be evaluated.

Figure 1. Study area with place names, reference or baseline (deep orange) and 8, 16 and 30 m isobaths. (Adapted from geomar.hidrografico.pt, accessed on 20 May 2023, using ESRI World Imagery).

The use of numerical models for predicting shoreline behaviour in the future is a widely used practice (e.g., [3]). Shoreline evolution models are simple but robust enough, when properly calibrated, to help analyse different coastal management options (e.g., [13,14]) or changes in the forcing mechanism (e.g., [15]). A recent review highlighting the capabilities and restrictions of one-line models is given by Chataigner et al. [16].

This paper aims at comparing different intervention strategies at Costa da Caparica beaches, comprising soft- and hard-engineering solutions (as detailed in Section 3.2.4), using a one-line shoreline evolution model. This provides a unique framework to assess possible coastal management interventions.

2. Study Area

The study area comprises the Atlantic ocean sandy beach immediately south of Lisbon (Portugal), which extends southward from *Cova do Vapor* to *Praia da Pipa* (Figure 1). This coastal stretch includes: (i) a northern sector—Costa da Caparica beaches—strongly intervened by artificial structures, protecting the town with the same name; (ii) a low-lying beach–dune system extending southwards from Costa da Caparica until *Fonte da Telha;* (iii) a sandy beach backed by soft (sandstone/clay stone) coastal cliffs; and (iv) the intermittently opened coastal lagoon *Lagoa de Albufeira*. South of the sandy beach, the littoral arch is completed by limestone cliffs until *Cape Espichel* (to the south of the region shown in Figure 1).

At the northern extreme, *Cova do Vapor* faces the Tagus river mouth, a tide-dominated estuary, with complex morphological interactions and evolution [17], which also affects the Costa da Caparica beach sediment transport fluxes and stability. Along the 5 km stretch to the south of *Cova do Vapor*, the coastline presents an alluvial plain with strong anthropogenic occupation [17].

Costa da Caparica beach (Figures 2 and 3) is an urban beach, with buildings constructed over foredunes, and part of the approximately 24 km low-elevation sandy beach sector, with a predominantly NNW–SSE orientation. Coastal defence works in the area include two groins (EV1 and EV2) at *Cova do Vapor*, delimiting the Tagus river entrance, and seven groins (EC1 to EC7) and one ~2.9 km long rock armour revetment at Costa da Caparica beaches (Figure 3). Figure 2 illustrates two photographs, one showing the beach to the north of Caparica, named *Praia de São João da Caparica*, and the other an urban beach, in front of the town, confined by groynes and a coastal revetment.

Figure 2. Images of Costa da Caparica beaches: *Praia de São João da Caparica* (**left**), and central urban beach, confined by groynes and an inland revetment (**right**).

According to Lira et al. [18], between 1958 and 2010, the entire sector presented a relative stability, with minor erosive tendency of -0.04 ± 0.03 m/year. However, the northern sector of Costa da Caparica exhibited a spatial maximum coastal retreat of -4.57 ± 0.2 m/year, whereas the sandy beaches at the south of the coastal arch presented accretion with a maximum of 1.20 ± 0.2 m/year. These data agree with the findings of Silva et al. [11], who report mean retreat rates of -3.09 ± 1.12 m/year at *São João da Caparica* beach (between groins EC7 and EV1; see location in Figure 3) and -1.69 ± 1.94 m/year at the urban beaches of Costa da Caparica, from the analysis of shoreline evolution data from 1958 to 2013. Accordingly, Pinto et al. [19] reported erosion over 200 m in the past 60 years at Costa da Caparica.

In order to counteract this erosive tendency, between the late 1950s and early 1970s, a groin field and a seawall/revetment were built in order to stabilize the shoreline and, tentatively, increase the beach width. However, coastline stabilization was not enough to reduce beach erosion and protect inland infrastructures. Between 1995 and 2003, several erosive episodes were reported north of Costa da Caparica, followed by emergency repair works (such as dune reinforcement) between 2003 and 2004, and strengthening and reconfiguration of the groin field and seawall from 2004 to 2006 [9,20]. Afterwards, according to the designed coastal management master plan [21], an extensive beach nourishment program was carried on by Portuguese coastal authorities between 2007 and 2019, comprising the placement of 4.5 million m^3 of sand (in five phases) in the sub-aerial beach, along the 3.8 km northern shoreline. Sand sources came from regular maintenance dredging performed by the Lisbon Port Authority (APL) in the outer Tagus estuary navigation channel (located 7 to 9 km from the project site) [19].

Figure 3. Groin's identification at Costa da Caparica beach.

According to Silva et al. [11], from the analysis of a 35-year long hindcast record, the area is frequently affected by moderate wave conditions with significant wave heights lower than 2 m with a dominant NNW–SSE direction. Raposeiro et al. [22] and Garzon et al. [23] indicate that the averaged significant wave height at the entrance of the Tagus river (at an approximate depth of 24 m Chart Datum, C.D.) is 1.22 m and 1.19 m, respectively. The averaged mean wave direction and mean wave period are 280° and 5.7 s, respectively. Sancho et al. [24] presented the wave climate spatial variation along the coastal arch, where an increase in the mean wave direction from north to south (from approx. 230° to 285°) is noticeable, as well as an increase in the mean significant wave height from half the coastal arch until the southern end. Dodet [25] reveals a high seasonal variability in the wave climate in front of *Lagoa de Albufeira*, with a maximum mean value in January (1.6 m) and a minimum mean value in August (0.8 m). Similarly, those results also exhibit a large seasonal variability in the significant wave height standard deviation, with larger values during the winter period (up to 0.4 m) than during the summer period (up to 0.1 m).

Regards to astronomical sea level variation, the study area is affected by semi-diurnal tides, with a mean tidal range of 2.20 m recorded in *Cascais* (west of Lisbon) tide gauge [11].

In front of Costa da Caparica beaches, the residual tidal currents intensities are much lower than at the river mouth, of the order of 0.2 m/s, directed northward [26].

The magnitude and net direction of the longshore sediment transport at Costa da Caparica and the coastal arch *Caparica-Espichel* has been subject to debate [24]. While some authors (e.g., [27]) advocate that the net alongshore yearly mean sediment transport is directed south to north over the entire sector, other results have pointed out that the net transport has the opposite orientation, for example, in front of *Lagoa de Albufeira* [25]. Recent results by Sancho [28] sustain that the net transport is indeed northward oriented in the northern sector of the coastal arch (in the first 5 km south of *Cova do Vapor*), and then the net direction changes southward, with increasing magnitudes towards the south. At Costa da Caparica beaches, the net transport is of the order of 50,000 m^3/year [28], resulting from the (in)balance of larger bulk transports, in either direction. At the groin-confined urban beaches, the bulk transports are lower than those at non-confined beaches, due to sand retention at the structures. It is worth noticing that, at *São João da Caparica*, Sancho [28] identified an inter-annual variability in the net alongshore sediment transport rates and direction, despite the long-term averaged value being northward directed. Finally, it is worth noting that the sediment dynamics at Costa da Caparica beaches is intimately connected with the Tagus river inlet dynamics, with complex sediment exchanges between the beaches and the inlet southern sand banks (e.g., the Cachopo Sul sand bank) [17]. Possible causes for erosion at the Caparica beaches have been debated, amongst others, by Veloso-Gomes et al. [9,21], Silva et al. [10], Taborda and Andrade [27], and Fortunato et al. [17]. These point towards a combination of anthropogenic factors (sand extraction and river regulation) and natural agents (sea level rise, wave energy increase and shoreline rotation), which are not entirely understood. The modelled alongshore sediment fluxes are further discussed in Section 5.2.

Beach sediments consist of well-sorted medium-to-coarse sand (median grain size between 0.2 mm and 0.7 mm) [29]. At Costa da Caparica beaches, the dominant grain size is 0.3 mm. According to Freire [30], this sand is composed, predominantly, by siliceous grains and bioclasts.

The local closure depth was estimated according to Hallermeier [31] formula, using the onshore wave climate estimated at 14 m depth [28]. According to the original formula [31], this depth depends on the non-breaking significant wave height that is exceeded, on average, 12 h per year. This resulted in a value of 9.0 m at *São João da Caparica*, which is in agreement with beach cross-shore profile data from the COSMO monitoring programme [32], at *Praia de São João* and *Praia do Tarquínio*, from 2018 to 2021, that shows minor bathymetric changes at depths greater than −11 m (MSL) (approx. −8.7 m C.D.). The estimated closure depth increases to the south, and thus, the unique value of −12 m (MSL) was adopted for the entire coastal sector.

3. Data and Methods

3.1. Data

In the present analysis, aerial photographs, orthophotographs and topo-hydrographical surveys from 1979 to 2018 were used. Table 1 contains information on the data type, date and time of photo/orthophoto, scale, spatial range and tidal level. Regarding the spatial range, it covers either the total continuous sandy beach, from "*Cova do Vapor*" to "*Praia da Pipa*", or just part of it, as indicated in each case. The total period was subsequently divided from 1978 to 2004 for model calibration, and from 2004 to 2018 for model performance verification.

These data were analysed with GIS-designated software, adopting the Official Portuguese reference system—PT-TM06/ETRS89—and the vertical Nautical Chart Datum, abbreviated as C.D., established 2.26 m below the present mean sea level (MSL) at Cascais [20]. Times of photos/orthophotos are referred to UTC (Coordinated Universal Time). Tidal levels were obtained using the "*Service Hydrographique et Océanographique de la Marine*" (https://maree.shom.fr/, accessed on 30 March 2023) tidal model estimates at Cascais.

Table 1. Aerial photographs, orthophotographs and topo-hydrographical surveys used in present study.

Data Type	Time and Date (dd/mm/yy)	Scale and Resolution (If Available)	Spatial Range	Tidal Level (m C.D.)
Aerial photo	7 April 1979; 11 h 00–12 h 00 (UTC + 1)	1:10,000	Entire sector	2.46–2.65
Topo-hydrograpic survey	September and October 1979	1:5000	*"Praia da Mata"* to *"Lagoa de Albufeira"*	-
Topo-hydrograpic survey	September 1980	1:2000	"Costa da Caparica"	-
Aerial photo	9 March 1989, 11 h 00–11 h 10 (UTC), 14 March 1989; 15 h 10 (UTC), 22 March 1989; 11 h 05 (UTC)	1:15,000	"Costa da Caparica" to *"Lagoa de Albufeira"*	0.74–0.85 1.67 1.66
Orthophotomap (False colour)	29 August 1995, (low-tide)	1:40,000 1 m resolution	Entire sector	1.0
Orthophotomap	9 November 2004, (high tide) 13 November 2004, (high tide) 14 November 2004, (high tide)	- 0.1 m resolution	Entire sector	3.2 3.7 3.5
Orthophotomap	6–7 September 2008, 15 September 2008, 10 October 2008	- 0.1 m resolution	Entire sector	-
Orthophotomap	24 October 2014, (high tide) 26 October 2014, (high tide)	- 0.1 m resolution	Entire sector	3.44 3.50
Orthophotomap	17 September 2018, (high tide)	- 0.03 m resolution	*"Cova do Vapor"* to *"Praia da Cornélia"*	2.7

3.2. Methods

3.2.1. Shoreline Identification

A mosaic for each collection of orthophotos/photos was constructed, and the shoreline was extracted. Several distinct shoreline indicators (e.g., [33]) can be obtained, but here, the use of a dynamic indicator was favoured, in line with the shoreline model assumptions, which was able to catch the annual and inter-annual variability. Moreover, not all photos allowed us to identify the high-tide water line. Hence, the wet/dry-sand line is the adopted proxy for the coastline or shoreline. In a recent work, Buccino et al. [34] used the instantaneous waterline. A single operator visually identified the shorelines for all data, thus avoiding any subjectivity associated with multiple operators.

An average beachface slope of 0.1 was assumed uniform for the whole sector, despite the occurrence of spatial and temporal variations [29], which would affect the shoreline position, albeit within the method's accepted error (see below). Using the tidal level data at the time of the photo and the above beachface slope, all shorelines were transferred landwards or seawards to the mean sea level position, if in low tide or high tide, respectively. Wave runup was not accounted for, but it was checked that it would only affect minimally the results, within the methodology uncertainty.

For comparison and data analysis purposes, shoreline data points were extracted every 50 m along the baseline, as depicted in Figure 1.

Regarding the shoreline determination, a specific uncertainty or error analysis was not carried out here. Nevertheless, since the present analysis follows the procedure of Silva et al. [35], who performed an extensive error analysis, and uses similar data sources, we assume an identical error in the shoreline identification method, of the order of 3 m, not factoring in the tidal level adjustment. Moreover, according to the same authors [35], an uncertainty of a similar magnitude (circa 4 m) is associated with georeferencing the aerial photo mosaics. Combining these uncertainties, which sum non-linearly (the total uncertainty is equal to square-root of the sum of the, partial, squared uncertainties), results in an overall positioning root-mean-square error of circa 5 m [35], and thus, generally, the shoreline evolution is of significance only at places where this value is exceeded. Note that although this error estimate may appear underestimated, given the methodological approximations and uncertainties, it is larger than that provided by other works using similar orthophotomaps (e.g., [34]), but it is similar to that obtained from air photos analysed by Hapke et al. [36]. Lastly, the correction for the tidal level adds an uncertainty, which sums linearly with the above. This is further discussed in Section 5.1.

3.2.2. Local Wave Climate

The wave climate at the closure depth (−12 m MSL) bathymetric contour, along the study area, was calculated using the SWAN (version 41.10) spectral wave model [37]. Sancho et al. [24] provide details of the model setting and application. In particular, the model was forced with spatial-uniform conditions at the domain western boundary, using the hindcast offshore wave climate time series at approximately 122 m water depth (hindcast data point located at 38°36′ N and 9°33′ W, WGS84 coordinate system), provided by Dodet et al. [38] and Dodet [25]. This time series consists of a 65-year hindcast data set, with data every 6 h, for the period from 1 January 1948 until 31 December 2012. Here, only the period from 1979 until 2012 was used.

The modelled results were compared with local measurements from a wave buoy placed at −22 m C.D. (38°37′33.6″ N and 9°23′16.8″ W, WGS84 coordinate system), at the mouth of Tagus estuary, for the period of 5 years, from 2007 until 2012. For a total of 4906 records, approximately equally distributed along the years, Sancho [28] obtained root-mean-square errors for the significant wave height, wave peak period and mean direction equal to 0.34 m, 2.08 s and 26°, respectively. The bias were −0.002 m, −0.222 s and −4.2°, respectively. These results are considered quite satisfactory, except for the wave direction mean-square error. Despite the latter appearing to be high, it is similar to the accuracy obtained for other wave model implementations (e.g., [17,39,40]), and thus, it was accepted. However, it cannot be ignored that ±25° errors in the wave direction can significantly affect the estimated longshore sediment transport fluxes and, presumably, the modelled shoreline evolution. According to Chataigner et al. [16], assuming that the one-line model is conceptually correct, a small wave angle bias (approximately 5–10°) has significant impacts on simulated long-term shoreline change trends. The lower bias obtained here for the wave direction indicates, however, that there should be no tendency along all the model results to underestimate or overestimate the wave direction (and alongshore sediment transport direction), and thus, one might expect that the modelling errors from this source should cancel out.

3.2.3. Shoreline Modelling

In the following, the shoreline evolution in the region of interest is modelled using the LITMOD model [41,42]. This is a classical one-line model, similar to the well-known GENESIS model [43], that allows us to easily define and evaluate scenarios, and to obtain medium-to-long-term (of the order of decades) estimates of the shoreline position for long (of the order of kilometres) coastal stretches. This model was designed to apply mostly to linear (or small-curvature) sandy beaches, where the shoreline evolution is mainly governed by the alongshore sediment transport and wave energy gradients, dependent on the local wave-breaking climate, sediment sources and sinks, sediment characteristics, cross-shore profile, coastal structures and boundary conditions. According to the model, at each time step, the shoreline moves parallel to itself, representing the movement of the entire beach profile (from the upper berm down to the closure depth), from the mass balance of the alongshore sediment fluxes and sediment sources/sinks. The model is presently prepared to deal with the effect of groins on the alongshore sediment fluxes, detached breakwaters, soft and hard coastal cliffs, and coastal revetments [41].

The wave-breaking characteristics are determinant to estimate the alongshore sediment fluxes. As the shoreline moves and rotates, the breaking wave properties (wave height and wave direction with the coast) alter iteratively, taking into account the breaking depth. The Kamphuis [44] formula was chosen to calculate the coastal alongshore sediment transport, with the calibration dimensional transport coefficient k equal to 70% of the original value ($k_{orig.} = 2.3 \times 10^{-3}$ $m^{1.25}$ $s^{-2.5}$), in order to better fit the data. All simulations were carried out ignoring the tidal water level fluctuations, using a constant water level equal to the mean sea level.

According to the available data [29], the present model application considered a variable median grain size along the study area, increasing from north to south, from 0.27 mm to 0.75 mm. The Dean [45] equilibrium cross-shore profile form was adjusted to some profiles extracted from the 1980 hydrographic survey (identified in Table 1). Hence, at each alongshore position, depending on the sediment size, the cross-shore profile is given by $h = Ay^{2/3}$, with $A = 0.21D_{50}^{0.48}$, where h represents the water depth and y is the seaward distance from the water line. For D_{50} = 0.32 mm, the beach slope at the breaking zone is approximately equal to 1:30. Thus, in order to properly account for wave transmission around the groins, a locally adjusted beach profile was used. From the available topographic data, the beach crest height is at +4 m (MSL), and the depth of closure is 12 m (MSL), for the alongshore-averaged cross-shore profile. Thus, the active beach volume (per unit length) that moves back and forth parallel to the shoreline is 16 m in height. In the model setup, this height is further adjusted from north to south, stretch by stretch, allowing for a variation of that value by 94% to 115%, respectively. Hence, according to the data, the active profile height at the northern sectors is lower than that at the southern sectors.

For the numerical model, a 24 km reference or baseline was established, oriented N–20°-W, with origin at *Praia da Pipa* (Figure 1). All shoreline positions are measured relative to this reference, and a 50 m cell length was adopted, yielding a total of 456 points along the coast. Note that the cell length equals the data extraction spatial step, allowing for a direct comparison, without further data interpolations. Twenty-nine local wave climate points, approximately uniformly distributed along the coast, were considered for the model simulations. The model time step is 1.2 h. Northern and southern model boundary conditions were set such that the shoreline orientation is constant at each extremity.

Finally, the application of the proposed model to specific situations, such as the present one, requires the numerical model calibration and validation. The model calibration data and setup is further discussed in Section 5.1.

3.2.4. Local Coastal Protection Strategies

Table 2 contains 16 possible intervention strategies, planned with the Portuguese coastal management Authority (Portuguese Environmental Agency), for a period of 25 years. The management goals are halting beach erosion at Costa da Caparica beaches and protecting the city (people, property and economy) from wave overtopping and coastal flooding. Two distinct groups were conceived: the first (letter A) includes intervention strategies without any further hard engineering interventions (except for maintenance works of the present structures); the second group (letter B) contains alternatives with rearrangements of the present coastal structures, alone or complemented with beach nourishment operations. It is assumed that the present coastal revetment remains in place for all interventions. The groins mentioned in Table 2 are identified in Figure 3.

Scenario A1 corresponds approximately to the present (last decade) coastal management strategy, where Costa da Caparica beaches are nourished every 5 years with 1×10^6 m^3 of sand, distributed along the coastal sector within $20{,}200 \leq x \leq 24{,}000$ m (see range in Figure 1). Scenario A2, albeit of minor intensity, is inspired by the strategy outlined by Duarte-Santos et al. [46], where it was recommended to perform a one-time, large volume, beach nourishment "shot" of 5×10^6 m^3. Here, a volume of 3×10^6 m^3, instead of 5×10^6 m^3, was suggested by the Portuguese Environmental Agency. Scenario A3 was defined mainly for comparison purposes, in terms of what would happen if no action were taken. Scenario A4 is similar to A1, except that the nourishment is confined to a shorter alongshore area, within groins EC1 to EC7 ($20{,}750 \leq x \leq 22{,}650$ m).

Table 2. Coastal intervention scenarios.

Scenario	Description
A	**Without hard engineering interventions**
A1	1×10^6 m^3 beach nourishments, every 5 years, distributed alongshore over 3800 m
A2	One large (3×10^6 m^3) beach nourishment "shot", at project beginning lifetime
A3	No intervention
A4	1×10^6 m^3 beach nourishments, every 5 years, distributed alongshore over 1900 m
B	**With hard engineering interventions**
B1	Similar to A1; removal of all groins from EC1 to EC7
B2	Similar to A1; removal of groins EC2, EC3, EC5, EC6 and EC7
B3.1	Similar to B1; EV1 groin length increase of 100 m
B3.2	Similar to B1; EV1 groin length increase of 200 m
B3.3	Similar to B1; EV1 groin length increase of 300 m
B4.1	Similar to A1; EV1 groin length increase of 100 m
B4.2	Similar to A1; EV1 groin length increase of 200 m
B4.3	Similar to A1; EV1 groin length increase of 300 m
B5.1	Similar to A3; EV1 groin length increase of 100 m
B5.2	Similar to A3; EV1 groin length increase of 200 m
B5.3	Similar to A3; EV1 groin length increase of 300 m
B6	Similar to A1; removal of groins EC1, EC2, EC4, EC6 and EC7; EV1 groin length increase of 300 m

The second group comprises strategy B1, which equals A1 in terms of beach nourishment, but accounts for the removal of all groins in front of Costa da Caparica, retaining the ones at *Cova do Vapor* (at the northern limit of the coastal arch), namely, groins EV1 and EV2. Hence, A1 versus B1 shall allow us to understand the capacity of the groins in retaining the added sand. Scenario B2 is similar to A1 and B1 regarding the nourishments, but it considers removing groins EC2, EC3, EC5, EC6 and EC7.

An extension of the groin at the northern end of this coastal sandspit has been debated for a long time [47] and is thus evaluated here, in comparison with other alternatives. Hence, simulated interventions B3.1, B3.2 and B3.3 are all similar to B1, but they include an extension of groin EV1 by 100, 200 and 300 m, respectively. Alternatives B4.1, B4.2 and B4.3 include, likewise, the same expansions of groin EV1, but they are otherwise similar to strategy A1 (present management option), i.e., all present structures would be maintained, and beach nourishment would be performed. The combination of no-nourishment (A3) with the EV1 groin expansions results in scenarios B5.1, B5.2 and B5.3. Finally, scenario B6 considers removing groins EC1, EC2, EC4, EC6 and EC7, beach nourishments equal to A1 and a 300 m increase in groin EV1.

All the above intervention scenarios were independently configured into LITMOD, and the shoreline evolution was simulated for a period of 25 years, starting from the position measured in 2018. The forecast-period wave climate in the model was the same as the wave climate series from 1979 to 2004. The comparison between the different interventions will focus in Costa da Caparica frontal and adjacent beaches, bounded by $20{,}000 \leq x \leq 24{,}000$ m. This is the region where present interventions take place and where coastal erosion and wave overtopping are problematic.

4. Results

4.1. Calibration and Verification

Figure 4 shows the simulated shoreline position at 2004 (after 25 years of simulation) for the entire study area, after model calibration, in comparison with the measured line. Vertical and horizontal scales are distorted. One generally notes good agreement over the entire coastal arch, except at the southern, first 1000 m, where the model predicts shoreline retreat, which is not observed in the data. The modelled–observed discrepancy is close to 150 m in that sector, but it is as low as 6 m in some of the northern groin-confined Costa da Caparica beaches.

The average measured shoreline movement for the entire study area, excluding the profiles where the difference between the two lines was less than 5 m (the method's precision), was 0.09 m/year, corresponding to accretion. This value matches satisfactorily the value of 0.14 m/year obtained by Silva et al. [48], using other shoreline indicators, for the period 1980–2005. Despite the overall slight accretion for the entire sandy stretch, erosion was significant at the northern 3 km-long stretch, reaching a localised maximum of approximately −100 m (−4 m/year).

Table 3 provides the average shoreline advance or retreat, measured and modelled, between 1979 and 2004, for six sectors (A to F; see location in Figure 4) along the studied sandy coast for both the measured and simulated results. The difference between data and model results are minimal and within the accepted data precision (see Section 3.2.1) for sectors C and F. It is also slightly above that value for sector D. Disparities in sectors B and E are around 30 m. Sector A, at the southern extremity, is the one where discrepancies are the greatest, most likely due to an inadequate southern boundary condition for the sediment dynamics at this study site. However, errors in the wave direction estimation and its effect on shoreline modelling cannot be neglected [16], but they also cannot be appraised since no wave monitoring data exist at the coast.

Table 3. Difference between final (2004) and initial (1979) shoreline positions in the calibration period (in meters), spatially averaged by coastal stretch (A to F, from south to north).

	A $1500 < x \leq 2000$ (m)	B $2000 < x \leq 7000$ (m)	C $7000 < x \leq 14{,}000$ (m)	D $14{,}000 < x \leq 17{,}000$ (m)	E $17{,}000 < x \leq 20{,}700$ (m)	F $20{,}700 < x \leq 24{,}000$ (m)
Simulated	−159.6	−29.8	3.5	11.5	−25.2	−14.7
Measured	−33.8	2.0	4.7	17.2	4.9	−19.4

The calibrated model was further verified against field data for the period 2004 to 2018. It is noted that in 2005–2006, the coastal structures at the northern sector (Costa da Caparica beaches) were rearranged relative to the previous configuration [21], so the new arrangement was set-up in the model. Furthermore, four nourishment operations (from 2007 to 2014, [19]) were introduced in the model, simulating as closely as possible the nourished volumes, locations and periods of interventions. The total added sand was 3.5×10^6 m^3. Since the wave climate had only been determined until 2012, and it remained relatively stable during the period 1979–2004 [28], i.e., without any noticeable tendency for long-term variation, it was decided that the wave time series used in the calibration period would be used in the verification period.

Figure 5 compares the shoreline position in 2018 using the previously calibrated model (left) with that obtained after a finer re-calibration setting (right), for the northern 4 km-long stretch (covering mainly sector F), where all past coastal works and beach nourishment interventions occurred. A better match is observed for the post-verification model configuration, particularly at the northern beach ("*Praia de São João*"). In that beach, the root-mean-square difference between modelled and measured coastlines were 50.1 m and 35.6 m for the post-calibration and post-verification model settings, respectively. The largest differences between modelled and measured shorelines are visible at the groin-confined beaches, where the model is unable to properly simulate the diffraction processes leeward of the groins (e.g., [49]), and generally predict a greater shoreline erosion than that measured. It is noted that the recalibration process was for the entire coastal arch, but the aim was to achieve a better match at the Costa da Caparica beaches (sector F of Table 3), where the different management strategies will be compared.

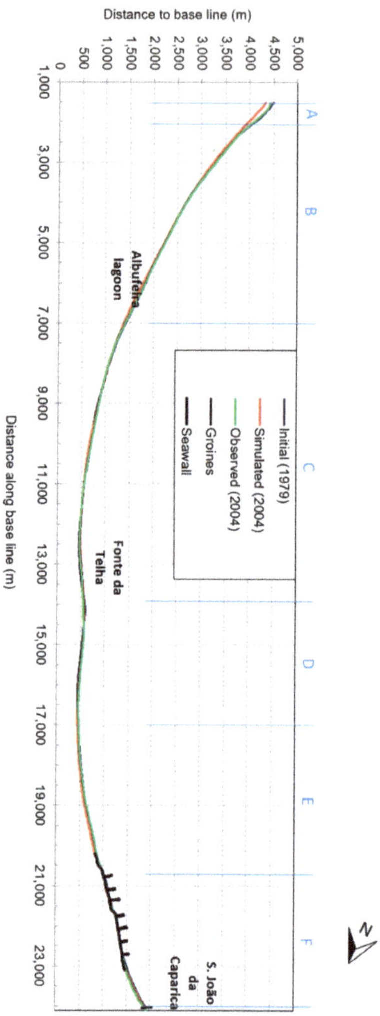

Figure 4. Modelled (calibrated; red) and measured (green) shoreline positions in 2004 over the entire study area. Initial shoreline configuration in 1979 (dark blue) and hard structures (groins and coastal revetment; black). The limits of the sectors referred to in Table 3 are market in light blue, at the top.

4.2. Coastal Intervention Scenario Simulations

4.2.1. Soft Solutions Plus Present Coastal Structures (Group A)

Figure 6 compares the predicted shoreline positions 20 years past the initial 2018 configuration, for all soft intervention strategies (group A in Table 2). Solution A3 ("no intervention") is, clearly, the one that allows for the largest coastal erosion at this sector. On the contrary, alternatives A1 and A4 induce slight coastline advances, particularly at some of the pocket beaches and *Praia de São João* (between EC7 and EV1). This is confirmed in Figure 7, which shows the time-evolution of the averaged shoreline advancement or retreat for these interventions, within the region $19{,}500 \leq x \leq 24{,}000$ m. Strategy A2 also induces overall shoreline retreat at this sector, which might be explained by the fact that this strategy adds to the system, in 25 years, less sediment (3×10^6 m^3) than strategies A1 and

A4 (totalizing 5×10^6 m^3). A second reason is related to the fact that, for A2, the volume is placed only once, at the project start, and thus, its long-term effect is less noticeable.

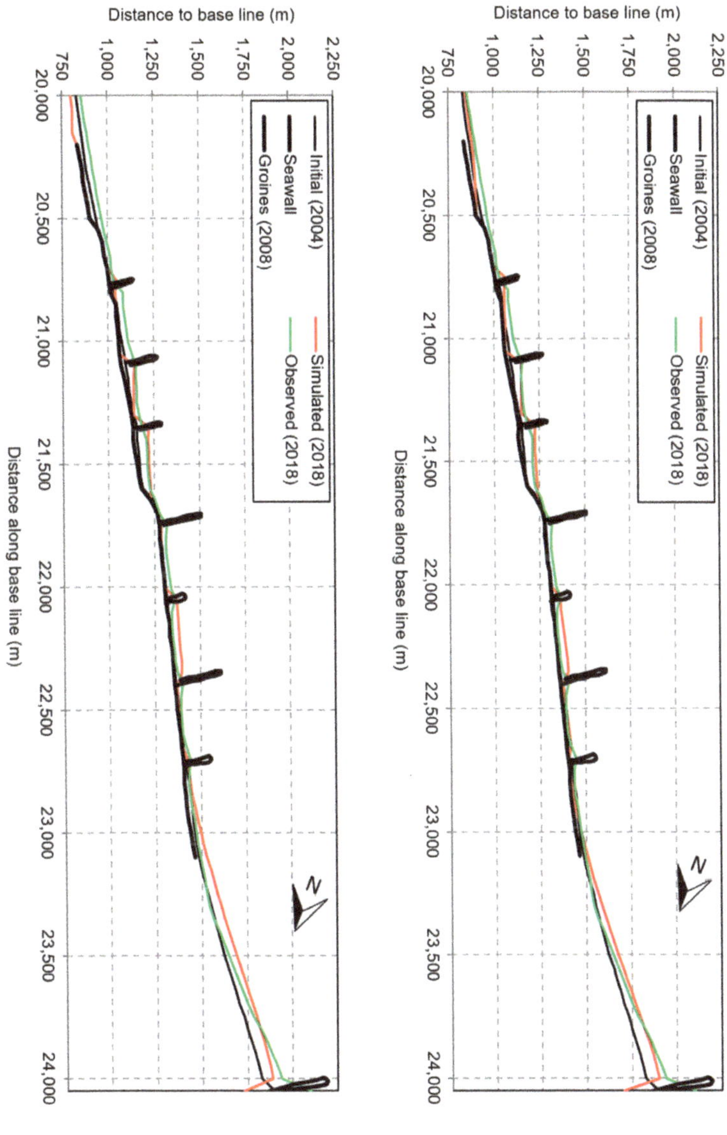

(**a**) (**b**)

Figure 5. Modelled (red) and measured (green) shoreline positions in 2018 at Costa da Caparica beaches. Initial shoreline configuration in 2004 (black thin line) and hard structures (groins and coastal revetment; black thick line). (**a**) Results from post-calibration simulation; (**b**) results from post-verification simulation.

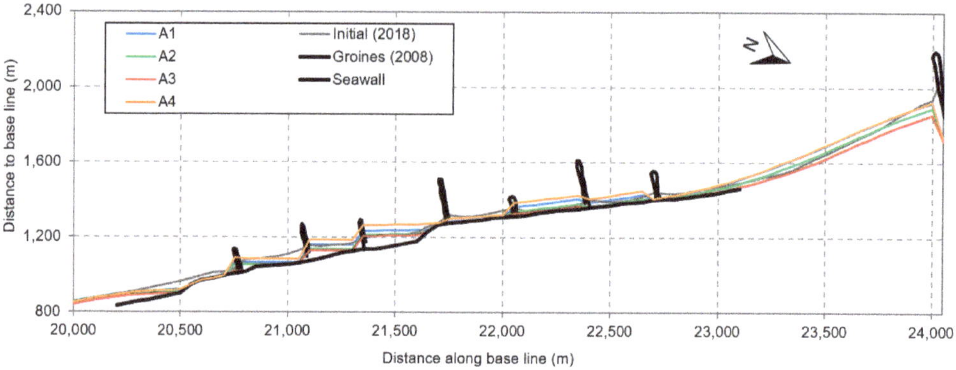

Figure 6. Modelled shoreline positions 20 years past initial configuration (2018) at Costa da Caparica beaches for intervention strategies A1, A2, A3 and A4.

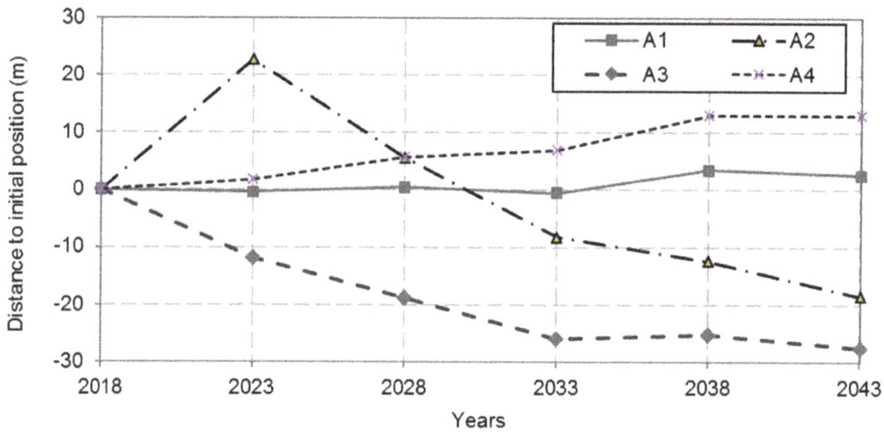

Figure 7. Time-evolution of mean shoreline position relative to initial position at Costa da Caparica beaches for intervention strategies A1, A2, A3 and A4.

4.2.2. Structural Interventions, with or without Soft Solutions (Group B)

Figure 8 shows the shoreline positions after a simulation of 20 years for all the group B intervention scenarios. The top sub-figure compares strategies B1, B2, B6 and A1, evidencing that the removal of several intermediate groins (e.g., strategies B1 and B2) reduces the sand accumulation capacity of the groin-confined beaches (comparing to strategy A1). Strategy B2 shows also that retaining groin EC1 (the southern-most groin) allows for removing EC2 without beach loss between EC1 and EC3. The results for strategy B6 show that the increase in EV1 provides a significant positive impact in retaining sand from EC6 to EV1 ($22{,}400 \leq x \leq 24{,}000$ m). For all these scenarios, the shoreline positions between EC4 and EC6 nearly coincide with the coastal revetment alignment, indicating thus a very narrow beach.

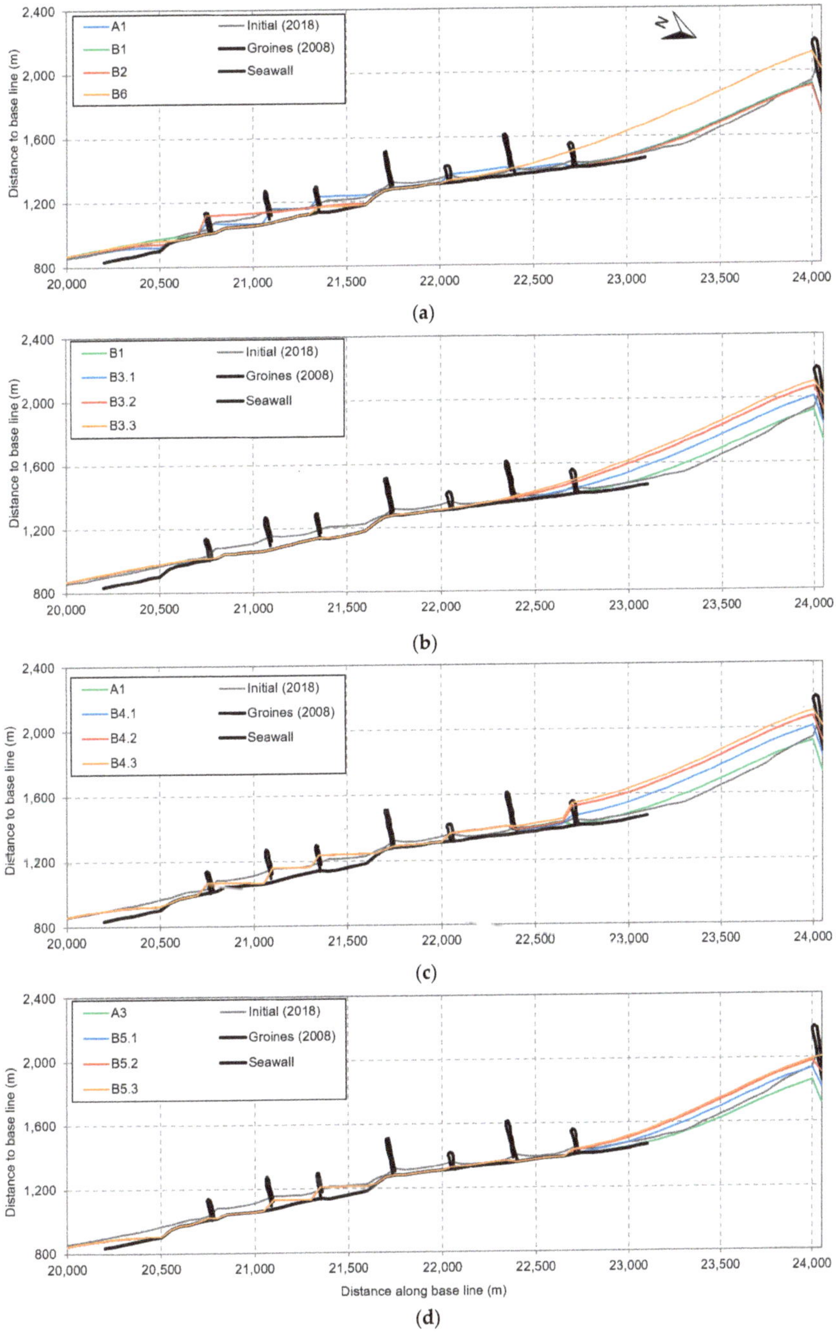

Figure 8. Modelled shoreline positions 20 years past initial configuration (2018) at Costa da Caparica beaches for intervention strategies: (**a**) A1, B2, B2 and B6; (**b**) B1, B3.1, B3.2 and B3.3; (**c**) A1, B4.1, B4.2 and B4.3; (**d**) A3, B5.1, B5.2 and B5.3.

The results for the scenarios considering the removal of all groins from EC1 to EC7 (B1 and B3) are given in Figure 8b. It is clear to observe the beneficial effect of increasing the terminal groin EV1 (simulations B3), which is able to trap the northward (down-drift)-directed sediment flux, proportionally to its length increase. This beach widening is limited, at best, to $x \geq 22{,}400$ m, that is, northward of the present position of groin EC6. Southward of this position, predictions show the shorelines attaching to the seawall, thus reducing the beach width to nearly zero. Figure 8c evidences again the positive effect of EV1 groin increase (scenarios B4), in addition to maintaining the present strategy (A1).

Figure 8d compares results without any intervention (A3) with a similar set but considering the length expansion of groin EV1 (B5). As in Figure 8c, differences between all modelled shorelines are only discernible at *Praia de São João*. Even without beach nourishment, the EV1 groin extension is effective in retaining sediments in that beach that would otherwise by-pass the present groin.

Figure 9 shows the time-evolution of the sector-averaged shoreline advance or retreat, within the sector $19{,}500 \leq x \leq 24{,}000$ m, for all group B interventions. Overall, strategies B3.3, B4.3 and B6, all corresponding to a 300 m increase in groin EV1 together with beach nourishment, are the ones that promote the most noticeable beach accretion. Strategy B5 does not generally allow for beach growth. For the same EV1 groin increase, strategy B4 is generally slightly more positive that B3; that is, retaining the present Costa da Caparica groins (EC1 to EC7) promotes more accretion than removing them. This is also observed by comparing results of A1 with B1 (Figure 9a).

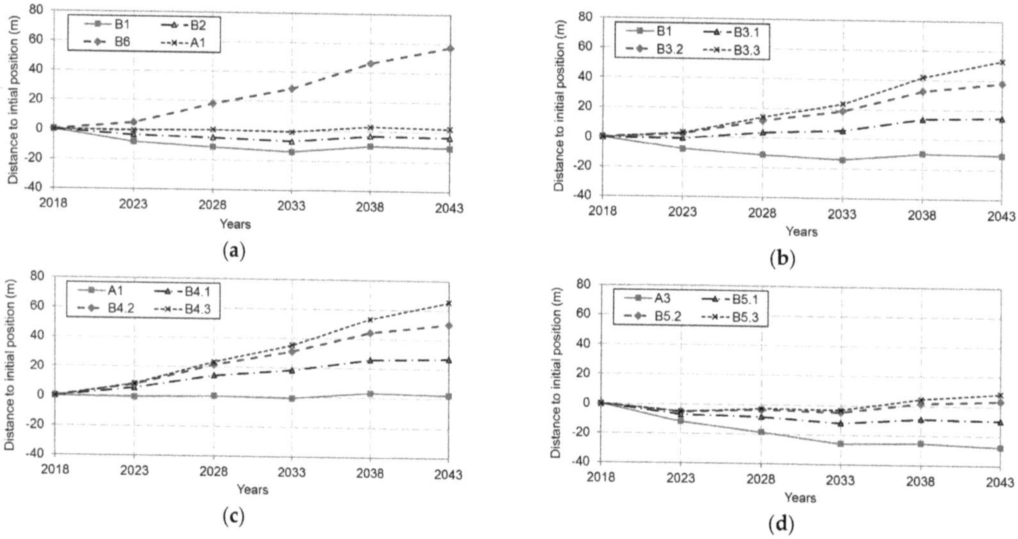

Figure 9. Time-evolution of mean shoreline position relative to initial configuration at Costa da Caparica beaches for intervention strategies: (**a**) A1, B1, B2 and B6; (**b**) B1, B3.1, B3.2 and B3.3; (**c**) A1, B4.1, B4.2 and B4.3; (**d**) A3, B5.1, B5.2 and B5.3.

5. Discussion

5.1. One-Line Model Calibration Data and Setup

A brief discussion on one-line model limitations is provided by Chataigner et al. [16], who highlight the role of the active profile depth, the assumption of a constant equilibrium profile and the use of k as a calibration parameter in the longshore transport rate. The one-line model approach depends on the quality of both the input data and the choice of model free parameters. Depending on the inherent complexity of the study site and the quality of the input data, the one-line longshore model may generate changes in the beach planform (e.g., reorientation) that are different from observations [16]. Hence, the issues of model implementation and limitations are discussed here.

The measured shoreline movement displayed in Figure 4 and characterised in Table 3 relate to the comparison between the positions in April 1979 and November 2004. As these dates refer to different maritime wave regimes (post-winter energetic versus post-summer low-energy seasons), the seasonal cross-shore profile-shaping processes can induce shoreline movements at this coastal sector of up to circa 20 m (this number results from the analysis of 2018–2021 cross-shore profile data, at *Praia de São João* and *Praia do Tarquínio*, available from the monitoring programme COSMO [32]). Hence, their absolute inter-comparison for estimating long-term shoreline movement rates could lead to erroneous interpretation. However, the seasonal behaviour of the coastline does not affect the model calibration, as the input-modelling conditions reproduce the wave time series and shoreline position from a given start date to a given end date, regardless of their seasons. This is also acknowledged by Chataigner et al. [16], who state that over long time-scales (e.g., interannual to decadal) longshore processes become increasingly important.

A shoreline determination precision of ±5 m was estimated in Section 3, based on the analysis carried out by Silva et al. [35]. The shoreline determination procedure included a correction for the tidal level, at the time of data acquisition, which adds more uncertainty to the above precision. The beach profile, particularly at the northern sector of the study site, changes its slope from the upper to the lower foreshore. Using data from the COSMO monitoring programme [32], at *Praia de São João* and *Praia do Tarquínio*, the average upper-foreshore (medium-to-high tide water line) slopes are circa 0.12 and 0.09, respectively. These numbers are in accordance with the average beachface slope of 0.1, used before. However, the average lower foreshore (medium-to-low tidal levels) slopes are 0.04 and 0.033, respectively, at the same profiles. Thus, the horizontal correction of the shoreline position needed to be larger if below mean tidal level at the time of data acquisition. This did not occur for the aerial photos and orthophotos of 1979, 2004 and 2018 (see Table 1), as all times corresponded to instants where the waterline was in contact with the upper foreshore. Hence, using the average high-tide level of the 2004 orthophoto (3.5 m C.D.), the error associated with the tidal correction is of the order of $((3.5 - MSL)/slope) - (3.5 - MSL)/0.1)$, where *slope* is the true upper beachface slope. Using *slope* = 0.12 yields an error of 2.5 m. A lower value is obtained for the 1979 dataset, as the tidal level was lower (but above mean sea level). This highlights the validity of the tidal-level shoreline correction procedure. We note that more care in the correction procedure would be needed in case of the 1989 and 1995 datasets, obtained at low-tide phases, but not used for the present model calibration.

The 1980 topo-hydrographic survey was chosen for model calibration, since this survey lies within the calibration period (1979–2004), and should, therefore, represent adequately the beach profile characteristics for that period. The data used herewith was extracted from 1 m-apart bathymetric contours, which are rather crude and thus of questionable quality. These data were thus compared with recent, 2018–2021, multi-beam, topo-hydrographic profiles collected at *Praia de São João*, from the COSMO programme [32]. The comparison indicates that the 1980 lower-resolution profiles can also represent the present beach profiles, the major difference being a slightly larger foreshore volume, which is explained by the beach being more robust, prior to the long-term erosion that occurred until the first beach nourishment operations.

The study site description in Section 2 indicates that the sandy stretch is somewhat non-homogeneous (e.g., the spatial grain size variation [29]) and long, possibly reducing the validity of the LITMOD (one-line) model application. However, the model parametrization and setup used here, namely, the alongshore variable grain size, variable beach profile and variable active beach volume (both varying sectorially, in a total of five sectors), allowed us to account for that non-homogeneity.

Further, in relation to the study site length and curvature, one-line models are valid to limited (small) shoreline curvatures, usually due to the decoupling of the wave refraction and shoreline movement computations [50]. In the present case, the one-line model baseline is closer to the shore-parallel configuration at the northern extremity than at the southern (Figure 1), and the shoreline presents a small curvature. This configuration ensures a potential low error due to those effects at the northern sector, where the coastal protection strategies are inter-compared. The greater model-data discrepancy found at the southern extremity in the model calibration phase (Figure 4) may also be due to this effect.

The data at the northern sector indicates that there is a clear long-term trend in shoreline behaviour. The assumption of a clear trend implies that the wave action producing longshore sand transport and boundary conditions are the major factors controlling long-term beach change [51]. A model-data comparison in the calibration stage (Figure 4) allowed us to conclude that the northern boundary condition was appropriate, unlike the southern one. Other southern boundary conditions were tested, providing poorer results. This is further discussed in the next section.

The Kamphuis [44] formula for the alongshore transport rate is a function of the (alongshore varying) breaking wave height and wave direction. A commonly used alternative transport formula is the CERC formula [51]. Most often, for both formulae, the transport coefficient k is used as a calibration parameter (e.g., [51–53]). For example, Hanson and Kraus [51] refer a common range of 75 to 100% of the original k value, for the CERC formula. Pilkey and Cooper [54] refer to the use of a much broader range, with k values from 2 to 300%. Considering the above, the value of k used in the present model configuration is quite acceptable and enabled us to obtain a satisfactory calibration. Since this affects the transport magnitude uniformly, the beach alongshore variability was accounted for through the other model configurations, already mentioned.

5.2. Alongshore Sediment Fluxes

The discrepancy of the predicted versus measured coastlines at the southern extreme of this coastal sector, in the calibration phase (Figure 4), relates to an over-estimation of the sediment flux at that boundary. In the calibration and verification phases, the averaged net transports at the model boundary, x = 1500 m, were 160×10^3 m^3/year and 200×10^3 m^3/year, respectively, directed southward. For the intervention scenarios, the flux resulted equal to 167×10^3 m^3/year for all simulations. These estimated fluxes are not corroborated by data and result from an ineffective model boundary condition. This may be due to the existing headland circa 650 m north of the sandy beach southern extreme, which is likely to act to stabilize the southern beach (*Praia da Pipa*), which was not included in the model setup. One other reason may be related to the fact that the sediment characteristic dimension (D_{50}) is much larger at that extreme (>0.75 mm), and despite the model accounting for a variable alongshore grain size, the used simplification could not exactly reproduce the reality. Nevertheless, the effect of this exaggerated net sediment flux at the southern extreme propagates only to approximately 5000 m, northward, in the model domain (Figure 10). Indeed, the net averaged sediment fluxes in the region x > 5000 m is limited to approximately $\pm 50 \times 10^3$ m^3/year (Figure 10). This net flux results, generally, from the balance between a 5-to-20-times-larger bulk transport, northward or southward directed, except at Costa da Caparica beaches (20,000 $\leq$ x $\leq$ 23,000 m), where the groins limit these fluxes. For x > 5000 m, it is also observed that the net flux varies alongshore from positive (southward) to negative (northward), and is negative at Costa da Caparica beaches.

The averaged net sediment transport at the northern extremity was -50×10^3 m^3/year and -116×10^3 m^3/year in the calibration and verification phases, respectively. These fluxes are within the known estimates of data-inferred alongshore values at *Praia de São João*, which are of the order of -200×10^3 to -275×10^3 m^3/year [55,56], and other model estimates [57].

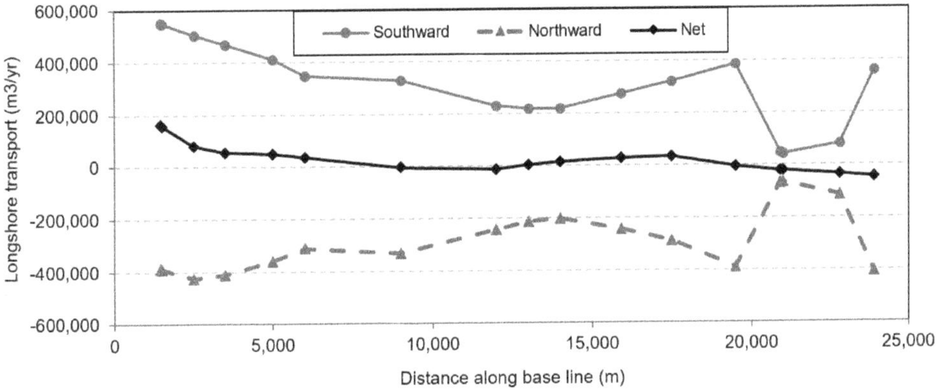

Figure 10. Alongshore annual mean sediment flux spatial distribution, for the post-calibration model simulations (1979–2004).

Figure 11 shows the annual mean net sediment fluxes for all intervention scenarios (see Table 2) at groin EV1 cross-section (x = 24,000 m) and immediately south of Costa da Caparica beaches (x = 19,500 m). For all soft-intervention simulations (group A), the predicted fluxes at EV1 ranged from -170×10^3 m^3/year to -185×10^3 m^3/year, except for scenario A3—no beach nourishment—where the flux was smaller due to the scarcity of sediments to transport. Figure 11a further evidences the expected reduction in the net alongshore sediment transport as the terminal groin EV1 length increases. Scenario B5.3 is the one with the least amount of mean annual sediment (-5.2×10^3 m^3/year) bypassing EV1, followed by scenarios B5.2 (-17.6×10^3 m^3/year) and B3.3 (-23.8×10^3 m^3/year). At x = 19,500 m, the annual-averaged sediment fluxes are much smaller, and, depending on the intervention strategy, the flux towards the coastal cell comprising Costa da Caparica beaches can be either positive or negative (Figure 11b). We note that this cross-section corresponds roughly to the one where the annual mean alongshore net sediment flux changes its direction, in the calibration simulation (see Figure 10). Figure 11b also shows that interventions B1, B2, B3 and B6 induce positive (southward) net fluxes, that is, promoting the removal of sediments from Costa da Caparica beaches. On the contrary, strategies A3 and B5 (both without beach nourishment) mobilise significant sediment fluxes from the south (northward), in an attempt to feed the starving beaches at the north. Intervention scenarios A1, A2, A4 and B4 are the ones with the smallest net fluxes, with A4 being preferable because it promotes sediment into Costa da Caparica cell.

5.3. Sediment Volume Budget

The accumulated sediment budget at the littoral cell of Costa da Caparica beach (between sections x = 19,500 m and x = 24,000 m) was determined from the computed alongshore sediment fluxes at the boundary sections and the overall projected sand nourishments, for the 25-year projection lifetime. Figure 12 (grey bars) compares these results for the 16 intervention scenarios (see Table 2). The total nourishment for each strategy is also presented (dashed-pattern bars) as well as the final alongshore average shoreline movement (orange line, positive corresponds to accretion). In relation to the overall sediment budget, strategy A4 is, amongst the non-structural interventions (group A in Table 2), the most efficient in retaining sediment. All the others (A2 and A3), except for A1, result in less sediment in the cell than at the initial condition. Strategies B1 and B2, which are similar to A1 but include the removal of some groins, also induce sediment losses. As shown previously, the lengthening of groin EV1 (see Figure 3) in strategies B3 and B4 contributes significantly to sediment accumulation. Accordingly, strategy B4.3 causes the largest sediment retention in the system, nearly preserving all the sand added to the system. Naturally, since the average shoreline displacement is proportional to the accumulated volume, those two curves follow the same tendency in Figure 12. The largest average beach growth is 65.6 m for scenario B4.3, and the continuation of the present strategy (A1) results in an accretion of only 2.6 m.

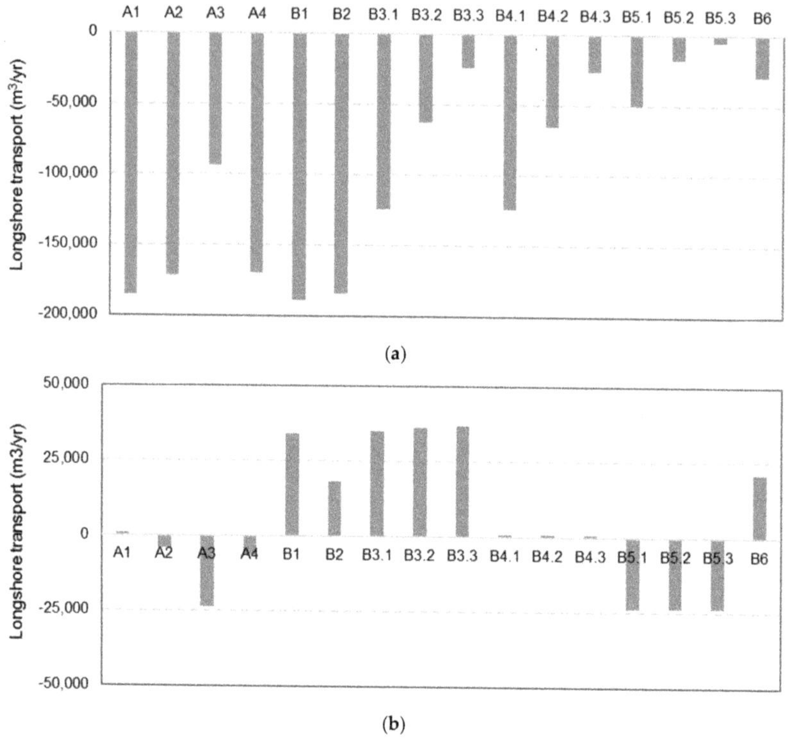

Figure 11. Annual mean net sediment fluxes for all intervention scenarios at: (**a**) x = 24,000 m; (**b**) x = 19,500 m.

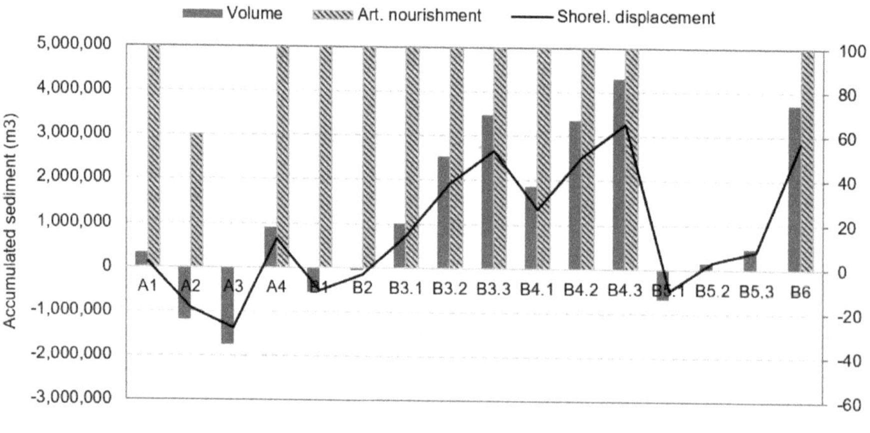

Figure 12. Accumulated sediment budget, artificial nourishments and average shoreline advance/retreat at Costa da Caparica cell (comprised within 19,500 ≤ x ≤ 24,000 m), for the various intervention strategies.

6. Conclusions

Costa da Caparica beach, in Portugal, has been subject to coastal erosion over the past 70 years (e.g., [10]). This chronic situation led, in the period 2004–2006, to reconstructing and reshaping the previously erected groins and to strengthening the seawall [9]. This action was followed by a succession of five beach nourishment interventions in 2007, 2008, 2009, 2014 and 2019, feeding to the beach a total of 4.5×10^6 m^3 of compatible sand [58,59]. This paper compares the present management strategy with 15 alternative ones (Table 2), also based on beach nourishment interventions, some combined with coastal structures interventions. These intervention scenarios were compared in terms of shoreline advance or recession for a period of 25 years, using a classical one-line shoreline evolution model [41,42].

The results indicate that the present management strategy (designated by A1) is effective in holding the shoreline position, although a similar strategy (A4), deploying the same overall sediment volume but over a shorter area, could lead to slightly better results (Figures 6 and 12). The best options, which are capable of promoting beach accretion at Costa da Caparica beaches, implicate the lengthening of groin EV1 (shown in Figure 3), north of *São João da Caparica*. The longer the increase in groin EV1, the greater the beach sediment volume gain, moved by littoral drift, at the Costa da Caparica beaches. However, such an intervention would need a detailed assessment of the hydro- and sediment dynamics changes at the Tagus river mouth and adjacent beaches.

All other assessed coastal engineering strategies (A2, A3, B1, B2 and B5) are not only incapable of increasing the beach width but also lead to even narrower beaches than the baseline solution (A1). Of all the intervention scenarios, the one associated with maintaining the present beach nourishment strategy and increasing the terminal groin EV1 by 300 m (B4.3) promotes the largest sediment retention in the upper beach and foreshore, nearly preserving all the sand nourishment provided to the system. Additionally, it is also concluded that groins EC1, EC2, EC4, EC6 and EC7 are not strictly necessary to the beach alongshore equilibrium, as intervention strategy B6 leads to a nearly similar beach accretion as strategy B4.3.

Lastly, one concludes that shoreline evolutions models can be used with success in comparing coastal management alternatives, provided they are properly calibrated with site data. Here, the results at Costa da Caparica beaches are in accordance with the present knowledge of the beach sediment dynamics.

Funding: This research was conducted within a contract coordinated by the Portuguese Environmental Agency and co-funded by the Portuguese "POSEUR—Programa Operacional Sustentabilidade e Eficiência no Uso de Recursos" (reference POSEUR-02-1809-FC-000039).

Data Availability Statement: Not applicable.

Acknowledgments: The author acknowledges the Portuguese Territorial General Directorate (DGT) for making available the orthophotos, the Portuguese Environmental Agency for supporting this study, and Dodet et al. (2010) for providing the hindcast offshore wave data. Discussions with Celso Pinto and Luís Portela were very fruitful in defining the intervention strategies evaluated here. Vitor Pisco helped with the model setup and runs. The data collection programme "Monitoring of the Portuguese continental coastal zone" (COSMO, [32]) was financed by the POSEUR funding programme ("Operational Programme for the Sustainability and Efficiency of the Resources Expenditure"). The manuscript revision by Óscar Ferreira and the anonymous reviewers is greatly appreciated.

Conflicts of Interest: The author declares no conflict of interest.

References

1. Pranzini, E. Coastal erosion and shore protection: A brief historical analysis. *J. Coast. Conserv.* **2018**, *22*, 827–830. [CrossRef]
2. Mentaschi, L.; Vousdoukas, M.I.; Pekel, J.; Voukouvalas, E.; Feyen, L. Global long-term observations of coastal erosion and accretion. *Sci. Rep.* **2018**, *8*, 12876. [CrossRef]
3. Vousdoukas, M.I.; Ranasinghe, R.; Mentaschi, L.; Plomaritis, T.A.; Athanasiou, P.; Luijendijk, A.; Feyen, L. Sandy coastlines under threat of erosion. *Nat. Clim. Chang.* **2020**, *10*, 260–263. [CrossRef]

4. Petropoulos, A.; Kapsimalis, V.; Evelpidou, N.; Karkani, A.; Giannikopoulou, K. Simulation of the Nearshore Sediment Transport Pattern and Beach Morphodynamics in the Semi-Enclosed Bay of Myrtos, Cephalonia Island, Ionian Sea. *J. Mar. Sci. Eng.* **2022**, *10*, 1015. [CrossRef]
5. Hinkel, J.; Lincke, D.; Vafeidis, A.T.; Perrette, M.; Nicholls, R.J.; Tole, R.S.J.; Marzeiong, B.; Fettweish, X.; Ionescu, C.; Levermann, A. Coastal flood damage and adaptation costs under 21st century sea-level rise. *Proc. Natl. Acad. Sci. USA* **2014**, *111*, 3292–3297. [CrossRef]
6. Gracia, A.; Rangel-Buitrago, N.; Oakley, J.A.; Williams, A.T. Use of ecosystems in coastal erosion management. *Ocean. Coast. Manag.* **2018**, *156*, 277–289. [CrossRef]
7. Sánchez-Arcilla, A.; García-León, M.; Gracia, V.; Devoy, R.; Stanica, A.; Gault, J. Managing coastal environments under climate change: Pathways to adaptation. *Sci. Total Environ.* **2016**, *572*, 1336–1352. [CrossRef]
8. Ferreira, J.C.; Cardona, F.S.; Jóia-Santos, C.; Tenedório, J.A. Hazards, vulnerability, and risk analysis on wave overtopping and coastal flooding in low-lying coastal areas: The case of Costa da Caparica, Portugal. *Water* **2021**, *13*, 237. [CrossRef]
9. Veloso-Gomes, F.; Costa, J.; Rodrigues, A.; Taveira-Pinto, F.; Pais-Barbosa, J.; Neves, L. Costa da Caparica artificial sand nourishment and coastal dynamics. *J. Coast. Res.* **2009**, *56*, 678–682.
10. Silva, R.; Veloso-Gomes, F.; Pais-Barbosa, J. Morphological Behaviour of Costa da Caparica Beaches Monitored during Nourishment Operations. *J. Coast. Res.* **2013**, *65*, 1862–1867. [CrossRef]
11. Silva, S.F.; Martinho, M.; Capitão, R.; Reis, T.; Fortes, C.J.; Ferreira, J.C. An index based method for coastal-flood risk assessment in low-lying areas (Costa de Caparica, Portugal). *Ocean. Coast. Manag.* **2017**, *144*, 90–104. [CrossRef]
12. Santos, Â.; Mendes, S.; Corte-Real, J. Impacts of the storm Hercules in Portugal. *Finisterra* **2014**, *XLiX 98*, 197–220. [CrossRef]
13. Brown, J.M.; Phelps, J.J.C.; Barkwith, A.; Hurst, M.-D.; Ellis, M.A.; Plater, A.J. The effectiveness of beach mega-nourishment, assessed over three management epochs. *J. Environ. Manag.* **2016**, *184*, 400–408. [CrossRef]
14. Stripling, S.; Panzeri, M.; Blanco, B.; Rossington, K.; Sayers, P.; Borthwick, A. Regional-scale probabilistic shoreline evolution modelling for flood-risk assessment. *Coast. Eng.* **2017**, *121*, 129–144. [CrossRef]
15. Pereira, C.; Coelho, C. Mapping erosion risk under different scenarios of climate change for Aveiro coast, Portugal. *Nat. Hazards* **2013**, *69*, 1033–1050. [CrossRef]
16. Chataigner, T.; Yates, M.L.; Le Dantec, N.; Harley, M.D.; Splinter, K.D.; Goutal, N. Sensitivity of a one-line longshore shoreline change model to the mean wave direction. *Coast. Eng.* **2022**, *172*, 104025. [CrossRef]
17. Fortunato, A.B.; Freire, P.; Mengual, B.; Bertin, X.; Pinto, C.; Martins, K.; Guérin, T.; Azevedo, A. Sediment dynamics and morphological evolution in the Tagus Estuary inlet. *Mar. Geol.* **2021**, *440*, 106590. [CrossRef]
18. Lira, C.P.; Nobre-Silva, A.; Taborda, R.; Andrade, C.F. Coastline evolution of Portuguese low-lying sandy coast in the last 50 years: An integrated approach. *Earth Syst. Sci. Data* **2016**, *8*, 265–278. [CrossRef]
19. Pinto, C.A.; Gomes, E.; Rodrigues, A. Dredging and beach nourishment: A sustainable sediment management approach in Costa da Caparica beach (Portugal). In Proceedings of the Dredging 2015 Conference Moving and Managing Sediments, Savannah, GA, USA, 19–22 October 2015.
20. Freire, P.; Fortunato, A.; Oliveira, F.S.B.F. *Modelação Para Apoio às Intervenções nas Praias da Costa da Caparica–Almada. Estudo IV–Avaliação do Comportamento e Longevidade das Alimentações Artificiais. Efeito do Forçamento*; Report 223/2019–DHA/NEC; LNEC: Lisbon, Portugal, 2019. (In Portuguese)
21. Veloso-Gomes, F.; Taveira-Pinto, F.; Pais-Barbosa, J. Rehabilitation study of coastal defense works and artificial sand nourishment at Costa da Caparica, Portugal. In Proceedings of the 29th International Conference of Coastal Engineering 2004, Lisboa, Portugal, 19–24 September 2004; pp. 3429–3440.
22. Raposeiro, P.D.; Fortes, C.J.E.M.; Capitão, R.; Reis, M.T.; Ferreira, J.C.; Pereira, M.T.S.; Guerreiro, J. Preliminary phases of the HIDRALERTA system: Assessment of the flood levels at S. João da Caparica beach, Portugal. In Proceedings of the 12th International Coastal Symposium 2013, Plymouth, UK, 9–12 April 2013; pp. 808–813. [CrossRef]
23. Garzon, J.L.; Ferreira, A.M.M.; Ferreira, Ó.; Fortes, C.J.; Reis, M.T. Beach State Report: Quarteira, Praia de Faro and Costa da Caparica. EW-Coast Project-Early Warning System for Coastal Risks Induced by Storms. 2020. ALG-LISBOA-01-145-FEDER-028657. Available online: https://www.cima.ualg.pt/EW-COAST/ (accessed on 16 April 2023).
24. Sancho, F.; Silva, J.; Neves, M.G. Avaliação e quantificação da intensidade da agitação marítima no arco Caparica-Espichel. In Proceedings of the 9as Jornadas Portuguesas de Engenharia Costeira e Portuária 2017, Lisboa, Portugal, 23–24 November 2017.
25. Dodet, G. Morphodynamic Modelling of a Wave-Dominated Tidal Inlet: The Albufeira Lagoon. Ph.D. Thesis, La Rochelle University, École Doctorale Sciences pour L'Environnement Gay-Lussac, La Rochelle, France, 2013; 181p.
26. Fortunato, A.B.; Baptista, A.M.; Luettich, R.L., Jr. A three-dimensional model of tidal currents in the mouth of the Tagus estuary. *Cont. Shelf Res.* **1997**, *17*, 1689–1714. [CrossRef]
27. Taborda, R.; Andrade, C. Morfodinâmica do Estuário Exterior do Tejo e Intervenção na Região da Caparica–v1. In *Report Contributo para o Grupo de Trabalho do Litoral*; Lisbon, Portugal, 2014. (In Portuguese)
28. Sancho, F. *Modelação Para Apoio às Intervenções Nas Praias da Costa da Caparica–Almada; Estudo II–Modelação da Evolução Morfológica para Diferentes Cenários de Intervenção. Aplicação do Modelo*; Report 111/2020–DHA/NEC; LNEC: Lisbon, Portugal, 2020. (In Portuguese)

29. Diogo, Z.S.; Silveira, T.M.; Sousa, H.; Carapuço, A.M.; Silva, A.N.; Taborda, R.; Andrade, C. *Estudo de Caso da Costa da Caparica; Caracterização da Variabilidade Morfodinâmica Sazonal e Pós-Temporal das Praias da Costa da Caparica–Entregável 2.2.c*; FCUL: Lisbon, Portugal, 2013; 119p. (In Portuguese)
30. Freire, M.E.F. A planície litoral entre Trafaria e a Lagoa de Albufeira-estudo de geomorfologia litoral. *Estudos (SNPRCN)* **1989**, *3*, 1–204.
31. Hallermeier, R. Uses for a calculated limit depth to beach erosion. In *Coastal Engineering Proceedings*; ASCE: Hamburg, Germany, 1978; Volume 1, p. 88. [CrossRef]
32. Programa COSMO. Available online: https://cosmo.apambiente.pt (accessed on 20 March 2023).
33. Boak, E.H.; Turner, I.L. Shoreline definition and detection: A review. *J. Coast. Res.* **2005**, *21*, 688–703. Available online: http://www.jstor.org/stable/4299462 (accessed on 30 March 2023). [CrossRef]
34. Buccino, M.; Di Paola, G.; Ciccaglione, M.C.; Del Giudice, G.; Rosskopf, C.M. A Medium-Term Study of Molise Coast Evolution Based on the One-Line Equation and "Equivalent Wave" Concept. *Water* **2020**, *12*, 2831. [CrossRef]
35. Silva, A.N.; Lira, C.; Matildes, R.; Andrade, C.; Taborda, R.; Freitas, M.C. *Utilização de Ortofotomapas e Fotografias Aéreas Para a Delimitação da Linha de Costa–Entregável 1.2.2.2.c*; FCUL: Lisbon, Portugal, 2013; 59p. (In Portuguese)
36. Hapke, C.J.; Himmelstoss, E.A.; Kratzmann, M.; List, J.H.; Thieler, E.R. *National Assessment 528 of Shoreline Change: Historical Shoreline Change Along the New England and Mid-Atlantic 529 Coasts. Open-File Report 2010-1118*; U.S. Geological Survey: Reston, VA, USA, 2010; p. 57.
37. Booij, N.; Ris, R.C.; Holthuijsen, L.H. A third-generation wave model for coastal regions, Part I, Model description and validation. *J. Geophys. Res.* **1999**, *104*, 7649–7666. [CrossRef]
38. Dodet, G.; Bertin, X.; Taborda, R. Wave climate variability in the North-East Atlantic Ocean over the last six decades. *Ocean. Model.* **2010**, *31*, 120–131. [CrossRef]
39. Rusu, L.; Bernardino, M.; Guedes-Soares, C. Wave forecast at the entrance of the Tagus estuary. In *Proceedings on the Third International Conference on the Application of Physical Modelling to Port and Coastal Protection, COASTLAB-2010*; Black Sea Coastal Research Association: Varna, Bulgaria, 2014.
40. Fortunato, A.B.; Nahon, A.; Dodet, G.; Pires, A.R.; Freitas, M.C.; Bruneau, N.; Azevedo, A.; Bertin, X.; Benevides, P.; Andrade, C.; et al. Morphological evolution of an ephemeral tidal inlet from opening to closure: The Albufeira inlet, Portugal. *Cont. Shelf Res.* **2014**, *73*, 49–63. [CrossRef]
41. Vicente, C.; Clímaco, M. *Evolução de Linhas de Costa. Desenvolvimento e Aplicação de um Modelo Numérico*; Report ICT-ITH 42; LNEC: Lisbon, Portugal, 2003. (In Portuguese)
42. Oliveira, F.S.B.F.; Freire, P.; Sancho, F.; Vicente, C.M.; Clímaco, C. Rehabilitation and protection of Colwyn Bay beach: A case study. In Proceedings of the 11th International Coastal Symposium 2013, Szczecin, Poland, 9–13 May 2011; pp. 1272–1276.
43. Hanson, H. Genesis: 'A generalized shoreline change numerical model'. *J. Coast. Res.* **1989**, *5*, 1–27.
44. Kamphuis, J.W. Alongshore Sediment Transport Rate. *J. Waterw. Coast. Ports Ocean. Eng.* **1991**, *11*, 624–640. [CrossRef]
45. Dean, R.G. *Equilibrium Beach Profiles: U.S. Atlantic and Gulf Coasts*; Ocean Engineering Technical Report No. 12; Department of Civil Engineering and College of Marine Studies, University of Delaware: Newark, DE, USA, 1977.
46. Duarte-Santos, F.; Mota-Lopes, A.; Moniz, G.; Ramos, L.; Taborda, R. *Grupo de Trabalho do Litoral: Gestão da zona Costeira. O desafio da Mudança*; Duarte Santos, F., Penha-Lopes, G., Mota Lopes, A., Eds.; Agência Portuguesa do Ambiente: Lisboa, Portugal, 2018; 368p, ISBN 978-989-99962-1-2. (In Portuguese)
47. Mota-Oliveira, I.B. Port of Lisbon improvement of the access conditions through the Tagus estuary entrance. *Coast. Eng. Proc.* **1992**, *1*, 2745–2757. [CrossRef]
48. Silva, A.N., Lira, C.; Sousa, H.; Silveira, T.M.; Andrade, C.; Taborda, R.; Freitas, M.C. Análise da Evolução da Linha de Costa nos Últimos 50 Anos–Caso Especial da Costa da Caparica–Entregável 1.2.2.2.b. FCUL: Lisbon, Portugal, 2013; 24p. (In Portuguese)
49. Canelas, S.T.; Sancho, F.; Trigo-Teixeira, A. A Coastal Defense Work Plan 40 Years Later: Review and Evaluation of the Espinho Case Study in Portugal. *J. Waterw. Port Coast. Ocean. Eng.* **2022**, *148*, 05022004. [CrossRef]
50. Hanson, H.; Larson, M.; Kraus, N.C. A new approach to represent tidal currents and bathymetric features in the oneline model concept. *Proc. Coast. Dyn.* **2001**, *1*, 172–181.
51. Hanson, H.; Kraus, N.C. *GENESIS-Generalized Model for Simulating Shoreline Change. Vol. 1: Reference*; Manual and Users Guide Technical Report CERC-89-19; Coastal Engineering Research Center, U.S. Army Corps of Engineers: Washington, DC, USA, 1989; 247p.
52. Antolínez, J.A.A.; Méndez, F.J.; Anderson, D.; Ruggiero, P.; Kaminsky, G.M. Predicting climate-driven coastlines with a simple and efficient multiscale model. *J. Geoph. Res. Earth Surf.* **2019**, *124*, 1596–1624. [CrossRef]
53. Mil-Homens, J.; Ranasinghe, R.; van Thiel de Vries, J.S.M.; Stive, M.J.F. Re-evaluation and improvement of three commonly used bulk longshore sediment transport formulas. *Coast. Eng.* **2013**, *75*, 29–39. [CrossRef]
54. Pilkey, O.; Cooper, A. Longshore transport volumes: A critical view. *J. Coast. Res.* **2002**, *36*, 572–580. [CrossRef]
55. Barceló, J.P. *Esporões Marítimos–Funcionamento Hidráulico de Unidades Fisiograficamente Independentes*; LNEC: Lisboa, Portugal, 1966; 115p. (In Portuguese)
56. TECHINT. *Estudo do Aproveitamento Turístico da Área da Cova do Vapor*; Compagnia Tecnica Internazionale: Castellanza, Italy, 1966. (In Portuguese)
57. Taborda, R.; Andrade, F.C.; Silva, N.A.; Silveira, M.T.; Lira, C.; Freitas, C.M.; Pinto, C. Caparica-Espichel Longshore Sediment Transport Model. In Proceedings of the IX Congresso Nacional de Geologia, Comunicações Geológicas, Porto, Portugal, 18–24

July 2014; Volume 101, pp. 641–644. Available online: http://www.lneg.pt/iedt/unidades/16/paginas/26/30/185 (accessed on 30 March 2023). (In Portuguese).

58. Pinto, C.; Silveira, T.; Teixeira, S. *Alimentação Artificial de Praias na Faixa Costeira de Portugal Continental: Enquadramento e Retrospetiva das Intervenções Realizadas (1950–2017)*; Relatório técnico; Departamento do Litoral e Proteção Costeira. Núcleo de Monitorização Costeira e Risco: Amadora, Portugal, 2018; 61p. (In Portuguese)
59. Pais, D.; Andrade, C.; Pinto, C. Evolution of a beach-dune system after artificial nourishment: The case of São João da Caparica (Portugal). In Proceedings of the Building Coastal Resilience 2022, Dune in Front of a Dike International Conference, KU Leuven Bruges Campus, Bruges, Belgium, 12–13 April 2022.

Journal of
Marine Science
and Engineering

MDPI

Article

3D Physical Modeling of an Artificial Beach Nourishment: Laboratory Procedures and Nourishment Performance

André Guimarães [1,*], Carlos Coelho [1], Fernando Veloso-Gomes [2] and Paulo A. Silva [3]

[1] RISCO & Civil Engineering Department, Campus Universitário de Santiago, University of Aveiro, 3810-193 Aveiro, Portugal; ccoelho@ua.pt
[2] Civil Engineering Department, Faculty of Engineering, University of Porto, 4200-465 Porto, Portugal; vgomes@fe.up.pt
[3] CESAM & Physics Department, Campus Universitário de Santiago, University of Aveiro, 3810-193 Aveiro, Portugal; psilva@ua.pt
* Correspondence: asaguimaraes@ua.pt

Abstract: Beach nourishment represents a type of coastal defense intervention, keeping the beach as a natural coastal defense system. Altering the cross-shore profile geometry, due to the introduction of new sediments, induces a non-equilibrium situation regarding the local wave dynamics. This work aims to increase our knowledge concerning 3D movable bed physical modeling and beach nourishment impacts on the hydrodynamics, sediment transport, and morphodynamics. A set of experiments with an artificial beach nourishment movable bed model was prepared. Hydrodynamic, sediment transport, and morphological variations and impacts due to the presence of the nourishment were monitored with specific equipment. Special attention was given to the number and positioning of the monitoring equipment and the inherent constraints of 3D movable beds laboratory tests. The nourishment induced changes in the beach dynamics, leading to an increase in the flow velocities range and suspended sediment concentration, and effectively increasing the emerged beach width. Predicting and anticipating the morphological evolution of the modeled beach has a major impact on data accuracy, since it might influence the monitoring equipment's correct position. Laboratory results and constraints were characterized to help better define future laboratory procedures and strategies for increasing movable bed models' accuracy and performance.

Keywords: wave basin; morphodynamic; hydrodynamic; sediment transport; cross-shore profile

Citation: Guimarães, A.; Coelho, C.; Veloso-Gomes, F.; Silva, P.A. 3D Physical Modeling of an Artificial Beach Nourishment: Laboratory Procedures and Nourishment Performance. *J. Mar. Sci. Eng.* **2021**, *9*, 613. https://doi.org/10.3390/jmse9060613

Academic Editor: Rodger Tomlinson

Received: 6 May 2021
Accepted: 29 May 2021
Published: 3 June 2021

1. Introduction

According to [1], from the 1980s onward, due to the failure of many "hard coastal engineering structures" (e.g., groins, breakwaters, or revetments), the "soft engineering" coastal protection interventions (e.g., artificial beach nourishments) became more appealing from the technical and economic point of view. However, to correctly identify the type of coastal protection to use, one must look at the causes of the problem (e.g., massive erosion and littoral drift sediment budget deficit), the morphological processes, and the safety, recreational, environmental, and economic aspects. Thus, evaluation and assessment criteria for the design and accreditation of nourishment projects are needed. Critical scientific and technical issues must be addressed prior to its development [2].

Physical modeling of coastal interventions, such as beach nourishment, helps in understanding the nourishment impacts on the beach morphodynamics and physical processes, thus assessing its performance [3]. There are several works that address 2D physical modeling in general [4,5], and more precisely, artificial nourishments [6,7]. These are usually specific to a certain sediment transport mechanism or morphological element of the beach (e.g., dune stability). However, the lack of quantitative studies related to 3D morphological evolution makes the validation of morphodynamical models and the understanding of the beach morphodynamic more difficult [8]. The present work describes a set of laboratory tests performed in the wave tank at the Faculty of Engineering of the University of Porto, in the laboratory of the Hydraulics, Hydraulic Resources, and Environment Section (SHRHA) between the 23rd September and the 23rd December 2016. It was intended to obtain hydrodynamic, sediment transport, and morphodynamic data concerning the performance of artificial nourishments (Scenario A).

This work also aims to increase knowledge related to 3D movable bed physical modeling, with special focus on the required setup and configuration of the experiments and monitoring procedures, in order to obtain tangible data that allow the proper assessment of the nourishment morphological, hydrodynamic, and sediment transport impact. A generic beach was modeled, taking into consideration the laboratory's available equipment and constraints. Sediment size, available sand quantities, existing monitoring equipment, wave basin dimensions, and the wave generators' capacity were the main aspects taken into consideration to characterize the beach and define the laboratory tests. Cross-shore and longshore velocities and suspended sediment concentrations were also monitored to assess the impact of the nourishment in an equilibrium beach. The main topics discussed are related to the model's hydrodynamic and morphodynamic behavior and constraints in preparing and testing a physical model with a movable bed. Constraints are mainly related to the monitoring of the morphological evolution of the bed, hydrodynamic phenomena in the wave basin, sediment transport quantification, and monitoring instruments' locations.

1.1. Artificial Nourishments

According to [9], beaches offer storm protection through a natural dynamic response to changing incoming waves and water levels. According to [7,10–12], shoreface nourishments, involving submerged mounds of sediments, appear to be a good alternative to hard engineering coastal defense structures (incorporating elements of efficiency, ecology, and operating cost). According to [13], the primary objective of beach nourishment is the protection of upland structures and infrastructures. It is often related to protection of seawalls and revetments against local scour, strengthening of dunes and beaches in order to maintain the coastline position, dune width extension for protection against floods, compensation of downdrift erosion due to coastal defense structures (e.g., groins), and enhancing recreation at a small scale [2,14–16]. To design an artificial nourishment, a cross-shore profile needs to be defined. A common assumption is that after nourishment, the profile shape will evolve towards its initial shape (provided that the same type of sediments has been used), so the final geometric result should be similar to the initial one before the filling. There is a need for maintenance (post-storm and periodic) to guarantee the desired protection level [13]. Usually, artificial nourishments are designed for a time scale of tens of years, considering a periodic (re)nourishment from 3 to 5 years after construction or renourishment. Longevity is primarily determined by the degree to which the placed sand volume addresses any pre-project profile volume deficit and the rate at which fill material is transported out of the project domain in the alongshore direction (lateral and cross-shore direction spreading losses). Wave-driven longshore sand transport processes are the major cause of lateral spreading. In addition, the fill itself creates a perturbation in the shoreline and beach orientation, particularly on the up and downdrift limits of the fill. In these transitions, local wave transformation patterns and consequently the longshore sand transport regime change. Changes lead to higher rates of fill loss from the longshore limits of the project.

1.2. Physical Modeling

According to [17], reduced-scale physical models replicate nearshore processes, without simplifying the governing equations that characterize the involved phenomena/processes, i.e., they overcome the inherent limits of deterministic fluid mechanics due to turbulence. Physical modeling has a reduced cost and presents fewer difficulties when compared to on-site data collection [18–21]. However, working at a reduced scale implies that all relevant variables will not be correctly simulated and even some phenomena that occur in nature will not be modeled. The laboratory itself may also have an impact on the results. Common laboratory effects are related to the inability to create realistic forcing conditions and the influence of the presence of physical boundaries like wave basin walls [17]. When trying to reproduce a specific beach scenario, these effects limit the extrapolation of the results to the real prototype dimensions [17,18,21]. Considering specific physical modeling tests to evaluate nourishment interventions, the work of [7] can be referred to. [7] analyzed shoreface and bar nourishments morphodynamic, hydrodynamic, and sediment grain size distribution under storm events using a 2D physical model. Results show important insights about the nourishment cross-shore behavior. An increase in the profile steepness was observed in all the nourishments, while the shoreface nourishment was the only scenario that presented a beach width increase after a full storm cycle. Physical tests concerning artificial nourishment performance using 3D physical modeling are scarce in the literature review. [22] studied the impact of several nourishment positions and nourishment frequencies using a distorted physical model. The study was performed to identify which nourishment scenario would have a positive impact in increasing Copacabana beach (Brazil) width. Other studies concerning nearshore nourishments on 3D physical models exist, such as [23,24].

2. Laboratory Tests

2.1. Model Setup

The modeled coastal stretch represents a linear beach without the presence of any coastal structures, before and after a construction of an artificial nourishment, where the main sediment transport mechanism is governed by the wave action.

SHRHA Hydraulic Laboratory wave basin has a longitudinal and cross extension of 28 and 12 m, respectively, with a maximum depth of 1.2 m. The laboratory is equipped with a multi-element, piston-type wave generation system that possesses an active reflection absorption system (HR Wallingford). Since the objective was to test the biggest possible beach extension, a longshore and cross-shore extension of 10 m was defined. The remaining 2 m of the total basin width was used to build a sediment collector at the downdrift limit of the basin with the objective to estimate the net longshore sediment transport volumes at the end of each tested scenario. The available sediments at the laboratory were characterized as quartz sand, with a specific mass of 2650 kg/m^3 and a median diameter (d_{50}) of 0.27 mm. Offshore significant wave height (H_S) mean period, (T_0) peak period (T_p), and wave length (L_0) were, respectively, 12.5 cm, 1.63 s, 2.13 s, and 3.97 m. Despite not trying to reproduce a specific prototype situation, these parameters were defined based on the Northwest Portuguese Coast characteristics (sediments and waves) considering a geometric scale of about 1:20, obtained through Froude similarity. A JONSWAP Spectrum was used (Equations (1) and (2), with f_p being the peak frequency and f representing the possible frequencies), with shape parameters γ and σ of 3.3 and 0.08, respectively, and $\alpha = 0.2189$ [20]. The mean water surface at the wave tank was kept constant during the laboratory tests.

$$E(f) = \alpha\, g^2\, (2\pi)^{-4}\, f^{-5} \exp(-1.25(f/f_p)^{-4})\, \gamma^r \tag{1}$$

$$r = \exp(-0.5\, ((f/f_p - 1)\, /\, \sigma)^2) \tag{2}$$

The initial cross-shore beach profile is characterized by an emerged slope (β) and length (l_e) of 0.03 and 4 m, respectively, and a height of 12 cm. The submerged part is described by Dean's [25] profile: $h = A \times y^m$. A value of 2/3 was chosen for the "m" parameter (intermediate beach), while the "A" parameter is obtained through [17] formulation. This formulation leads to an average steeper profile at the surf zone, but it allows respecting the 10 m cross-shore profile extension. The initial cross-shore profile geometry influences the initial rate of morphologic beach changes, since the further from equilibrium the initial beach geometry is, the more changes it will suffer. At the prototype scale, the cross-shore profile geometry would have been characterized by the same "A" and "m" parameters, since the model is designed to be geometrically non-distorted.

The profile has a submerged length (l_s) of 6 m that develops towards a total depth of 0.4 m (defined through Roberts' [26] formulation 27), which corresponds to the closure depth (d_c) and the adopted maximum offshore depth, giving the profile a total height (h_t) of 52 cm (Figure 1), with a slope (β) of 0.03. The total cross-shore extension of the beach was of 10 m.

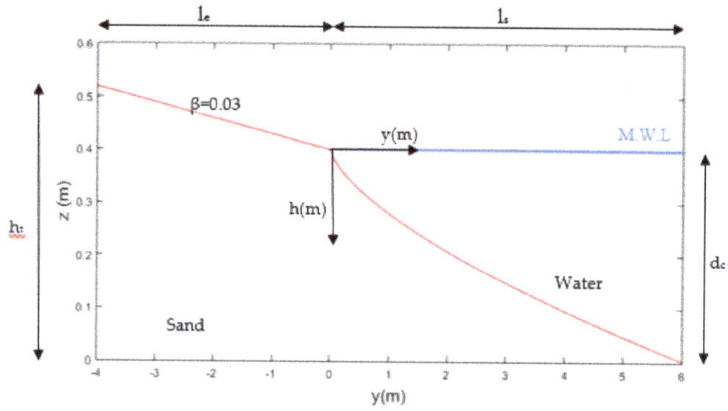

Figure 1. Initial cross-shore profile of the physical model (full extension).

The modeled beach was built over a dissipative beach to reduce the amount of sediment necessary for the model and to dissipate the wave energy propagating and breaking over the model landward limit (Figure 2). The initial low sand compaction affects the rate of the morphological cross-shore changes. An initial liquefaction of the sand was observed when filling the wave basin with water, which led to an initially higher value of the sediment transport. Additionally, no pumps were used to match the longshore current to prevent large circulation eddies.

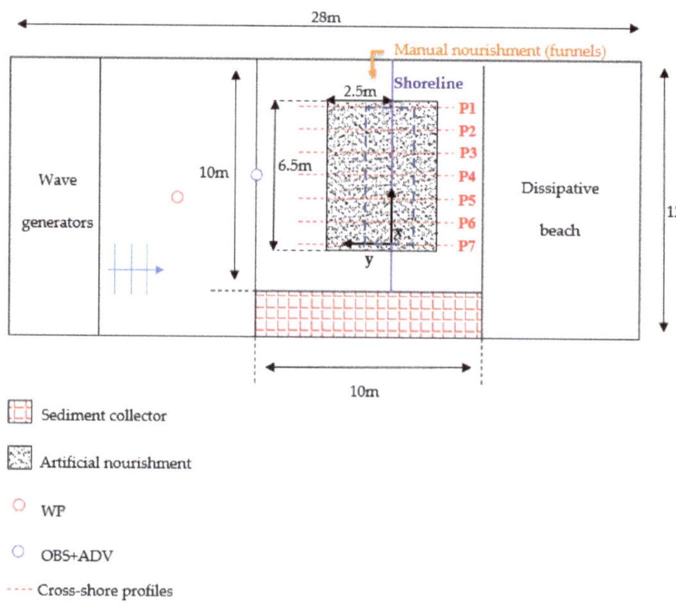

Figure 2. Beach plan scheme at the wave basin. The area represented by a blue dashed rectangle, above the nourishment, represents the monitored cell.

2.2. Tested Scenarios

Two sets of experiments were performed: A.1, which aimed to find a dynamic equilibrium or near-equilibrium state of the natural beach, and A.2, where an artificial nourishment is built on the pre-existing beach (final beach of scenario A.1). Beach configuration is presented in Figures 2 and 3. Scenario A.1 had a total duration of 26.5 h (from t = 0 min to t = 1590 min), while scenario A.2 had a duration of 20 h (from t = 1590 min to t = 2790 min). During the first 20 h (1200 min), scenario A.1 was subjected to wave action in order to guarantee that an equilibrium or near-equilibrium (static) configuration was reached by the beach and that the initial sand at the beach would be compacted. During this interval, only the initial and final profiles, at t = 0 min and t = 1200 min, respectively, were monitored. Throughout the rest of the tests, morphological evolution of the beach was regularly monitored. The rest of scenario A.1 consisted in subjecting the beach to wave action during 6.5 h (from t = 1200 min to t = 1590 min), while in the A.2 scenario, the beach was subjected to the already referred 20 h of wave action.

(a)

(b)

Figure 3. Initial beach configurations for both scenarios (the dashed lines represent the top limits of the nourishment emerged dimensions): (**a**) Initial beach before nourishment (Scenario A.1); (**b**) Initial nourished beach (Scenario A.2).

To simulate the littoral drift, sediments were continuously added at the updrift beach boundary to create a perturbation in the bathymetry that could induce a longshore sediment transport due to an alteration in the shoreline and bathymetric orientation with the wave direction. Sediments were added to the basin near the updrift wall, by manual nourishment, using 5 funnels placed at different distances from the initial shoreline position (Table 1 and Figure 3). Each funnel's opening was calibrated to feed the same amount of sediment at a constant flow rate of 48 kg/h. This value was estimated accordingly to [20] in order to simulate the littoral drift. At t = 1200 min, the beach reached an equilibrium with no significant changes in the shoreline position being recorded during the last hour of tests.

Table 1. Scenario A.1, funnels' position. Cross-shore distances to the initial shoreline.

Funnels	1	2	3	4	5
Distance to the initial shoreline (m)	2.92	2.42	1.92	1.42	0.42

The funnels used to simulate the littoral drift represent a discrete supply of sediments. Despite being evenly spaced, they do not cover the entire active cross-shore extension and are focused on supplying the most energetic area where wave breaking is observed. This reproduces neither a real cross-shore distribution of the longshore sediment transport in natural conditions nor the spatial continuity of the littoral drift.

Regarding the downdrift sediment losses, accumulated in the sediment trap, those volumes were only measured at the end of each scenario, since in order to measure them, the wave tank had to be emptied. Therefore, those values were only used to estimate the total average longshore sediment transport rates in each scenario.

During the last 6.5 h of scenario A.1 (from t = 1200 min to t = 1590 min) and over the full duration of scenario A.2 (20 h, from t = 1590 min to t = 2790 min), the wave generator system and monitoring devices were stopped periodically to measure the bathymetry using the HR Wallingford profiler. The stopping instants are represented by dashed lines in Figures 4–6 and the following ones, being more frequent at the early stages of each set of tests. There was no need to empty the tank, since the profiler had a mechanical probe.

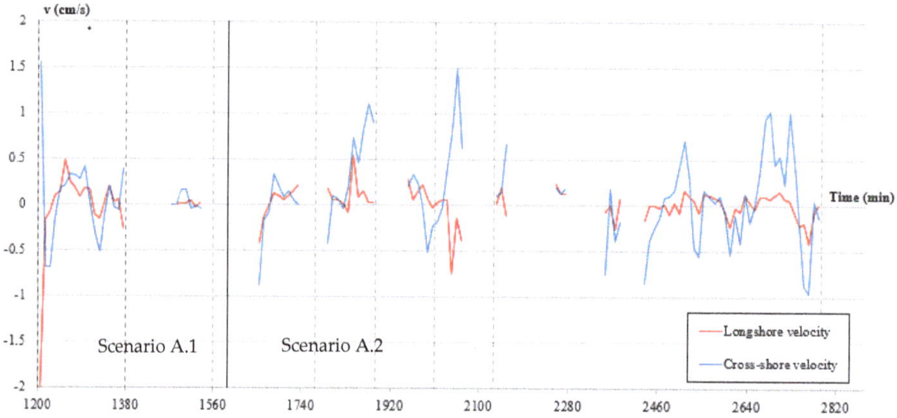

Figure 4. Recorded data timeline for the longshore and cross-shore velocity for the A.1 and A.2 scenarios. The vertical black line represents the transition between scenarios A.1 and A.2. The ver-tical black dashed lines represent the instants when the equipment was stopped.

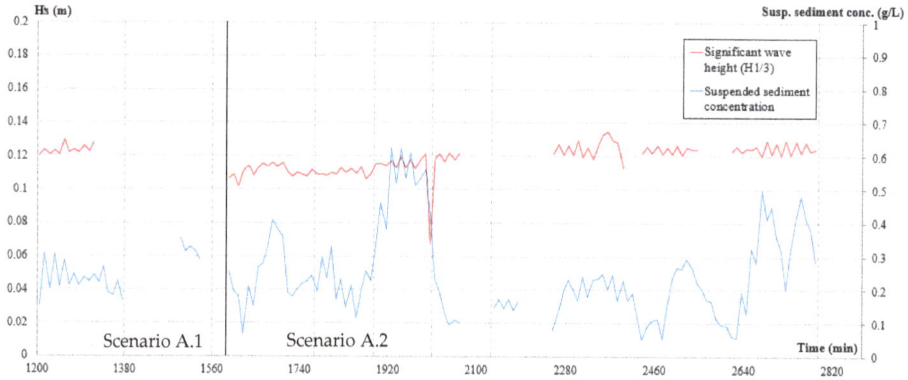

Figure 5. Recorded data timeline for significant wave height and suspended sediment concentration for the A.1 and A.2 scenarios. The vertical black line represents the transition between scenarios A.1 and A.2. The vertical black dashed lines represent the instants when the equipment was stopped.

Figure 6. Timeline for the estimated suspended longshore and cross-shore sediment transport. The vertical black line represents the transition between scenarios A.1 and A.2. The vertical black dashed lines represent the instants when the equipment was stopped.

The nourishment was designed to increase the beach width by 2.5 m, with a flat emerged surface intercepting the existing beach slope, with a longshore extension of 6.5 m (see Figures 2 and 3). The nourishment configuration placed over the modeled beach led to the use of a total amount of 9080 kg of sand for its construction.

2.3. Equipment

One acoustic Doppler velocimeter (SONTEK 10 MHz ADV), one optical backscatter (OBS 3+), and one wave probe (WP) were used to monitor flow velocity components, suspended sediment concentrations, and wave heights and periods, respectively. The location of these instruments in the wave basin is shown in Figure 2. An HR Wallingford profiler was used to monitor 7 cross-shore profiles (P1 to P7), spaced 1 m apart (middle profile, P4, was centered on the beach longshore extension). Due to its limited length, the profiler monitored 2 m of the cross-shore extension of the profile. Thus, a total longshore and cross-shore extension of 6 and 2 m, respectively, was monitored.

Preparation of physical tests with movable beds presents some specific procedures. One of the main struggles in preparing the laboratory tests was related to the definition of the instrument's location. Due to the morphodynamic evolution during the tests, the shoreline position over time will change and the equipment will measure different locations on the profile. Thus, the relative positions of the instruments vary with increasing or decreasing cross-shore profile width. Additionally, bed accretion could cause misreading due to the risk of the monitored sample not being positioned in the water column.

Due to the morphological evolution uncertainty, plus the fact that an artificial nourishment was built to increase the beach width, it was chosen to place the ADV and the OBS at the offshore limit, aligned with the middle cross-shore profile (P4). The measured sample, for both pieces of equipment, was initially positioned 10 cm above the model bed. The WP was placed in the middle of the wave basin to measure the incoming wave's characteristics. Sampling frequencies for the OBS, ADV, and WP were 5, 25, and 25 Hz, respectively.

Since the OBS received signal is influenced by the size, composition, and shape of the suspended sediments, it was previously calibrated using sediments from the laboratory. ADV data quality is defined by the signal to noise ratio (SNR) and correlation. The lowest possible values for SNR and correlation were selected based on the instrument's manual [27] and on the works of [28,29]. The values of 15 dB and 70% were chosen for the lowest admissible limit of SNR and correlation, respectively.

The profiler allowed the collection of only 2 m of the profile cross-shore extension (6 m). This increases the difficulty in the data analysis, since it was not possible to precisely track the cross-shore sediments' movement over time, over the full extension of the beach for both longshore and cross-shore directions. In the situation that the cross-shore profile increases its length towards offshore (e.g., due to an undertow current), leading to an increase in the beach extension, the bathymetric data reduces its representability regarding the beach morphological evolution.

3. Results

3.1. Hydrodynamic and Sediment Transport

The hydrodynamic and sediment transport collected data, for both scenarios A.1 and A.2, are presented in Figures 4–6. Figure 4 shows the longshore and cross-shore flow velocity components, obtained by the ADV (negative values are considered in the direction of the sediment collector and towards offshore, respectively, for longshore and cross-shore velocity components). Significant wave height (based on WP records) and suspended sediment concentrations (OBS records) are presented in Figure 5, while Figure 6 shows the suspended longshore and cross-shore sediment transport rates estimated based on flow velocities and suspended sediment concentrations. The total longshore suspended sediment transport was estimated by the combination of the longshore velocity rates and suspended sediment concentrations. The concentration and velocity values were assumed constant throughout the water column (0.4 m) and over the cross-shore extension of the active profile (a total extension of 8 m) to estimate the net transport. For the cross-shore transport, the cross-shore velocities were used instead and were assumed constant over the water column (0.4 m) and longshore extension of the beach (10 m).

The values represent an average over 10 min of collected data. Gaps in the velocities and suspended sediment concentration data series, with a duration inferior to 2 s, were filled by linear interpolation, and those with a duration superior to 2 s were kept empty. Since a qualitative analysis was intended, these gaps might still allow the overall analysis of the hydrodynamic and sediment transport evolution throughout the tests. Wave climate records were not subjected to any kind of filling. All the data were processed and analyzed using MATLAB.

Though it was intended for the significant wave height to be 12.5 cm for both experiments, the significant wave height in scenario A.2, during the first 540 min, is on average about 10.5% smaller than 12.5 cm, corresponding to 11.2 cm. This was due to the beach nourishment morphology evolution and thus a correction of the gain factor in the wave generation and active dissipation system was performed during this time interval to achieve the expected wave heights.

Over the full duration of the laboratory tests, several rip-currents were observed; however, their position was not monitored. The presence of rip-currents' presence combined with the water recirculation and local bathymetric changes (near the ADV) are responsible for the registered alteration of velocities sign/direction.

Longshore velocities present a range between −0.75 and 0.55 cm/s, while cross-shore velocities present a larger range, from −0.95 to 1.55 cm/s. The average longshore and cross-shore velocities were approximately −0.05 and 0.08 cm/s for A.1 scenario and 0.01 and 0.11 cm/s in scenario A.2, respectively.

Suspended concentration time series shows a nearly constant value of 0.25 g/L during A.1, while in A.2 a larger variability is present, with higher values of concentration from t = 1890 min to t = 2010 min and from t = 2610 min to t = 2790 min. Due to the nourishment, in scenario A.2, a higher average value of the suspended sediment concentration than scenario A.1 was recorded. The suspended sediment concentration peak was not registered immediately after the nourishment, as would be expected due to the large perturbation and the lack of sand compaction. The highest value of suspended sediment concentration of 0.60 g/L was recorded in scenario A.2 between t = 1890 min and t = 2010 min, about 300 min after the start of the laboratory tests for scenario A.2. This 300 min delay in the recorded suspended sediment concentrations peak during scenario A.2 might be explained by the distance of the OBS to the shoreline and the nourishment area (and the eventual influence of the re-circulation currents inside the basin).

Over the last 6.5 h of scenario A.1 (from t = 1200 min to t = 1590 min), an average value of 0.23 g/s in the direction of the sediment collector (with an absolute maximum peak of 9.92 g/s) and 0.35 g/s in the offshore direction (with an absolute maximum of 14.39 g/s) was obtained for the longshore and cross-shore suspended transport rates, respectively. During scenario A.2, the maximum value of longshore and cross-shore suspended sediment transport rate of 7.68 and 17.36 g/s, respectively, was obtained between 360 t = 2130 min and t = 2790 min (last 11 h hours of tests during scenario A.2). In scenario A.2, the mean suspended longshore transport rate inverted its direction (in the funnel direction) and decreased its magnitude to 0.08 g/s, while the cross-shore suspended transport rate increased to 0.98 g/s (towards offshore). Therefore, while the suspended cross-shore transport is, on average, towards offshore, between scenarios A.1 and A.2, the average longshore transport changes direction (Table 2).

Table 2. Suspended sediment transport rates (g/s).

	Maximum Positive		Maximum Negative		Average	
Scenario	**Longshore**	**Cross-Shore**	**Longshore**	**Cross-Shore**	**Longshore**	**Cross-Shore**
A.1	4.46	11.42	−9.92	−14.39	−0.23	0.35
A.2	7.68	17.36	−6.66	−15.49	0.08	0.98

The total longshore net sediment transport value was estimated considering the accumulated sediments collected downdrift. At the end of scenarios A.1 and A.2, a total value of about 7630 and 7560 kg of sand was collected downdrift, respectively. The value obtained for scenario A.1 takes into consideration all the 1590 min of tests. A mean net transport rate of approximately 288 kg/h (80 g/s) for scenario A.1 and 378 kg/h (105 g/s) for scenario A.2 was obtained from the total sand volumes accumulated downdrift. Estimated sediment transport rates are much higher than the 48 kg/h added by the funnels. Although these values indicate a general retreat of the shoreline, they do not correctly characterize the sediment transport, since the initial conditions of the sand compaction do not reproduce the ideal condition to be modeled. By extrapolating the obtained results from the offshore monitoring point to the full extension of the beach (including the surf zone), it can be concluded that the bedload transport dominates the sediment transport for the modeled conditions.

3.2. *Morphodynamics*

Seven cross-shore profiles spaced 1 m apart and with an extension of 2 m were monitored. Figure 7 presents a digital model terrain of the modeled beach that was obtained by triangulation using a linear interpolation method and Figure 8 presents the corresponding erosion/accretion maps. The horizontal coordinate for y = 0 m represents the initial shoreline position, which corresponds to the interception of the water level at z = 0.4 m with the modeled beach. Figure 7 shows a shoreline retreat and slope decrease between the initial and final conditions in both A.1 and A.2 scenarios.

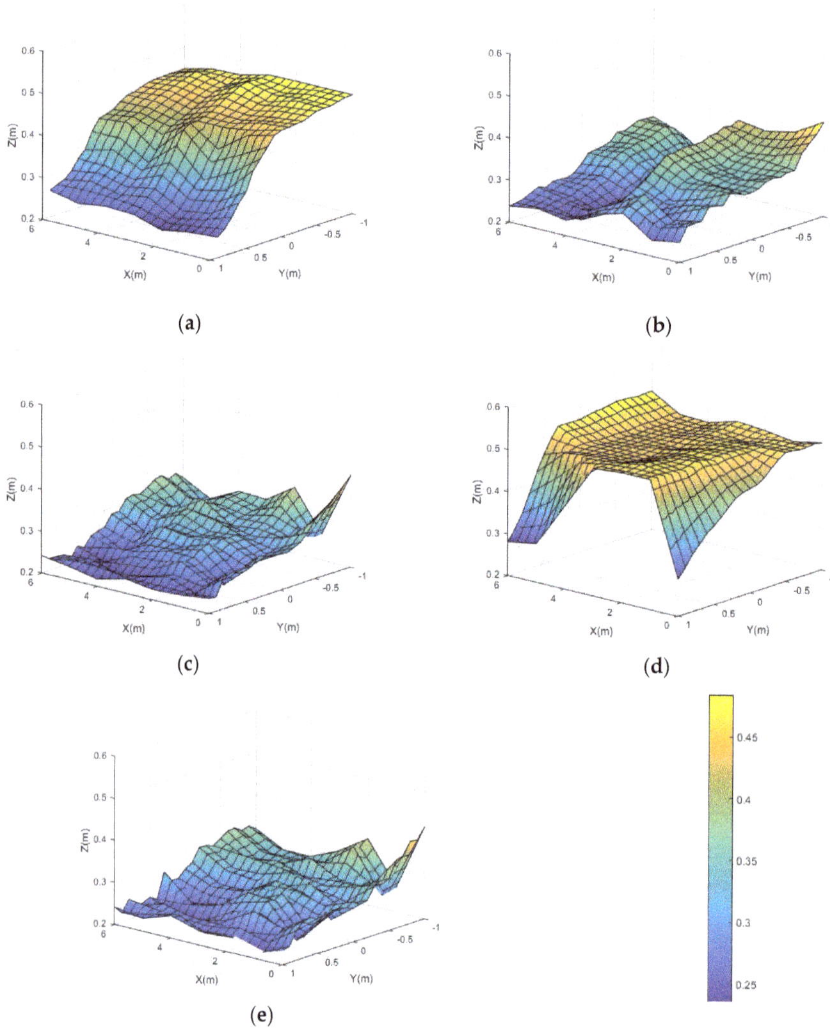

Figure 7. Bathymetric and topographic digital model terrain of the beach over the laboratory tests: (**a**) t = 0 min; (**b**) t = 1200 min; (**c**) t = 1590 min; (**d**) t = 1590 min (A.2); (**e**) t = 2790 min.

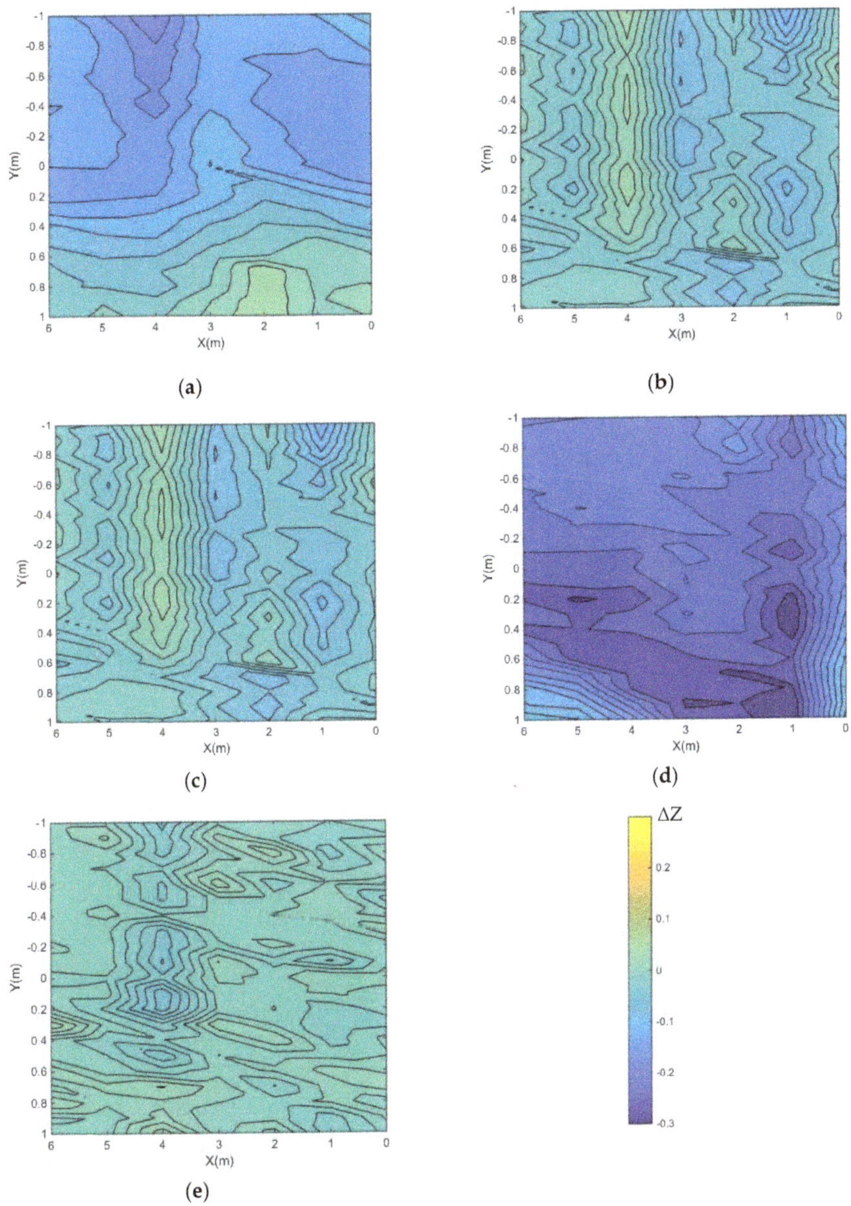

Figure 8. Erosion and accretion maps of the monitored area: (**a**) comparison between t = 0 min and t = 1200 min; (**b**) comparison between t = 1200 min and t = 1590 min; (**c**) comparison between t = 1590 min (A.1) and t = 1590 min (A.2); (**d**) comparison between t = 1590 min (A.2) and t = 2790 min; (**e**) comparison between t = 1590 min (A.1) and t = 2790 min.

During the first 1200 min of scenario A.1, there was a large erosion of the emerged beach, with sediments moving towards offshore, considering the lower compaction and initial adaptation of the beach profile (Figure 8). This tendency was also observed between 1200 and 1590 min. The deepening of the cross-shore profile at x = 4 m (P5) may arise due to a strong Z (m) rip-currents observed at this location. From the bathymetric data acquired during scenario A.2, a total erosion of the nourishment volume was registered and a bed morphological configuration similar to the end of scenario A.1 was observed. Erosion and accretion volumes' variation, presented in detail in Table 3, show that every scenario recorded mainly erosion, despite the artificial nourishment construction, as seen in Figure 8c. Comparing Figure 8c,d, it can be observed that the artificial beach nourishment volume disappeared, since the range of vertical accretion (defined by the color range) is symmetric between both figures, and the contour line shapes are similar. A few differences are shown in Figure 8e between the end of scenario A.1 and scenario A.2, with a total difference in the beach volume of -0.03 m^3. When comparing the final beach of each scenario, a similar bottom configuration is observed (Figure 7c,e). The results show that the beach tends to an equilibrium that is independent of the nourishment performed and allows the estimation of the time scale of the nourishment existence.

Table 3. Erosion and accretion volumes at the monitored area.

Time Interval (min)	Erosion (m^3)	Accretion (m^3)	Balance (m^3)
0–1200	1.06	0.03	−1.03
1200–1590 (A.1)	0.37	0.02	−0.35
1590 (A.1)–1590 (A.2)	0.00	2.56	+2.56
1590 (A.2)–2790	2.58	0.00	−2.58
1590 (A.1)–2790	0.06	0.03	−0.03
Total	4.07	2.64	−1.43

Figure 9 represents the position of the bathymetric contour lines of the P4 profile, z = 0.4 m and z = 0.3 m over time. The contour for z = 0.4 m corresponds to the shoreline and mean water lever (MWL); the z = 0.3 m contour is represented, since the profiler was not able to monitor the complete movement of the shoreline throughout the laboratory tests. Due to the lack of available data concerning the shoreline position, it is assumed that the shoreline presents a similar behavior to the evolution of the z = 0.3 m contour. Thus, the retreat of the z = 0.3 m contour line was used to estimate the average retreat rates for each profile. At some instants, profile P4 intercepted the z = 0.3 m contour twice (from t = 1020 min till the end of scenario A.1), which indicates the presence of a bar. Different colors represent different contour lines, while the dashed lines represent a second profile interception at the same level. All the lines show a general retreat of the cross-shore profile.

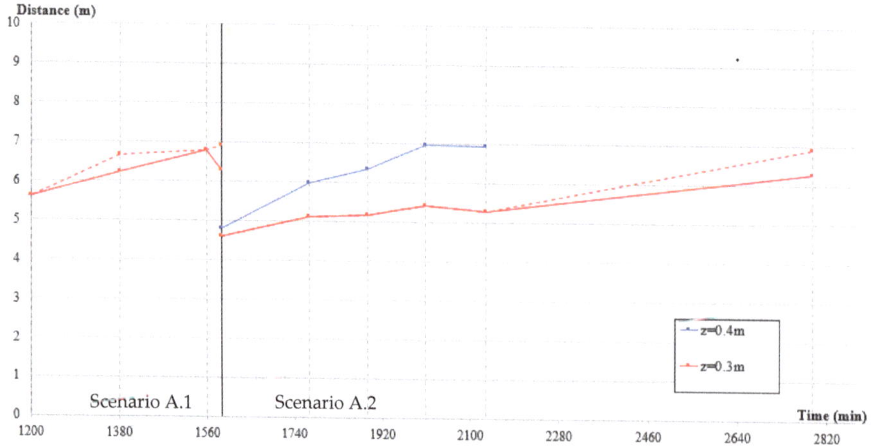

Figure 9. Timeline for profile P4 contour lines distance from the ADV. The vertical black line represents the transition between scenarios A.1 and A.2. The vertical black dashed lines represent the instants when the equipment was stopped.

Looking at scenario A.2, the contour that corresponds to the shoreline (z = 0.4 m) shows a higher retreat rate than the contour for z = 0.3 m, as expected due to the morphological perturbance introduced by the nourishment. According to Table 4, on average, all the profiles presented a retreat of the 0.3 m contour for both scenarios (between 1200 and 1590 min, in A.1 and between 1590 and 2790 min, in A.2). On average, and considering all the profiles, scenario A.2 has a higher retreat rate than scenario A.1. It was expected for the nourishment scenario to possess higher retreat rates than scenario A.1, due to an initial loss of sand and the fact that the new profile was further away from the equilibrium or near-equilibrium configuration [8,30]. Nevertheless, not all the profiles in scenario A.2 recorded a higher retreat than the respective profiles for scenario A.1 (Table 4); on the profiles other than P2 and P3, the average rate is lower for A.2 scenario. The more updrift of the nourishment the profiles are, the higher the shoreline retreat rate is. The results in Figure 9 also show that the position of the shoreline in scenario A.2 is not located as offshore as in scenario A.1, which can be interpreted as a direct benefit from the nourishment performed.

Table 4. Average retreat rates (cm/min) for z = 0.3 m contour.

Profile	Scenario A.1	Scenario A.2
P1	0.21	0.14
P2	0.02	0.26
P3	0.06	0.29
P4	0.18	0.14
P5	0.21	0.04
P6	0.00	0.06
P7	0.24	0.06
Average	0.13	0.14

Considering the geometric evolution of the seven cross-shore profiles, an estimation of beach volume and beach volume variation between these seven profiles was made (Figure 10), with the coastal cell volume being defined as the volume between the horizontal contour of z = 0 m and the upper limit of the profile geometry, through a cross-shore extension of 2 m and a longshore extension of 6 m. The red line represents the total sediments' volume in the measured control volume, while the blue line represents the volume variation between time instants in the monitored control volume. An almost constant erosion state is present in the monitored cell. This analysis is in accordance with the Table 3 and Figure 8 results.

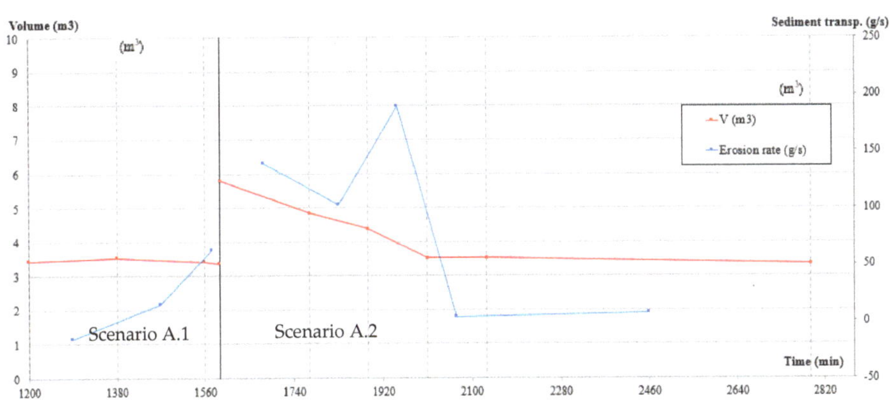

Figure 10. Timeline for beach volume variation and sediment transport volumes rate. Vertical black line represents the transition between scenarios A.1 and A.2. The vertical black dashed lines represent the instants when the equipment was stopped.

After the beginning of scenario A.2, the artificial nourishment effect in the monitored cell had a duration of 420 min. At the end of this time interval, the monitored cell volume reaches a value close to the volume determined at the end of scenario A.1 and the average volume variation rate goes from 0.323 m^3/h, during the first 420 min, to an average value of 0.015 m^3/h onwards. The volumes considered in Figure 10 can be assumed to demonstrate the nourishment impact/efficiency in the emerged beach width value, since the control volume incorporates the shoreline position after the nourishment.

Figure 11 contains the shore face slope evolution over time. The shoreface was defined as the zone between the shoreline (z = 0.4 m) and z = 0.2 m, since there is little or no data available to the geometry of the swash zone and deeper than z = 0.2 m. To determine the slope angle, only the top and bottom profile limits between the z = 0.4 m and z = 0.2 m were used. Using the classification proposed by [31] and [32], it was concluded that the mean slope of the beach described a dissipative state for scenario A.1, while for scenario A.2, despite during the tests an intermediate state was registered, the beach is also moving toward a dissipative state (slope values close to the final values for scenario A.1). For t = 1590 min, the beach is characterized as reflective.

Figure 11. Timeline for mean beach shoreface slope. The vertical black line represents the transition between scenarios A.1 and A.2. The vertical black dashed lines represent the instants when the equipment was stopped.

Wave breaking mechanisms (spilling, plunging, and oscillatory) influence the cross-shore distribution of the longshore sediment transport and may consequently influence the beach morphological evolution geometry and rates [33]. During scenario A.1, the wave breaking was only characterized as spilling breaking. Scenario A.2 registered plunging breaking during the first 300 min and spilling breaking afterward.

4. Discussion

4.1. Sediment Transport Dynamics

The observed variation of the cross-shore velocity in Figure 4 signal may be explained by undertow currents and bathymetric changes that change the wave breaking location and the relative position of the ADV on the cross-shore profile [34]. Bottom elevation variation near the ADV location may cause the ADV sample to be placed in a different vertical relative position in the velocity profile over time, capturing either lower and upper layers in the water column, leading to variations of the signal's direction and intensity of the recorded velocities. Scenario A presents an average suspended cross-shore sediment transport toward offshore. This effect can be observed in Figure 12, where the grid near the wave generator system, destined for emptying the basin, was initially set about 3 m away from the cross-shore profile limit at the beginning of the tests, while at the end, the cross-shore profile offshore limit was placed right before that same grid.

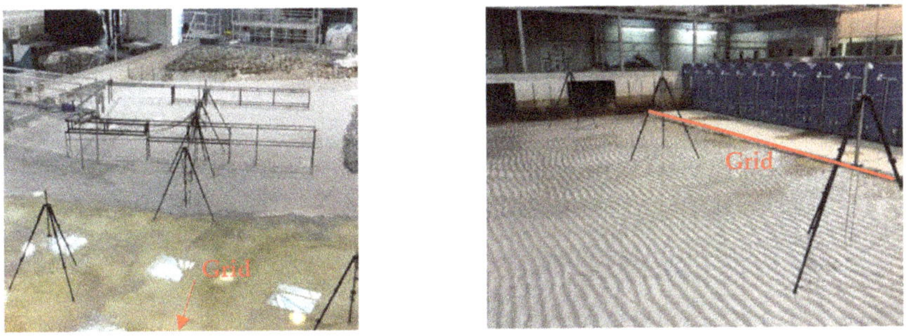

(**a**) (**b**)

Figure 12. Offshore beach limits for scenario A: (**a**) Initial offshore profile limit (A.1); (**b**) Final offshore profile limit (A.1).

Recirculation of water inside the basin might be the cause for the presence of a longshore current toward the updrift limit of the tank, as observed at some instants. This might be due to the physical boundaries that reflect the downdrift currents that are generated in the surf zone. Note that the ADV is in depths where longshore currents should be weak, and thus the recirculation currents can change the flow direction.

Suspended sediment concentration peaks in scenario A.2, registered from t = 1890 min to t = 2010 min and from t = 2610 min to t = 2790 min, are justified by the introduction of sediments through the nourishment, which also, consequently, led to a distancing of the dynamic equilibrium, or near-equilibrium, beach state. As referred to by [8], beach profiles far from the equilibrium present an increase in the beach dynamics. Just after the placement of the nourishment, there is a quick decrease of the sediment's volumes in the controlled cell, as shown in the topo-bathymetric variations (Figure 7) and the volumes' variation describing the durability of the nourishment (Figure 10). For scenario A.1, the total amounts of sediment that left the control volume that were fed at the updrift limit and collected in the downdrift limit were 2086, 1008, and 7632 kg, respectively (in 26.5 h), while the same values, respectively, for scenario A.2, were 3948, 958, and 7560 kg (in just 1200 min). The net erosion rates estimated from the total amount of sediments that left the control volume, obtained by the bathymetric surveys, are 79 kg/h for scenario A.1 and 197 kg/h for scenario A.2. Scenario A.2 registered higher erosion rates leading to a greater loss of sediment volume from the control cell than for scenario A.1. However, the volume of sediments collected at downdrift was greater for scenario A.1. This behavior may indicate that there was a movement of the nourished sediment toward the upper part of the beach profile, or to deeper areas, which were not monitored. Nevertheless, looking at the net sediment transport, estimated from the monitored coastal cell volumes (Figure 10), the initial transport rate in scenario A.2 is superior to scenario A.1.

Apart from the morphodynamic response to the nourishment, the volumes in the control cell, the position of 0.3 contour, and the beach slope tend to equilibrium values that are analogous to the ones when there is no nourishment, as depicted in Figures 9–11. Due to the initially loose sand placed as nourishment, there was a fast decrease of the shoreface slope (Figure 11) during the first moments of scenario A.2, which indicates that the profile quickly re-adapted to the new morphological conditions [8,30]. With the sudden drop in the slope, the beach developed into an intermediate state showing a decreasing beach slope tendency, developing toward a dissipative state like the beach at the end of scenario A.1. The instant when the control cell volume reaches the previous volume before the nourishment would represent the timing required for a renourishment (nourishment lifetime concept [31]). Although the monitored coastal cell does not represent the entire beach and sediments leaving the monitored area cannot be completely tracked, it was assumed that the artificial nourishment had a lifespan of about 420 min (7 h).

4.2. Laboratory Tests Constraints

Over this work, there were several difficulties in the interpretation of the obtained laboratory results. These issues were mainly due to the low spatial resolution and small monitoring area of the bathymetric surveys, the low sand compaction, and the low representativity of the velocities and concentrations results, since only one point was monitored. The number of pieces of monitoring equipment available at the laboratory is of major concern. One of the main constraints during the laboratory tests was the lack of available instruments, limiting the characterization of the beach morphodynamics in terms of space and time resolution. Thus, it was not possible to fully comprehend and analyze the profile behavior and the hydrodynamic and sediment transport processes.

The end of scenario A.1 was defined by visual observations of the beach morphology. When the beach presented no significant shoreline retreat, it was assumed to be in a near-equilibrium or dynamic equilibrium state with the updrift incoming sediments. This is confirmed by the stability registered in the longshore and cross-shore velocities. However, the uncertainty of the equilibrium state, due to the lack of spatial resolution, may lead to important misinterpretations [35,36]. To guarantee the dynamic equilibrium state of the beach, one should define a quantitative criterion to determine that same state. One example is the work of [8] that registered that with a decrease in bed velocities the beach is most likely tending to a steady state.

The performed nourishment had a direct feeder effect, since there was an onshore sediment movement partially restoring the beach width. Sediments moving offshore might have fed the bar, producing a secondary effect, which results in further away wave energy dissipation through breaking and a consequent protection of the upper profile. However, due to the physical limitations of the profiler (in its length) and since, for scenario A.2, the emerged nourishment effect (increased beach width) disappeared rapidly, it is not possible to quantify the amount of sediment feeding the bar nor the upper shoreface.

Low spatial resolution (6 by 2 m monitored coastal cell) does not allow full comprehension of the existence of bathymetric gradients. Despite the apparent shoreline orientation, parallel to the wave generator, there is little information about the bathymetric contours that are responsible for the wave refraction and breaking, which influence the longshore sediments transport. The reflective effect of the basin walls might have led to some updrift sediment movement, which may explain the shoreline orientation despite the updrift sediments added by the funnels. The physical barriers of the wave basin might have influenced the spread rate of the nourishment, which can affect the retreat rates. By representing a physical barrier, it is most likely that recorded rates were smaller than for a scenario where no lateral barrier exists.

Initial conditions for both scenarios were affected by the lack of sand compaction, which led to initially higher sediment transport rates, making it difficult to correctly determine the time interval needed for both beaches' scenarios to reach a dynamic equilibrium or near-equilibrium state and for the nourishment effect to dissipate.

It was observed that the cross-shore and longshore sediment transport rates are much higher than predicted and intended for the laboratory tests. The nourishment dimensions and geometric configuration, combined with the re-circulation inside the wave basin, might have accelerated the morphological evolution of the modeled beach. Due to the high morphological changes to the beach over time, the relative position of the instruments to the shoreline changed greatly (the instruments measured different relative profile position and depths over time). The relative distance was influenced by the retreating shoreline and the advancing offshore limit of the profile.

5. Conclusions

Two sets of laboratory tests were conducted over a movable bed physical model, using an irregular wave climate. The first stage aimed to obtain a beach at a near dynamic equilibrium state, while the second stage introduced a nourishment containing 9080 kg of sand (induced perturbance) to alter the morphological, hydrodynamic, and sediment transport processes over the previously obtained equilibrium beach. Placing an artificial nourishment in the modeled beach led to an increase in the beach width and variations in its dynamics. A total value of 1.43 m^3 of sediments left the monitored cell over the full duration of the test (scenarios A.1 and A.2). Although some areas of the cell experienced accretion, the volume balance is negative at all instants with exception of the instant when the artificial nourishment was introduced. The final beach for scenario A.2 evolved toward a similar volume and configuration to the beach at the end of scenario A.1. Although the size of the monitored coastal cell does not allow the observation of the accurate movement of the sediments, it was possible to determine the volume variation of the coastal cell and the nourishment efficiency: the lifetime of the artificial nourishment was about 420 min.

In both scenarios, the cross-shore profiles have retreated, with scenario A.1 presenting a higher average retreat rate (0.13 cm/min) than scenario A.2 (0.11 cm/min). Despite the initial non-compacted sediments in the artificial nourishments, the nourishment reduced the shoreline retreat rate by about 15.4% when compared to the conditions presented in scenario A.1. The artificial nourishment presence also led to an increase in the hydrodynamic activity: the highest longshore and cross-shore velocity components and suspended sediment concentrations occurred during the first 540 min of scenario A.2. These results and the fast morphodynamic evolution of the nourishment and beach go according to [8] and [30], noting that when introducing a perturbation on the beach, such as an artificial nourishment, there is an increase in the beach dynamics.

The suspended sediment transport values computed from the collocated ADV and OBS are much lower than the total sediment transport estimated from volume balance, which suggests that the bed load is the dominant mode of transport. However, it is difficult to characterize sediment transport with only one point monitoring velocities and suspended sediment concentrations over the full beach extension, whose relative position in the surf zone changes due to cross-shore profile evolution (when located closer to the surfzone, the equipment would register higher velocities and suspended sediment concentrations). Due to the spatial variation of the sediment transport characteristics (e.g., cross-shore distribution of the longshore transport), more points all over the beach, with an ADV coupled with an OBS, should be monitored to correctly characterize sediment transport. Moreover, additional monitoring points would decrease the difficulties in adequately locating the measuring instruments due to the bed evolution and morphological evolution along the tests. To mitigate these issues, a similar system to the one described by [37] could be adopted.

To avoid some problems during laboratory tests with sand movable bed physical models, it is suggested that after the construction of the initial beach profile, before experiments start, the wave generators should run over a certain amount of time. This should be done to assess if the initial profile geometry is close to the equilibrium and to compact the sand to avoid its liquefaction. During this interval, it should be assessed if the profile/bathymetry is having an impact on the wave height (e.g., through reflection) and corrected if necessary. The instruments' location should be defined with precision after the sand is compacted, despite the difficulties in placing a support for a similar profiler measuring device.

The difficulties in predicting and anticipating the morphological evolution of the modeled beach create constraints in correctly locating the measuring equipment and obtaining accurate data. Future definition of laboratory procedures should account for the hereby described conditions to improve the model's accuracy and performance.

Author Contributions: Conceptualization, A.G. and C.C.; methodology, A.G.; software, A.G.; validation, A.G., C.C., F.V.-G., and P.A.S.; formal analysis, A.G.; investigation, A.G.; resources, C.C., F.V.-G., and P.A.S.; data curation, A.G. and C.C.; writing—original draft preparation, A.G.; writing—review and editing, C.C., F.V.-G., and P.A.S.; visualization, A.G.; supervision, C.C., F.V.-G., and P.A.S.; project administration, A.G. and C.C.; funding acquisition, A.G. and C.C. All authors have read and agreed to the published version of the manuscript.

Funding: This research received no external funding.

Institutional Review Board Statement: Not applicable.

Informed Consent Statement: Not applicable.

Data Availability Statement: Not applicable.

Acknowledgments: A.G. is supported by the Foundation for Science and Technology PhD Grant SFRH/BD/103694/2014. Thanks to the Portuguese Foundation for Science and Technology (FCT) through the financial support to CESAM (national UID/AMB/ 50017/2013 and FEDER funds, within the PT2020 Partnership Agreement and Compete 2020). This work was financially supported by the project "Sandtrack-Beach nourishment: An integrated methodology for coastal management support", POCI-01-0145-FEDER-031779, funded by FEDER, through COMPETE2020-POCI, and by national funds (OE), through FCT/MCTES.

Conflicts of Interest: The authors declare no conflict of interest.

References

1. Capobianco, M.; Hanson, H.; Larson, M.; Steetzel, H.; Stive, M.J.F.; Chatelus, Y.; Karambas, T. Nourishment design and evaluation: Applicability of model concepts. *Coast. Eng.* **2002**, *47*, 113–135. [CrossRef]
2. Hanson, H.; Brampton, A.; Capobianco, M.; Dette, H.H.; Hamm, L.; Laustrup, C.; Spanhoff, R. Beach nourishment projects, practices, and objectives—A European overview. *Coast. Eng.* **2002**, *47*, 81–111. [CrossRef]
3. Bottin, R.R.; Earickson, J.A. Buhne Point, Humboldt Bay, California, Design for the Prevention of Shoreline Erosion, Hydraulic and Numerical Model Investigation. In *Technical Report CERC84-5*; Coastal Engineering Research Center, U.S. Army Engineer Waterways Experiment Station: Vicksburg, MS, USA, 1984; p. 340.
4. Grasso, F.; Michallet, H.; Barthélemy, E.; Certain, R. Physical modelling of intermediate cross-shore beach morphology: Transients and equilibrium states. *J. Geophys. Res.* **2009**, *114*, 15.
5. Grasso, F.; Michallet, H.; Barthélemy, E. Experimental simulation of shoreface nourishments under storm events: A morphological, hydrodynamic, and sediment grain size analysis. *Coast. Eng.* **2011**, *58*, 184–193. [CrossRef]
6. Dette, H.H.; Larson MMurphy, J.; Newe, J.; Peters, K.; Reniers, A.; Steetzel, H. Application of prototype flume tests for beach nourishment assessment. *Coast. Eng.* **2002**, *47*, 137–177. [CrossRef]
7. Grasso, F.; Michallet, H.; Barthélemy, E. Sediment transport associated with morphological beach changes forced by irregular asymmetric, skewed waves. *J. Geophys. Res.* **2011**, *116*, 12. [CrossRef]
8. Michallet, H.; Castelle, B.; Barthélemy, E.; Berni, C.; Bonneton, P. Physical modeling of three-dimensional intermediate beach morphodynamics. *J. Geophys. Res. Earth Surf.* **2013**, *118*, 1045–1059. [CrossRef]
9. Headland, J.; Smith, W.G.; Kotulak, P.; Alfageme, S. Coastal Protection Methods. In *Handbook of Coastal Engineering*; J. B. Herbich & McGraw-Hill: London, UK, 1999; pp. 1–66.
10. Tondello, M.; Ruol, P.; Sclavo, M.; Capobianco, M. Model tests for evaluating beach nourishment performance. In Proceedings of the 26th International Conference on Coastal Engineering, Copenhagen, Denmark, 22–24 June 1998.
11. Van Duin, M.; Wiersma, N.; Walstra, D.; Van Rijnand, L.; Stive, M. Nourishing the shoreface: Observations and hindcasting of the Egmond case, The Netherlands. *Coast. Eng.* **2004**, *51*, 813–837. [CrossRef]
12. Grunnet, N.; Ruessink, B.; Walstra, D. The influence of tides, wind and waves on the redistribution of nourished sediment at Terschelling, The Netherlands. *Coast. Eng.* **2005**, *52*, 617–631. [CrossRef]
13. USACE. *V-4 Beach Fill Design*; USACE: Vicksburg, MS, USA, 2008; Volume 1100, p. 113.
14. van Heuvel, T. *Sand Nourishment: A Flexible and Resilient, Adaptive Coastal Defence Measure*; Climate of Coastal Cooperation: Amsterdam, The Netherlands, 2011; 7p.
15. González, M.; Medina, R.; Losada, M. On the design of beach nourishment projects using static equilibrium concepts: Application to the Spanish coast. *Coast. Eng.* **2010**, *57*, 227–240. [CrossRef]
16. Verhagen, H.J. Method for Artificial Beach Nourishment. In Proceedings of the 23rd Coastal Engineering Proceedings, Venice, Italy, 4–9 October 1992.
17. Hughes, S.A. *Physical Models and Laboratory Techniques in Coastal Engineering*; World Scientific: Singapore, 1993; 570p.
18. Bayram, A.; Larson, M.; Miller, H.C.; Kraus, N.C. Cross-shore Distribution of Longshore Sediment Transport: Comparison between Predictive Formulas and Field Measurements. *Coast. Eng.* **2001**, *44*, 79–99. [CrossRef]
19. Wang, P.; Ebersole, B.A.; Smith, E.R. *Longshore Sand—Initial Results from Large-Scale Sediment Transport Facility, No. ERDC/CHL-CHETN-II-46*; Engineering and Development Center: Vicksburg, MS, USA, 2002; 12p.
20. Silva, R. Avaliação Experimental e Numérica de Parâmetros Associados a Modelos de Evolução Da linha de Costa. Ph.D. Thesis, Civil Engineering Department, University of Porto, Porto, Portugal, 2010.
21. Guimarães, A.; Lima, M.; Coelho, C.; Silva, R.; Veloso-Gomes, F. Groin impacts on updrift morphology: Physical and numerical study. *Coast. Eng.* **2016**, *109*, 63–75. [CrossRef]
22. Vera-Cruz, D. Artificial Nourishment of Copacabana Beach. In Proceedings of the 13th Intl. Conference on Coastal Engineering, Vancouver, BC, Canada, 10–14 October 1972.

23. Johnson, B.D.; Smith, E.R. Material placement in the nearshore: Laboratory and numerical model investigation. In Proceedings of the 33rd Conference on Coastal Engineering, Santander, Spain, 1–6 July 2012.
24. Smith, E.R.; Mohr, M.C.; Chader, S.A. Laboratory experiments on beach change due to nearshore mound placement. *Coast. Eng.* **2017**, *121*, 119–128. [CrossRef]
25. Dean, R.G. Equilibirum Beach Profiles: Characteristics and Applications. *J. Coast. Res.* **1991**, *7*, 53–84.
26. Roberts, T.M.; Wang, P.; Kraus, N.C. *Limits of beach and dune erosion in Response to Wave Runup Elucidated from SUPERTANK*; Coastal Sediments '07; ASCE: Reston, VA, USA, 2007.
27. Sontek. *Sontek ADVField Acoustic Doppler Velocimeter-Tecnhical Documentation*; Sontek: San Diego, CA, USA, 2001.
28. Cea, L.; Puertas, J.; Pena, L. Velocity measurements on highly turbulent free surface flow using ADV. *Exp. Fluids* **2007**, *42*, 333–348. [CrossRef]
29. Carrilho, A. Morfodinâmica e Transporte Sedimentar Longitudinal na Praia de Mira. Master's Thesis, Civil Engineering Department, University of Aveiro, Aveiro, Portugal, 2013.
30. Marino-Tapia, I.; Russell, P.; O'Hare, T.; Davidson, M.; Huntley, D. Cross-shore sediment transport on natural beaches and its relation to sandbar migration patterns: 1. Field observations and derivation of a transport parameterization. *J. Geophysical. Res.* **2007**, *112*, C03001. [CrossRef]
31. Klein, A.H.F.; Miot da Silva, G.; Ferreira, Ó.; Dias, J.A. Beach sediment distribution for a headland bay coast. *J. Coast. Res.* **2004**, *SI*, 285–293.
32. Bosboom, J.; Stive, M. *Coastal Dynamics I*; Lecture Notes CIE4305 Division of Hydraulic and Geotechnical Engineering Delft; Hydraulic Engineering Section, Delft Academic Press: Delft, The Netherlands, 2015.
33. Wang, P.; Ebersole, B.A.; Smith, E.R. Beach-Profile Evolution under Spilling and Plunging Breakers. *J. Waterw. Port Coast. Ocean Eng.* **2003**, *129*, 41–46. [CrossRef]
34. Aagaard, T.; Greenwodd, B.; Hughes, M. Sediment transport on dissipative, intermediate and reflective beaches. *Earth-Sci. Rev.* **2013**, *124*, 32–50. [CrossRef]
35. Crosato, A.; Desta, F.B.; Cornelisse, J.; Schuurman, F.; Uijttewaal, W.S. Experimental and numerical findings on the long-term evolution of migrating alternate bars in alluvial channels. *Water Resour. Res.* **2012**, *48*, 14. [CrossRef]
36. Crosato, A. *Issues in Laboratory Experiments of River Morphodynamics*; RCEM: Padova, Italy, 2017.
37. Molfetta, M.G.; Bruno, M.F.; Pratola, L.; Rinaldi, A.; Morea, A.; Preziosa, G.; Pasquali, D.; Di Risio, M.; Mossa, M. A Sterescopic System to Measure Water Waves in Laboratories. *Remote Sens.* **2020**, *12*, 2288. [CrossRef]

Journal of
Marine Science and Engineering

MDPI

Review

Does Sand Beach Nourishment Enhance the Dispersion of Non-Indigenous Species?—The Case of the Common Moon Crab, *Matuta victor* (Fabricius, 1781), in the Southeastern Mediterranean

Dov Zviely [1,*], Dror Zurel [2], Dor Edelist [3], Menashe Bitan [3] and Ehud Spanier [3]

1 Faculty of Marine Sciences, Ruppin Academic Center, Emek-Hefer 4025000, Israel
2 Israel Ministry of Environmental Protection, Marine and Coastal Protection Division, 15-a Pal-Yam. St., Haifa 3309519, Israel; DrorZ@sviva.gov.il
3 Department of Maritime Civilizations and the Leon Recanati Institute for Maritime Studies, The Leon H. Charney School for Marine Sciences, University of Haifa, Mount Carmel, Haifa 3498838, Israel; blackreefs@gmail.com (D.E.); bitanmen@netvision.net.il (M.B.); spanier@research.haifa.ac.il (E.S.)
* Correspondence: dovz@ruppin.ac.il; Tel.: +972-52-5805-758

Abstract: Sand beach nourishment (BN) is one of the commonest "soft solutions" for shore protection and restoration. Yet it may have ecological consequences. Can this practice enhance the introduction and dispersal of non-indigenous species (NIS)? There has been little research on the impacts of nourishment on NIS, especially in the southeastern Mediterranean, a region considered most affected by invading biota. However, so far only one study referred to the possible interaction between BN and the success of invading species. It reports increasing numbers and densities of the aggressive, omnivorous Indo-Pacific moon crab, *Matuta victor* (Fabricius, 1781) in Haifa Bay (northern Israel) between 2011 and 2017. This research suggests a possible role of anthropogenic disturbance in the outbreak of *M. victor* and blames the Israel Ministry of Environmental Protection for authorizing a (rather small scale) BN in Haifa Bay in 2011 as an alleged cause for this outbreak. Circumstantial indirect evidence is not sufficient to establish the role of nourishment in promoting the establishment and dispersal of NIS. There are plenty of examples of successful settlement and rapid and large-scale distribution of NIS (including another member of the genus *Matuta*), especially in the eastern Mediterranean, without any BN in the region. Furthermore, the location where the *M. victor* specimens were sampled was exposed to more prevailing and frequent anthropogenic marine stressors than BN, such as eutrophication, pollution, fishing activities and particularly port construction. To firmly establish an assumed role of nourishment in the invasion of NIS, assessments must be based on solid and orderly planned scientific research to be designed well before the beginning of any BN. It is suggested that direct communication between environmental regulators and scientists is crucial for improving both scientific research and environmental management policies.

Keywords: invasive species; Lessepsian migration; coastal processes; dredging; Levant; Haifa Bay

Citation: Zviely, D.; Zurel, D.; Edelist, D.; Bitan, M.; Spanier, E. Does Sand Beach Nourishment Enhance the Dispersion of Non-Indigenous Species?—The Case of the Common Moon Crab, *Matuta victor* (Fabricius, 1781), in the Southeastern Mediterranean. *J. Mar. Sci. Eng.* **2021**, *9*, 911. https://doi.org/10.3390/jmse9080911

Academic Editors: Carlos Daniel Borges Coelho and Rodger Tomlinson

Received: 23 June 2021
Accepted: 18 August 2021
Published: 23 August 2021

Publisher's Note: MDPI stays neutral with regard to jurisdictional claims in published maps and institutional affiliations.

1. Introduction

Beach nourishment (BN), mainly by sand, is one of the commonest "soft solutions" for beach restoration and shore protection [1–4] and is considered more environmentally acceptable than coastal defenses such as seawalls, revetments and detached breakwaters [5]. It is widely applied as a soft coastal measure because of its reduced ecological impact relative to hard coastal protection [6]. However, the nourished sand may be eroded after a relatively short period, and this costly practice must be repeated periodically [7,8]. In recent years, coastal ecosystems have been severely threatened by climate change due to changes in sea level, storms and wave regimes, flooding, altered sediment budgets and the loss of coastal habitat [9,10]. Because sandy beaches form the single largest coastal ecosystem on Earth, covering 70% of all continental margins [11], these threats are global. Sandy beaches have a multitude of ecosystem functions as they are important habitat for a variety of biota and are concurrently of enormous economic, social and cultural importance to humans [11–14]. Thus, BN may affect natural ecosystems as well as human societies.

BN has further environmental effects in the imported site (i.e., borrow area) as well as on the nourished beach [15–19]. The ecological consequences of the nourishment on coastal biota may be short- or long-term. The environmental impacts may lead to sedimentation and turbidity that affect light penetration and filtering organisms. It may cause burial of organisms that reside in the nourished area, and the effects of heavy equipment used in the nourishment operation may injure, kill or affect the behavior and physiology of the native biota. BN can change the nature of the local habitat (e.g., altering the grain size and type, or change hard substrate to a soft one). Changing the sediment composition may alter the types of organisms that inhabit the nourished beach [8,20–24]. BN may also displace native biota, but, does this activity enhance the introduction and dispersal of non-indigenous marine species? The goal of the present review is to reveal if there is solid evidence for this assumption.

Invasive species are considered one of the four major threats to the world oceans together with terrestrial and marine pollution sources, overutilization of marine resources and physical changes and destruction of coastal and marine habitats [25]. Shipping, via ballast water, is considered the most important marine invasion pathway, but other anthropogenic factors, such as man-made canals between previously separated biogeographic marine provinces, transport of marine species in aquaculture, research projects and aquarium trade, fishing activities and global warming also have important roles in invasion processes [26].

2. Research on Beach Nourishment and Non-Indigenous Species

Localized negative impacts of BN on many coastal species of fauna and flora are well documented [26]. Burial, habitat alteration (e.g., changing wet habitats to dry ones and vice versa) and habitat reduction may have harmful implications for many coastal populations of plants and invertebrates in the short term. They can be impacted not only by burial itself but also by sand compaction, which affects gas, nutrient and water availability in interstitial spaces [13,27].

Yet there has been little research on the consequences of BN on the impacts of invasive plants and animal species [13]. Non-indigenous species (NIS), also known as non-native, alien, or exotic organisms, are species that have been introduced outside of their natural previous or present range by human activities. They might survive and subsequently reproduce in the new environment. If these species become established and their spread threatens the local biodiversity or causes economic damage, they are known as invasive species ([28] and references therein).

Lessepsian migration, the influx of Indo-Pacific species into the Mediterranean, is the largest marine bioinvasion on the planet (e.g., [29,30]). The Suez Canal (Figure 1) is the primary vector for transfer of Indo-Pacific biota into the Mediterranean Sea, and climate change and warming of Mediterranean waters are known as the main protagonists exacerbating this process [31]. In the last decade, the species richness of marine organisms in the Mediterranean Sea has been reported to have reached $\sim$17,000 taxa, among which some 820 can be considered NIS [32–34]. The southeastern Mediterranean is considered the region most affected by invading biota [35], especially Indo-Pacific species termed Lessepsian migrants. Israel has by far the highest number and percentage of earliest records of NIS in the Mediterranean Sea [30], of which, over 70% were subsequently recorded in other Mediterranean countries [33–36]. Thus, Israel may indeed be an ideal location to study the effect of BN on establishment and dispersal of NIS.

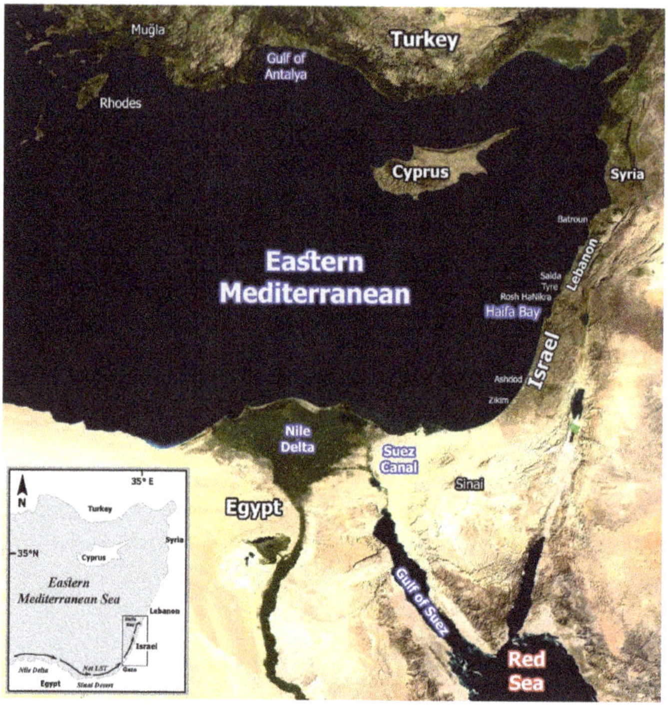

Figure 1. Eastern Mediterranean. The Nile littoral cell net longshore sand transport (LST) direction (black arrows) (bottom inset). Background: part of "Blue Marble: Land Surface, Shallow Water, and Shaded Topography". NASA Goddard Space Flight Center Image by Reto Stöckli, Robert Simmon and MODIS Groups; https://visibleearth.nasa.gov/images/57752/blue-marble-land-surface-shallow-water-and-shaded-topography (accessed on 11 February 2002).

To date, the only study that refers to an alleged effect of BN on species invasion in the southeastern Mediterranean is that of Innocenti et al. [37], which referred to nourishment activities carried out in Haifa Bay, northern Mediterranean coast of Israel (Figure 1). This study, however, lacks any before-after-control-impact (BACI) methodology.

3. Study Area

3.1. Haifa Bay Physical Setting

The Mediterranean coastline of Israel extends about 195 km from Zikim near the border with Gaza Strip in the south, to Rosh HaNikra near the Lebanese border in the north (Figure 1). It is generally a smooth coastline open to the west that gradually changes in orientation from northeast to almost north, except for Haifa Bay, the Mount Carmel headland and a few small rocky promontories [38].

From a sedimentological perspective, the Israeli coast and its inner shelf (i.e., from the shore to maximum water depth of about 30 m), can be divided into two main provinces. The Southern Province, stretching 175 km from Zikim to the Akko promontory (northern Haifa Bay) [39] (Figure 2), is considered the northern flank of the Nile littoral (sedimentary) cell [40,41]. This coastal compartment is mainly composed of Nile-derived quartz fine sand, transported from the Nile Delta eastward to the northern Sinai [41–46], then northeastward to Gaza Strip and the Israeli coasts by longshore currents [47–52] (Figure 1: bottom inset). These currents are generated by the radiation stress of breaking waves and shear stress of local winds [53,54]. Wave-induced and wind-induced longshore currents occur in both directions. However, the long-term net longshore sand transport has drifted northward up to Haifa Bay [50–52], the northernmost final depositional sink of the Nile littoral cell, in the last 7900–8500 years [55,56]. The Northern Province (i.e., the western Galilee coast) however, is a small, isolated and rocky littoral cell, partly covered with locally carbonated coarse sand [40,47,57,58].

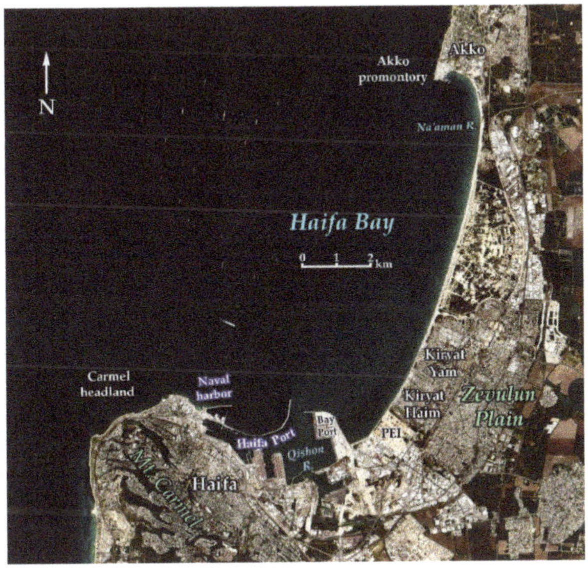

Figure 2. Haifa Bay area. Background: part of Sentinel-2 L2A, True color imagery,365XB https://www.sentinel-hub.com (accessed on 15 April 2021).

Haifa Bay is bordered by the Carmel headland to the south, the Akko promontory to the north, and the Zevulun Plain to the east. Two rivers traverse the Zevulun Plain: the Na'aman River in the north and the Qishon River in the south (Figure 2). Both transport silt and clay to the bay's offshore region [59]. The bay's coastline is crescent-shaped and includes a large port area, about 6 km long on its southern part, and about 12 km of sandy beaches along its eastern part. Haifa Bay's floor (0–30 m deep) consists of three sub-areas [52]: (1) a smooth seabed that extends from the shore to a maximum water depth of 10 to 20 m along the bay's coasts, and is mainly covered by Nile-derived quartz fine sand [56,60]; (2) a submerged calcareous sandstone (locally termed Kurkar) ridge area [61,62], which covers about two-thirds of the bay's floor at a water depth of 10 to 25 m; (3) a smooth seabed that is located in the bay's deep area (25–30 m deep) and covered by silty sand.

3.2. Haifa Port Main Breakwater Morphological Impact

During the last 4000 years, the sea level along the Mediterranean coast of Israel was relatively stable, and Nile-derived sand was continuously transported to Haifa Bay by longshore currents. As a result, the Zevulun Plain dried up and the bay's shoreline shifted 3 to 4.8 km to its present location [55,56]. This long-term trend ceased in 1932 after completion of the Haifa Port main breakwater, which became a large trap for sand transported to the bay.

Between 1978 and 1980, a container quay (Eastern Quay) was built in the eastern part of Haifa Port (Figure 3). To protect this terminal, the main breakwater was extended by 600 m to a total length of 2810 m, and the breakwater's head was located at a water depth of 13.5 m.

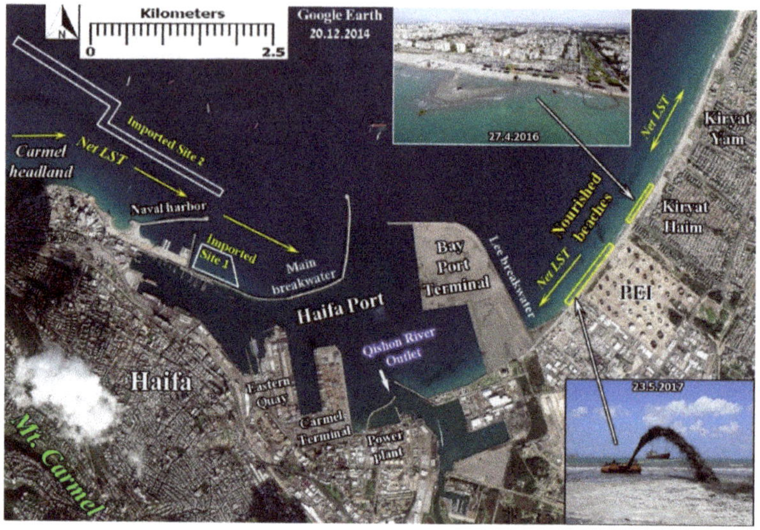

Figure 3. South Haifa Bay sand beach nourishment projects (May 2016–June 2017). Aerial photograph of Kiryat Haim bathing beach (Yehudit Naot Beach) during the first nourishment operation (top inset). Nourishment rainbowing operation via discharge pipe south of Petroleum and Energy Infrastructure Ltd. (PEI) (23 May 2017) (bottom inset). Background: Google Earth image, 20 December 2014 (after [39]).

The total amount of sand trapped along the main breakwater between 1929 and 2004 has been estimated at 5 million m^3 or an average of 66,000 m^3/year. Only a small amount of sand (8000–10,000 m^3/year) bypassed the breakwater's head and drifted to the eastern coast [52,55,63]. Analysis of shoreline migration between 1928 and 2006 shows that most of the bay's coasts were in a steady state, with seasonal fluctuations of less than about $\pm$20 m and slight erosion [63].

During the years 2005–2008, another container quay (Carmel Terminal) was built in Haifa Port, and the last container quay (Bayport Terminal) was built in the last 5 years. This huge terminal is designed to handle Class EEE container vessels [39]. To protect the Bayport Terminal, the Haifa Port main breakwater once again was extended to 3682 m, and its head was located at a water depth of 20 m.

3.3. Haifa Bay Sand Beach Nourishment Activity

In December 2010, an extremely strong storm with significant wave heights (Hs) above 7 m occurred along the coast of Israel and caused severe erosion in many places, including in Haifa Bay [39]. To restore part of the eroded bathing beaches in southeast Haifa Bay, a small amount of sand (50,000 m^3) was dredged along the Haifa port main breakwater and used to nourish the Kiryat Haim beaches via a discharge pipe at a water depth of 3 m (Figure 3). The limited nourishment was inefficient and the Kiryat Haim beaches remained eroded. In February 2015, another extreme storm with Hs > 7 m hit the Israeli coast, and for the first time a severe erosion developed along the southern part of the PEI fence coast (Figure 3). This fence starts 730 m northeastwards of the Bayport lee breakwater and runs 800 m along the southeastern coast of Haifa Bay.

In March 2015, construction of the Bayport lee breakwater began. From then on, the erosion along the PEI fence coast continued to develop further north, including along Kiryat Haim bathing beaches, which extend for about 1 km north of the fence. To cope with the coastal erosion and prevent further damage to the PEI fence and the bathing beach infrastructure, several sand BN activities were carried out during 2016 and 2017 [39]. The first nourishment was carried out between April and May 2016, when a sand volume of

70,000 m^3 was dredged along the port's main breakwater (Figure 3: Imported Site 1) and deposited between the coastline and water depth of 2 m, in two sites: (1) a coastal section of 450 m along the Kiryat Haim bathing beaches (45,000 m^3); and (2) a coastal section of 250 m along the southern part of the PEI fence (25,000 m^3). The second nourishment was carried out in October 2016, when a sand volume of 100,000 m^3 was dredged north of the new Naval harbor main breakwater (Figure 3: Imported Site 2) and nourished a coastal section of 500 m along the PEI fence. Unfortunately, at the beginning of December 2016, a storm with Hs = 5.2 m eroded the nourished beach. The third nourishment was carried out between May and June 2017, when a sand volume of 185,000 m^3 was also dredged north of the naval harbor and used to nourish the same coastal section along the PEI fence again. When the nourishment along the PEI fence was completed, the coast had been widened by 40 m on average. However, by mid-January 2018 the beach was eroded again by storm waves [39]. The fourth nourishment was carried out along the Kiryat Haim bathing beaches, in May 2019. This operation was conducted in two stages. First, a sand volume of 28,000 m^3 was excavated from the beach near the Bayport lee breakwater and dumped directly on the shores, and a month later, a sand volume of 45,000 m^3 was dredged north of the Naval harbor and used to nourish the Kiryat Haim beaches via a discharge pipe.

4. The Erythraean Moon Crab *Matuta victor* and Beach Nourishment

The second-most abundant group of NIS in the Mediterranean Sea are crustaceans, among which more than 90 species belong to the order Decapoda [64]. Nearly 400 decapods have been recorded in the Mediterranean [65], of which about 24% are NIS [64]. Most recent records of decapod NIS in the Mediterranean include species of Indo-Pacific and Red Sea origin such as the moon crab, *Matuta victor* (Fabricius, 1781) from Turkey and Greece (Rhodes Island) [66,67] (Figure 1). *Matuta* species (moon crabs) belong to superfamily Calappoidae, which inhabit sandy beach areas [68]. Matutid crabs can bury themselves very quickly into sand [69]. *M. victor* is an omnivorous predator and a voracious scavenger, exhibiting intraspecific feeding competition and aggressive behavior [37]. It preys mainly on crustaceans and mollusks; smaller individuals prey on small, soft-shelled species while larger individuals eat slow-moving invertebrates such as anomurans, bivalves and gastropods [64]. It is widely distributed in the Indo-Pacific region, including the Red Sea, Southeast Asia to Fiji and New Hebrides [70], where it is caught by local communities in nets, by hand or beach seines. The establishment success of *M. victor* is further assisted by its reproductive plasticity and the production of numerous eggs [68].

Innocenti et al. [37] studied the distribution, feeding habits and behavior of the invasive common moon crab *M. victor*, mainly in Haifa Bay. Two separate adult specimens of this species were first recorded in Haifa Bay in 2012 [71]. One male was caught by trammel net in 10 m water depth, about 1 km north of the Qishon River outlet (Figure 2), on 31 October 2012, and one female *M. victor* was caught by seine net in 5 m water depth off the coast of Kiryat Yam (Figures 2 and 3), on 20 November 2012. This initial report was followed by reports of several dozens of *M. victor* in Tyre and Saida, (Figure 1: South Lebanon) in 2014 [72], a record of a single individual off Antalya (Turkey) in August 2015 [73] and later multiple observations in Israel, including on Ashdod bathing beach [38], southern Mediterranean coast of Israel (Figure 1), in June 2021 (Figure 4). It is interesting to note that this beach, which is located between the Ashdod Marina and Ashdod Port, has never been a sand nourishment site [39].

In the autumn of 2013, Innocenti et al. [37] observed the largest densities of *M. victor* ever recorded in Israel (1–10 crabs/10 m^2) in water shallower than 30 cm near Kiryat Yam (Figure 2). Their surveys of the Haifa Bay coast (Figure 1) in 2013 and 2015 revealed the presence of numerous specimens at its northern end, near the Na'aman River estuary (Figure 2) (up to 27 specimens/m^2 in 2013). By 2016 they recorded *M. victor* along the Israeli coast from Ashdod to Akko (Figure 1). They also reported that in May 2017, *M. victor* densities in Kiryat Yam reached 2 specimens/m^2. There was no description of the survey methodology used by Innocenti et al. [37], but they did mention that bait was used to attract the *M. victor* crabs, which might explain the sudden high densities observed.

Figure 4. A specimen of the common moon crab *Matuta victor* at Ashdod beach, southern Israel, (photographed by Shani Tubul, 9 June 2021).

In view of these findings, Innocenti et al. [37] suggested a possible role of anthropogenic disturbance in the outbreak of *M. victor*. They blamed the Israel Ministry of Environmental Protection (MOEP) for authorizing the November 2011 "massive" BN scheme depositing 50,000 m^3 of sand along a 2 km stretch of Kiryat Haim and Kiryat Yam beaches (Figures 2 and 3), at a depth of 3 m. They also accused MOEP of continuous operations "ever since" in 2016. They suggested that the most obvious and direct effect of BN was the obliteration of the resident benthic fauna (which they previously described as "almost barren"). They pointed out that anthropogenic disturbances such as eutrophication, pollution and the physical disturbance had been considered risk factors, reducing native diversity and increasing invader dominance. Innocenti et al. [37] accused MOEP of allegedly not considering the long-term consequences of providing suitable habitat for Erythraean NIS when authorizing BN in the polluted and eutrophic Haifa Bay. They stated: "It seems likely that the rapid increase in the population of *M. victor* in the newly "disturbed" Kiryat Yam beach next to the highly eutrophic Qishon River estuary, and soon after, next to the similarly polluted Na'aman River estuary, established the beachhead for its subsequent spread". They concluded by suggesting that to reduce the risk of the spread of NIS already present in the Mediterranean Sea and the establishment of new introductions, it was essential to lessen coastal eutrophication, pollution and physical disturbance.

5. The Possible Implications of the Appearance of *M. victor* in Haifa Bay

Innocenti et al. [37] point at eutrophication, pollution and BN as the main stressors affecting the biota of Haifa Bay. Sadly, this is not correct. The site of the *M. victor* study is located less than 2 km from a major marine stressor (i.e., Haifa Port—one of Israel's two busiest international seaports) recognized as a hotspot for NIS establishment [26]. Additionally, since 2005 the port has undergone several changes increased turbidity levels in the bay, including reclamation and construction of two new container terminals (i.e., the Carmel Port and the Bayport projects) and dredging and deepening of the entrance canal and the inner port basin. Moreover, the limited amount of sand (50,000 m^3) deposited in November 2011 and termed by Innocenti et al. [37] as "massive beach nourishment" was less than the natural annual average volume of sand deposited along the bay's coast before the construction of the port and after its completion [52,55]. It should also be mentioned that before 2011, no BN was carried out along the Israeli coast, and due to technical issues at that time, the sand was not deposited directly on the beach as planned, but at a depth of 3 m, some 200 m offshore. According to Innocenti et al. [37], a 2 km stretch was completely covered by deposited sand in 2011 and this had been repeated again and again ever since. The sand of the 2011 BN was deposited in four different point locations, 500 m apart, creating local sand spots at 3 m water depth, which were later dispersed by waves. There was no further deposition of sand at 3 m water depth in the bay. The sand of the 2016 BN was deposited directly on the beach [37]. This information appears in technical documents at the MOEP and is open and available to the public upon request.

The conclusions of Innocenti et al. [39] regarding the increase in *M. victor* sightings imply that these authors may not be fully familiar with the process of BN and the habitat it creates. In their paper, these authors provided references that showed that BN obliterated the benthic fauna at the nourished site, which is to be expected when a benthic submerged habitat becomes a dry beach. Yet, they failed to acknowledge that BN can also provide habitat for resident biota such as the native sand crab, the tufted ghost crab, *Ocypode cursor*, or even endangered nesting sea turtles (e.g., [74]). Both species are protected in Israel.

Beach nourishment activities are aimed at enlarging an eroded stretch of beach, and sand is thus deposited from the swash zone towards offshore. The nourished area does not remain a part of the subtidal; the main nourished area becomes a dry land and only the edge of the nourished area is submerged, becoming the new tidal zone, into which some motile fauna can migrate from the previous shoreline that is now covered. We do not dispute the claim that sand deposition may promote marine invasions, but in the case of *M. victor*, solid evidence for this assumption was not provided.

Additionally, the bay intertidal community is under constant fishing pressures, with clear effects on the shallow intertidal crab population. In September 2015, MOEP's marine pollution inspectors found a large number of dead crabs on the Akko South Beach, located in north Haifa Bay (Figure 5). According to local fishermen, intertidal crabs are often caught as bycatch by local fishing boats and illegal beach seine operations which are active in the bay. Due to lack of demand, they are discarded dead back into the sea or on the beach. Dredging and depletion from fishing activities in the intertidal habitat are much more likely candidates facilitating *M. victor*'s invasion into the bay than a single BN project.

More than 50 crustacean species of Indo-Pacific origin have colonized the shallow Levantine continental shelf before *M. victor*, displacing many indigenous species from both soft and hard sediments [31]. Tracing the establishment of a specific species to a single localized BN in a highly bio-invaded littoral that has been undergoing massive sedimentary changes for almost a century as well as repeated exploitation, eutrophication and pollution, mandates more rigorous supporting research. For example, a genetic study can compare specimens from Israel with those from Lebanon [71] and nearly 1000 km away in Turkey [73,75], in areas where no BN projects have been carried out. A genetic study may tell us whether a bottleneck effect is observed (i.e., a single invasion event has occurred) or if there were multiple introductions, in which case BN is much less likely to have been a protagonist. Such a genetic study was done, for example, in the marine gastropod *Strombus (Conomurex) persicus* (Swainson, 1821) [75], where genetic analyses were performed in specimens from the Persian Gulf and the Mediterranean coasts of Israel and Turkey.

Figure 5. Large amount of discarded crabs, apparently a bycatch of intertidal fishing, Akko beach, north Haifa Bay (photographed by Gidi Bettelheim, 18 September 2015).

Finally, Innocenti et al. [37] suggested that the Israeli MOEP neglected to consider the long-term consequences of providing suitable habitat for Indo-Pacific NIS when authorizing BN in the polluted and eutrophic Haifa Bay. There are several considerations that MOEP takes into account when authorizing BN, including the social impact of the loss of beaches to the human population of Haifa Bay and the possibility of marine pollution due to damage to existing infrastructure located at the back of the eroding beaches, such as the oil terminal. However, the issue of invasive species was not overlooked during the 2016 nourishment activities and was a major component in the infauna monitoring programs and regulations regarding BN projects. For example, the use of imported sand from Turkey in a major BN project in southern Israel was not authorized by MOEP due to the discovery of *Caulerpa taxifolia*, a known invasive seaweed species, in the dredging borrow area [76]. A strain of the species bred for use in aquariums has established non-native populations in waters of the Mediterranean Sea, the United States, and Australia and altered the structure of native biotic communities (e.g., [77]). Soft bottom invasive species are also monitored by Israel's national monitoring program, funded by MOEP since the early 1980s. Hard bottom invasive species in Haifa Bay have been monitored every 3 months since 2015 as part of the Environmental Monitoring and Management Program (EMMP) for construction of the Bayport.

6. Discussion and Conclusions

Circumstantial indirect evidence is insufficient to undeniably establish the role of artificial BN in promoting the establishment and dispersal of NIS. There are many examples of successful settlement and rapid and long-range proliferation of NIS especially in the eastern Mediterranean without any BN in the region [30,35,36]. *M. victor* is not the first of its family to be recorded in Haifa Bay. Three decades earlier, confamilial *M. banksii* was identified in the bay [78] with no prior BN. Firmly establishing an assumed role of BN in invasion by NIS must be based on solid and orderly, planned scientific research (e.g., BACI). Such a study should be designed well before the beginning of any BN. It has to include test sites to be artificially nourished, as well as similar control sites with no artificial BN. Systematic quantitative sampling of benthic biota (e.g., using transects or quadrats) has to be performed before the beginning of the nourishment activities and for a long period after their completion. These types of scientific quantitative samplings should be performed in equal patterns and by the same team and in the same seasons in both the impact and control sites. Unfortunately, there are very few examples of this approach. One of these rare studies is an experimental research project dealing with the role of BN in the success of an invasive Asiatic sand sedge, *Carex kobomugi Ohwi* [79]. Dune communities were subjected to 5 burial depth treatments in ~15 cm increments ranging from 0 (control) to 60 cm burial. Growth responses were monitored by quantifying emergent individuals and by harvesting all aboveground benthic biomass at the end of the season. Physiological responses were evaluated using an infrared gas analyzer to quantify photosynthesis rates. Burials lead to a reduction in community diversity and native species biomass, while favoring the invasive species. In addition, seaside goldenrod, *Solidago sempervirens*, within invaded communities exhibited significantly lower photosynthesis rates than those individuals in non-invaded communities [79]. Their results suggest that nourishment will promote Asiatic sand sedge invasion to the detriment of native dune species.

Kondylatos et al. [67] discuss the spread of *M. victor* in the eastern Mediterranean after it was first recorded from Haifa Bay in 2012 [71]. Later it was reported from Lebanon, at Batroun in 2013 and Tyre and Saida in 2014 [72], from the Mediterranean coasts of Turkey, at Phaselis, Gulf of Antalya, in 2015 [73], and from the southeastern Aegean waters of Turkey, at İztuzu Beach, Muğla (Figure 1), in 2017 [66]. These authors suggest that the occurrence of *M. victor* was expected in Greek waters [80], and they add that its record in Rhodes Island is not surprising because the last finding came from the area of Muğla (Turkey), close to Rhodes [67] (Figure 1). In view of this spreading pattern, Kondylatos et al. [67] propose that, since its initial discovery in Haifa Bay in 2012 [71], *M. victor* followed a north and westward expansion, on a "classic" route for NISs entering the Mediterranean Sea through the Suez Canal. In 2019, *M. victor* was also reported for the first time in Cyprus [81].

It is interesting to note that one of the coauthors of the article by Innocenti et al. [37], M. Mendelson, in a recent article (in Hebrew) in a local newspaper (https://haipo.co.il/item/277966 accessed on 1 June 2021) entitled: "When the sea 'swallows' the beach" supports BN. Referring to BN carried out on 20 June 2021 in Kiryat Haim beach (Figure 2) with clean sand, he praises this activity. He reports that it enriches the coastal environment with an abundance of harmless native species (in contrast to Lessepsian species) that were brought with the nourished sand, including indigenous mollusks, echinoderms and crustaceans. Governmental environmental agencies, including the Israeli MOEP, welcome criticism from the scientific community and often consult with scientists before approval of large projects. MOEP personnel are frequently approached by scientists for data. We believe that direct communication between regulators and scientists is beneficial to both sides. Information held by MOEP, such as technical documents on BN and prior knowledge of future nourishment projects, can be very useful when designing a study targeted at measuring the effects of nourishment on marine habitats. Had Innocenti et al. [37] approached MOEP before or during their study, their field surveys could have been better designed to target the actual effects of the Haifa BN projects on the local and invasive biota. Like many scientists, they based their conclusions regarding BN mostly on information found on the MOEP website. Not all types of data and documents gathered by the public sector are made available on the internet, due either to lack of public interest or to a lack of manpower needed to upload it, but these are available upon request. We conclude that direct communication between scientists and the public sector is crucial for the advancement of both environmental research and policy.

Author Contributions: Conceptualization, D.Z. (Dov Zviely) and D.Z. (Dror Zurel); investigation D.Z. (Dov Zviely) and M.B.; writing—original draft preparation, D.Z. (Dov Zviely) and E.S.; writing—review and editing, D.Z. (Dror Zurel), D.E. and M.B.; visualization, D.Z. (Dov Zviely). All authors have read and agreed to the published version of the manuscript.

Funding: This research received no external funding.

Institutional Review Board Statement: Not applicable.

Informed Consent Statement: Not applicable.

Data Availability Statement: The study did not report any data.

Conflicts of Interest: The authors declare no conflict of interest.

References

1. Hanson, H.; Brampton, A.; Capobianco, M.; Dette, H.H.; Hamm, L.; Laustrup, C.; Lechuga, A.; Spanhoff, R. Beach nourishment projects, practices, and objectives—A European overview. *Coast. Eng.* **2002**, *47*, 81–111. [CrossRef]
2. Özhan, E. *Coastal Erosion Management in the Mediterranean*; UNEP, MAP, Priority Actions Programme, Regional Activity Centre, Ankara/Split: Ankara, Turkey, 2002.
3. Waterman, R.E. *Integrated Coastal Policy Via Building with Nature*; Opmeer BV: The Hague, The Netherlands, 2008.
4. Anthony, E.J.; Sabatier, F. France. In *Coastal Erosion and Protection in Europe*; Pranzini, E., Williams, A.T., Eds.; Routledge: London, UK, 2013; Chapter 12; pp. 227–253.
5. Semeoshenkova, V.; Newton, A. Overview of erosion and beach quality issues in three southern European countries: Portugal, Spain and Italy. *Ocean Coast. Manage* **2015**, *118*, 12–121. [CrossRef]
6. Vanden Eede, S.; Van Tomme, J.; De Busschere, C.; Vandegehuchte, M.L.; Sabbe, K.; Stienen, E.W.; Degraer, S.; Vincx, M.; Bonte, D. Assessing the impact of beach nourishment on the intertidal food web through the development of a mechanistic-envelope model. *J. Appl. Ecol.* **2014**, *51*, 1304–1313. [CrossRef]
7. Pranzini, E.; Wetzel1, L.; Williams, A.T. Aspects of coastal erosion and protection in Europe. *J. Coast. Conserv.* **2015**, *19*, 445–459. [CrossRef]
8. Dean, R.G. *Beach Nourishment: Theory and Practice*; World Scientific Publishing Company: Singapore, 2003; Volume 18.
9. Harley, C.D.G.; Hughes, A.R.; Hultgren, K.M.; Miner, B.; Sorte, C.J.B.; Thornber, C.S.; Rodriguez, L.F.; Tomanek, L.; Williams, S.L. The impacts of climate change in coastal marine systems. *Ecol. Let.* **2006**, *9*, 228–241. [CrossRef] [PubMed]
10. Jones, A.; Gladstone, W.; Hacking, N. Australian sandy-beach ecosystems and climate change: Ecology and management. *Aust. Zool.* **2007**, *34*, 190–202. [CrossRef]
11. McLachlan, A.; Brown, A.C. *The Ecology of Sandy Shores*; Academic Press: Burlington, MA, USA, 2006.
12. Speybroeck, J.; Bonte, D.; Courtens, W.; Gheskiere, T.; Grootaert, P.; Maelfait, J.-P.; Provoost, S.; Sabbe, K.; Stienen, E.W.M.; Van Lancker, V.; et al. The Belgian sandy beach ecosystem: A review. *Mar. Ecol.* **2008**, *29*, 171–185. [CrossRef]
13. Defeo, O.; McLachlan, A.; Schoeman, D.S.; Schlacher, T.A.; Dugan, J.; Jones, A.; Lastra, M.; Scapini, F. Threats to sandy beach ecosystems: A review. *Estuar. Coast. Shelf. Sci.* **2009**, *81*, 1–12. [CrossRef]
14. Schlacher, T.A.; Schoeman, D.S.; Dugan, J.; Lastra, M.; Jones, A.; Scapini, F.; McLachlan, A. Sandy beach ecosystems: Key features, sampling issues, management challenges and climate change impacts. *Mar. Ecol.* **2008**, *29*, 70–90. [CrossRef]
15. Dean, R.G. Compatibility of borrow material for beach fills. In *Proceedings of the 14th International Conference on Coastal Engineering*; ASCE: Copenhagen, Denmark, 1974; pp. 1319–1333.
16. Dean, R.G. Principles of beach nourishment. In *Handbook of Coastal Processes and Erosion*; Komar, P.D., Ed.; CRC Press: Boca Raton, FL, USA, 1983; Chapter 11; pp. 217–232.
17. Anthony, E.J. The status of beaches and shoreline development options on the French Riviera: A perspective and a prognosis. *J. Coast. Conserv.* **1997**, *3*, 169–178. [CrossRef]
18. Aragonés, L.; Garcia-Barba, J.; García-Bleda, E.; López, I.; Serra, J.C. Beach nourishment impact on *Posidonia oceanica*: Case study of Poniente Beach (Benidorm, Spain). *Ocean Eng.* **2015**, *107*, 1–12. [CrossRef]
19. Danovaro, R.; Nepote, E.; Martire, M.L.; Ciotti, C.; De Grandis, G.; Corinaldesi, C.; Carugati, L.; Cerrano, C.; Pica, D.; Di Camillo, C.G.; et al. Limited impact of beach nourishment on macrofaunal recruitment/settlement in a site of community interest in coastal area of the Adriatic Sea (Mediterranean Sea). *Mar. Pollut. Bull.* **2018**, *128*, 259–266. [CrossRef]
20. Naqvi, S.; Pullen, E. *Effects of Beach Nourishment and Borrowing on Marine Organisms*; Miscellaneous Report No. 82–14; U.S. Army Corps of Engineers, Coastal Engineering Research Center: Fort Belvoir, VA, USA, 1982.
21. Van Dolah, R.F.; Wendt, P.H.; Martore, R.M.; Levisen, M.V.; Roumillat, W.A. *A Physical and Biological Monitoring Study of the Hilton Head Beach Nourishment Project. Unpublished*; South Carolina Wildlife and Marine Resources Department for Town of Hilton Head Island: Hilton Head Island, SC, USA, 1992.
22. Van Dolah, R.F.; Martore, R.M.; Lynch, A.E.; Levisen, M.V.; Wendt, P.H.; Whitaker, D.J.; Anderson, W.D. *Final Report: Environmental Evaluation of the Folly Beach Nourishment Project*; U.S. Army Corps of Engineers, Charleston District: Charleston, SC, USA, 1994.
23. Peterson, C.H.; Laney, T.R.W. Biological impacts of beach nourishment. In *Workshop on the Science of Beach re Nourishment*; Pine Knoll Shores: Carteret County, NC, USA, 2001.
24. Rice, T. The big picture: An overview of coastal resources and federal projects. In *Coastal Ecosystems & Federal Activities Technical Training Symposium*; US Fish and Wildlife Service: Gulf Shores, AL, USA, 2001.
25. Lakshmi, E.; Priya, M.; Achari, V.S. An overview on the treatment of ballast water in ships. *Ocean. Coast. Manag.* **2021**, *199*, 105296. [CrossRef]
26. Minchin, D.; Gollasch, S.; Cohen, A.N.; Hewitt, C.L.; Olenin, S. Characterizing vectors of marine invasion. In *Biological Invasions in Marine Ecosystems*; Rilov, G., Crooks, J.A., Eds.; Ecological Studies 204; Springer-Verlag: Berlin Heidelberg, Germany, 2009; pp. 109–116.
27. Franks, S.J.; Peterson, C.J. Burial disturbance leads to facilitation among coastal dune plants. *Plant. Ecol.* **2003**, *16*, 13–21. [CrossRef]

28. Rotter, A.; Klun, K.; Francé, J.; Mozetič, P.; Orlando-Bonaca, M. Non-indigenous species in the Mediterranean Sea: Turning from pest to source by developing the 8Rs model, a new paradigm in pollution mitigation. *Front. Mar. Sci.* **2020**, *7*, 178. [CrossRef]
29. Spanier, E.; Galil, B.S. Lessepsian migration—A continuous biogeographical process. *Endeavour* **1991**, *15*, 102–106. [CrossRef]
30. Edelist, D.; Rilov, G.; Golani, D.; Carlton, J.T.; Spanier, E. Restructuring the sea: Profound shifts in the world's most invaded marine ecosystem. *Divers. Distrib.* **2013**, *19*, 69–77. [CrossRef]
31. Galil, B.S. The alien crustaceans in the Mediterranean Sea: An historical review. In *the Wrong Place—Alien Marine Crustaceans: Distribution, Biology and Impacts*; Galil, B.S., Clark, P.F., Carlton, J.T., Eds.; Springer: Dordrecht, The Netherlands, 2011; pp. 377–401.
32. Katsanevakis, S.; Moustakas, A. Uncertainty in marine invasion science. *Front. Mar. Sci.* **2018**, *5*, 38. [CrossRef]
33. Galil, B.S.; Marchini, A.; Occhipinti-Ambrogi, A. East is east and West is west? Management of marine bioinvasions in the Mediterranean Sea. *Estuar. Coast. Shelf Sci.* **2018**, *201*, 7–16. [CrossRef]
34. Zenetos, A.; Cinar, M.E.; Crocetta, F.; Golani, D.; Rosso, A.; Servello, G.; Shenkar, N.; Turon, X.; Verlaque, M. Uncertainties and validation of alien species catalogues: The Mediterranean as an example. *Estuar. Coast. Shelf Sci.* **2017**, *191*, 171–187. [CrossRef]
35. Galil, B.S.; Mienis, H.K.; Hoffman, R.; Goren, M. Non-indigenous species along the Israeli Mediterranean coast: Tally, policy, outlook. *Hydrobiologia* **2021**, *848*, 2011–2029. [CrossRef]
36. Ben Rais Lasram, F.; Tomasini1, J.A.; Guilhaumon, F.; Romdhane, M.S.; Do Chi, T.; Mouillot, D. Ecological correlates of dispersal success of Lessepsian fishes. *Mar. Ecol. Prog. Ser.* **2009**, *363*, 273–286. [CrossRef]
37. Innocenti, G.; Stasolla, G.; Mendelson, M.; Galil, B.S. Aggressive, omnivorous, invasive: The Erythraean moon crab Matuta victor (Fabricius, 1781) (Crustacea: Decapoda: Matutidae) in the eastern Mediterranean Sea. *J. Nat. Hist.* **2017**, *51*, 2133–2142. [CrossRef]
38. Bitan, M.; Zviely, D. Lost value assessment of bathing beaches due to sea level rise: A case study of the Mediterranean coast of Israel. *J. Coast. Conserv.* **2019**, *23*, 773–783. [CrossRef]
39. Bitan, M.; Zviely, D. Sand beach nourishment: Experience from the Mediterranean coast of Israel. *J. Mar. Sci. Eng.* **2020**, *8*, 273. [CrossRef]
40. Nir, Y. Offshore artificial structures and their influence on the Israel and Sinai Mediterranean beaches. In *Proceedings of the 18th International Conference on Coastal Engineering*; ASCE: Cape Town, South Africa, 1982; pp. 1837–1856.
41. Inman, D.L.; Jenkins, S.A. The Nile littoral cell and man's impact on the coastal zone of the Southeastern Mediterranean. In Proceedings of the 19th International Conference on Coastal Engineering; ASCE: Houston, TX, USA, 1984; pp. 1600–1617.
42. Stanley, D.J. Sediment transport on the coast and shelf between the Nile Delta and Israeli margin as determined by heavy minerals. *J. Coast. Res.* **1989**, *5*, 813–828.
43. Frihy, O.E.; Fanos, A.M.; Khafagy, A.A.; Komar, P.D. Patterns of sediment transport along the Nile Delta, Egypt. *Coast. Eng.* **1991**, *15*, 409–429. [CrossRef]
44. Sharaf El Din, S.H.; Mahar, A.M. Evaluation of sediment transport along the Nile Delta coast, Egypt. *J. Coast. Res.* **1997**, *13*, 23–26.
45. Frihy, O.E.; Deabes, E.A.; El Gindy, A.A. Wave climate and nearshore processes on the Mediterranean coast of Egypt. *J. Coast. Res.* **2010**, *26*, 103–112. [CrossRef]
46. Khalifa, M.A. Adoption of recent formulae for sediment transport calculations applied on the Egyptian Nile delta coastal area. *J. Coast. Conserv.* **2012**, *16*, 37–49. [CrossRef]
47. Emery, K.O.; Neev, D. Mediterranean beaches of Israel. *Israel Geol. Surv. Bull.* **1960**, *26*, 1–24.
48. Goldsmith, V.; Golik, A. Sediment transport model of the southeastern Mediterranean coast. *Mar. Geol.* **1980**, *37*, 135–147. [CrossRef]
49. Rohrlich, V.; Goldsmith, V. Sediment transport along the southeast Mediterranean: A geological perspective. *Geo-Mar. Lett.* **1984**, *4*, 99–103. [CrossRef]
50. Carmel, Z.; Inman, D.; Golik, A. Directional wave measurements at Haifa, Israel, and sediment transport along the Nile littoral cell. *Coast. Eng.* **1985**, *9*, 21–36. [CrossRef]
51. Perlin, A.; Kit, E. Longshore sediment transport on the Mediterranean coast of Israel. *J. Waterw. Port. Coast. Ocean. Eng.* **1999**, *125*, 80–87. [CrossRef]
52. Zviely, D.; Kit, E.; Klein, M. Longshore sand transport estimates along the Mediterranean coast of Israel in the Holocene. *Mar. Geol.* **2007**, *237*, 61–73. [CrossRef]
53. Kit, E.; Sladkevich, M. Structure of offshore currents on sediment Mediterranean coast of Israel. In Proceedings of the 6th Workshop on Physical Processes in Natural Waters, Girona, Spain, 27–29 June 2001; Casamitjana, X., Ed.; 2001; pp. 97–100.
54. Kunitsa, D.; Rosentraub, D.; Stiassnie, M. Estimates of winter currents on the Israeli continental shelf. *Coast. Eng.* **2005**, *52*, 93–102. [CrossRef]
55. Zviely, D. Sedimentological Processes in Haifa Bay in Context of the Nile Littoral Cell. Ph.D. Thesis, Department of Geography and Environment Studies, University of Haifa, Haifa, Israel, 2006. (In Hebrew, English abstract)..
56. Zviely, D.; Sivan, D.; Ecker, A.; Bakler, N.; Rohrlich, V.; Galili, E.; Boaretto, E.; Klein, M.; Kit, E. The Holocene evolution of Haifa Bay area, Israel, and its influence on the ancient human settlements. *Holocene* **2006**, *16*, 849–861. [CrossRef]
57. Pomerancblum, M. The distribution of heavy minerals and their hydraulic equivalents in sediments of the Mediterranean shelf of Israel. *J. Sedim. Petrol.* **1966**, *36*, 162–174. [CrossRef]
58. Almagor, G.; Gill, D.; Perath, I. Marine sand resources offshore Israel. *Mar. Georesources Geotech.* **2000**, *18*, 1–42. [CrossRef]
59. Sandler, A.; Herut, B. Composition of clays along the continental shelf off Israel: Contribution of the Nile versus local sources. *Mar. Geol.* **2000**, *167*, 339–354. [CrossRef]

60. Nir, Y. *Recent Sediments of Haifa Bay*; Rep. MG/11/80; Geological Survey of Israel: Jerusalem, Israel, 1980; p. 8.
61. Bakler, N. *Gross Lithology of Drilling and Laboratory Data, Haifa Bay, Tel-Aviv and Caesarea*; Final summary 1, UNDP-GSI offshore dredging project, ISR/71/682; Geological Survey of Israel: Jerusalem, Israel, 1975; p. 23.
62. Hall, J.K. *Seismic Studies, Haifa Bay, Summary Report*; UNDP-GSI offshore dredging project ISR/71/522. Geological Survey of Israel: Jerusalem, Israel, 1976; Volume 1/76, p. 36.
63. Zviely, D.; Kit, E.; Rosen, B.; Galili, E.; Klein, M. Shoreline migration and beach-nearshore sand balance over the last 200 years in Haifa Bay (SE Mediterranean). *Geo Mar. Lett.* **2009**, *29*, 93–110. [CrossRef]
64. Galil, B.S.; Froglia, C.; Noel, P. Looking back, looking ahead: The CIESM Atlas, Crustaceans. *Manag. Biol. Invasions* **2015**, *6*, 171–175. [CrossRef]
65. Coll, M.; Piroddi, C.; Steenbeek, J.; Kaschner, K.; Lasram, F.B.R.; Aguzzi, J.; Ballesteros, E.; Bianchi, C.N.; Corbera, J.; Dailianis, T.; et al. The biodiversity of the Mediterranean Sea: Estimates, patterns, and threats. *PLoS ONE* **2010**, *5*, e11842. [CrossRef]
66. Gökoğlu, M.; Julian, D.; Dailianis, T.; Akyol, O.; Babali, N.; Bariche, M.; Crocetta, F.; Gerovasileiou, V.; Chanem, R.; Gökoğlu, M.; et al. Occurence of *Matuta victor* (Crustacea: Decapoda) in Turkey [p.619–620]. In New Mediterranean biodiversity records. *Medit. Mar. Sci.* **2016**, *17*, 608–626.
67. Kondylatos, G.; Corsini-Foka, M.; Perakis, E. First record of the isopod Idotea hectica (Pallas, 1772) (Idoteidae) and of the brachyuran crab *Matuta victor* (Fabricius, 1781) (Matutidae) in the Hellenic waters. *Medit. Mar. Sci.* **2018**, *19*, 656–661. [CrossRef]
68. Hanim, N.; Wardiatno, Y.; Perwitasari, D.; Suman, A.; Parlindungan, D.; Farajallah, A. Distribution of Matuta purnama J. C. Y. Lai and Galil, 2007 (Brachyura: Matutidae) outside type locality. *IOP Conf. Ser. Earth Environ. Sci.* **2021**, *744*, 012023. [CrossRef]
69. Bellwood, O. The occurrence, mechanics and significance of burying behavior in crabs (Crustacea: Brachyura). *J. Nat. Hist.* **2002**, *36*, 1223–1238. [CrossRef]
70. Ng, P.K.L. Crabs [pp. 1045–1155]. In *FAO Species Identification Guide for Fishery Purposes. The Living Marine Resources of the Western Central Pacific. Volume 2: Cephalopods, Crustaceans, Holothurians and Sharks*; Carpenter, K.E., Niem, V.H., Eds.; Food and Agriculture Organization of the United Nations: Rome, Italy, 1998; pp. 687–1396.
71. Galil, B.S.; Mendelson, M.A. Record of the moon crab *Matuta victor* (Fabricius, 1781) (Crustacea; Decapoda; Matutidae) from the Mediterranean coast of Israel. *BioInv. Rec.* **2013**, *2*, 69–71. [CrossRef]
72. Crocetta, F.; Bariche, M.; Crocetta, F.; Agius, D.; Balistreri, P.; Bariche, M.; Bayhan, Y.; Çakir, M.; Ciriaco, S.; Corsini-Foka, M.; et al. Six new records from Lebanon, with general implications for Mediterranean alien fauna [p. 696–698]. In New Mediterranean biodiversity records. *Medit. Mar. Sci.* **2015**, *16*, 682–702. [CrossRef]
73. Ateş, S.; Katağan, T.; Sert, M.; Özdilek, S.Y. A new locality for common box crab, *Matuta victor* (Fabricius, 1781), from the eastern Mediterranean Sea. *J. Black Sea Medit. Env.* **2017**, *23*, 191–195.
74. Crain, D.A.; Boltenu, A.B.; Bjorndal, K.A. Effects of beach nourishment on sea turtles: Review and research initiatives. *Rest. Ecol.* **1995**, *3*, 95–104.
75. Lubinevsky, H. Ecological, Genetic and Morphological Aspects of Migrant Gastropod Strombus (Conomurex) Persicus. Ph.D. Thesis, Department of Geography and Environment Studies, University of Haifa, Haifa, Israel, 2011. (In Hebrew, English abstract).
76. Çinar, M.E.; Bilecenoğlu, M.; Yokeş, M.B.; Öztürk, B.; Taşkin, E.; Bakir, K.; Doğan, A.; Açik, Ş. Current status (as of end of 2020) of marine alien species in Turkey. *PLoS ONE* **2021**, *16*, e0251086. [CrossRef] [PubMed]
77. Parreira, F.; Martínez-Crego, B.; Afonso, C.M.L.; Machado, M.; Oliveira, F.; dos Santos Gonçalves, J.M.; Santos, R. Biodiversity consequences of *Caulerpa prolifera* takeover of a coastal lagoon. *Estuar. Coast. Shelf Sci.* **2021**, *107344*, 1–7.
78. Galil, B.S.; Golani, D. Two new migrant decapods from the Eastern Mediterranean. *Crustaceana* **1990**, *58*, 229–236.
79. Daneshgar, P.P.; Phillips, L.B.; James, D.P.; Mickley, M.G.; Bohackyj, A.M.; Rhoads, L.J.; Bastian, R.B.; Wootton, L.S. The role of beach nourishment on the success of invasive Asiatic Sand Sedge. *Northeast. Nat.* **2017**, *24*, 110–120. [CrossRef]
80. Karachle, P.K.; Corsini Foka, M.; Crocetta, F.; Dulčić, J.; Dzhembekova, N. Setting-up a billboard of marine invasive species in the ESENIAS area: Current situation and future expectancies. *Acta Adriatica* **2017**, *58*, 429–458. [CrossRef]
81. Kleitou, P.; Doumpas, N. Cyprus: The alien moon crab *Matuta victor* is reported for the first time from the entire country (First record of the moon crab *Matuta victor* (Fabricius, 1781) (Crustacea; Decapoda; Matutidae) from Cyprus Periklis and Nikolaos Doumpas. *Medit. Mar. Sci.* **2019**, *20*, 244.

MDPI
St. Alban-Anlage 66
4052 Basel
Switzerland
Tel. +41 61 683 77 34
Fax +41 61 302 89 18
www.mdpi.com

Journal of Marine Science and Engineering Editorial Office
E-mail: jmse@mdpi.com
www.mdpi.com/journal/jmse